Model-based Health Monitoring of Hybrid Systems

Danwei Wang · Ming Yu
Chang Boon Low · Shai Arogeti

Model-based Health Monitoring of Hybrid Systems

Danwei Wang
Ming Yu
Chang Boon Low
Nanyang Technological University
Singapore

Shai Arogeti
Ben-Gurion University of the Negev
Beer-Sheva
Israel

ISBN 978-1-4899-9059-4 ISBN 978-1-4614-7369-5 (eBook)
DOI 10.1007/978-1-4614-7369-5
Springer New York Heidelberg Dordrecht London

Softcover re-print of the Hardcover 1st edition 2013

Printed on acid-free paper

Springer is part of Springer Science+Business Media (www.springer.com)

Preface

As the complexity of industrial systems increases, fault diagnosis and failure prognosis become more and more important since they are crucial means to maintain system safety and reliability. Many manmade systems, such as printers, converters, and automobiles, can be modeled as hybrid systems. These systems consist of interacting continuous and discrete parts. Monitoring of hybrid system requires estimation of continuous state variables as well as tracking discrete states, and more difficulties are involved due to the mode-varying nature of hybrid system. In the past two decades, there were several efforts for diagnosis of hybrid systems; however, to the best of our knowledge, no results have been reported for failure prognosis of hybrid systems. Subsequently, a framework and effective techniques are required for sophisticated analysis, reliable design, and efficient implementation of hybrid system diagnosis and prognosis algorithms.

As a multidisciplinary modeling language, bond graphs were introduced in 1959 by Professor Henry Paynter of Massachusetts Institute of Technology. Recently, it has shown its great power in different engineering applications, such as control engineering, process monitoring, and system identification, etc. For the purpose of modeling of hybrid systems, hybrid bond graph was developed in 1995 by Professor Gautam Biswas and his student Pieter J. Mosterman of Vanderbilt University. Since then, applications of hybrid bond graph to various hybrid systems have received intensive research.

This book explains theoretical and experimental works related to diagnosis and prognosis of hybrid systems using hybrid bond graph modeling. It offers an overview of the fundamentals of diagnosis, prognosis, and hybrid bond graph modeling. This book describes a framework of hybrid bond graph-based quantitative fault detection, isolation, and estimation. Moreover, it also presents strategies to track the system mode and predict the remaining useful life under multiple fault condition. A real-world complex hybrid system—a mobile robot steering control system is studied using the developed fault diagnosis method to exhibit practical significance. The material of this book is solely based on the result of several own research projects and our personal interactions with many leading figures in this area have motivated us to write this book.

In this book, the latest contributions to the area of model-based fault diagnosis and prognosis using hybrid bond graph are presented to understand the health monitoring process better and help the researchers in their future research. The book is written for undergraduate and graduate students with electrical engineering, mechanical engineering, or computer science background. It also serves as an introduction material for practicing engineers as well as academic researchers. With these aims, the scope of the book covers the following subjects: modeling of hybrid systems, quantitative hybrid bond graph-based fault detection and isolation, fault parameter identification, mode tracking techniques, application of real-time fault diagnosis to a vehicle steering system, and multiple failure prognosis.

Singapore, February 2013

Danwei Wang
Ming Yu
Chang Boon Low
Shai Arogeti

Contents

Abbreviations

AGA	Adaptive Genetic Algorithm
AGARR	Augmented Global Analytical Redundancy Relation
AHPSO	Adaptive Hybrid Particle Swarm Optimization
AI	Artificial Intelligence
ARR	Analytical Redundancy Relation
BDE	Binary Differential Evolution
BG	Bond Graph
BPSO	Binary-Valued Particle Swarm Optimization
CBM	Condition-Based Maintenance
DE	Differential Evolution
DHBG	Diagnostic Hybrid Bond Graph
EA	Evolutionary Algorithm
FDI	Fault Detection and Isolation
FSM	Fault Signature Matrix
GA	Genetic Algorithm
GARR	Global Analytical Redundancy Relation
HBG	Hybrid Bond Graph
HDE	Hybrid Differential Evolution
HMM	Hidden Markov Models
HPSO	Hybrid Particle Swarm Optimization
IMC	Initial Mode Coefficients
IMM	Interacting Multiple Model
MAHPSO	Multiple Adaptive Hybrid Particle Swarm Optimization
MAT	Mode AGARRs Table
MCSM	Mode Change Signature Matrix
MD-FSM	Mode-Dependent Fault Signature Matrix
MGY	Modulated Gyrator
MHDE	Multiple Hybrid Differential Evolution
MTF	Modulated Transformer
ODE	Ordinary Differential Equation
PCA	Principal Component Analysis
PHM	Prognostics and Health Management
PSO	Particle Swarm Optimization

QBG	Qualitative Bond Graph
QHBG	Quantitative Hybrid Bond Graph
QSIM	Qualitative Simulation
RDE	Real Differential Evolution
RPSO	Real-Valued Particle Swarm Optimization
RUL	Remaining Useful Life
SCAP	Sequential Causality Assignment Procedure
SCAPH	Sequential Causality Assignment Procedure for Hybrid Systems
SOC	State Of Charge
SVC	Support Vector Classifier
SVM	Support Vector Machine
TCG	Temporal Causal Graph

Chapter 1
Health Monitoring of Engineering Systems

1.1 Condition Based Maintenance

In general, a fault refers to an abnormal condition that may lead to reduction or loss of the capability of a system or its component to perform a required function. On the other hand, a failure means the inability of a system or its component to perform its required functions within specified performance requirements. A loosen belt is an example of fault in a mechanical system where the belt is still working but the transmission efficiency is decreased. However, a broken belt is a failure since the belt is not working anymore and must be replaced.

System health monitoring is a key feature for failure prevention and Condition Based Maintenance (CBM). A health monitoring system needs to detect a fault or failure in a timely manner so that and the faulty components can be replaced effectively to ensure system's normal operations. In the past few decades, maintenance strategies have evolved from early reactive maintenance, to age-based preventive maintenance, then to condition-based maintenance. Reactive maintenance is usually performed after system breakdown. In order to prevent catastrophic failures which cause emergency shutdowns, age-based preventive maintenance is introduced. This policy is carried out based on system operating time regardless of the health condition. Age-based preventive maintenance may sometimes reduce unexpected failures, but it is not cost effective and cannot eliminate major failures. These conventional maintenance strategies do not satisfy the demands of high reliability in modern engineering systems. Fortunately, CBM can be an effective alternative, and it tries to avoid unnecessary maintenance by taking maintenance actions when there is evidence of abnormality in a monitored system [1]. The monitoring is based on sensor measurements and does not interrupt normal operation. It attempts to avoid excessive or insufficient maintenance and ultimately results in higher system availability.

In general, CBM includes three key steps: data collection, data processing and decision-making. These steps are shown in Fig. 1.1. Data collection step is to obtain data related to system condition. Data processing is about handling and analyzing the data or signals collected for better understanding and interpretation. The purpose

D. Wang et al., *Model-based Health Monitoring of Hybrid Systems*,
DOI: 10.1007/978-1-4614-7369-5_1,

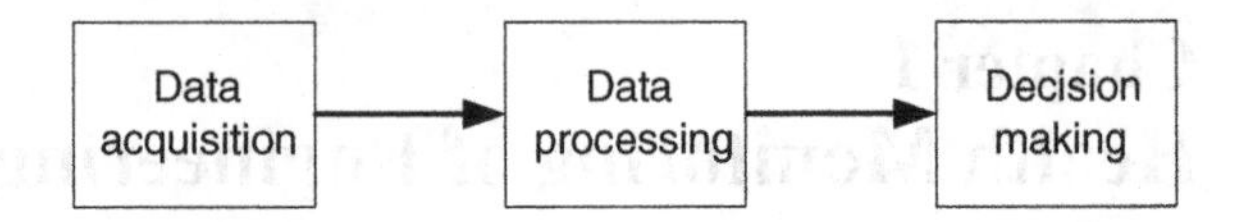

Fig. 1.1 Three steps of CBM

of decision-making is to recommend efficient maintenance strategies. Diagnosis and prognosis are two critical factors in CBM, and they are complementary tasks since diagnosis is a "static" indicator whereas prognosis is a "dynamic" indicator. The objective of diagnosis is to indicate whether or not a fault has occurred and at the same time provide some information about the severity of the fault [2]. Prognosis tries to track fault degradation and predict the Remaining Useful Life (RUL) of a faulty component or subsystem. Prognosis in CBM has received more attention in recent years and many efforts have been made for different applications. Prognosis is often much more efficient than diagnosis to achieve zero-downtime performance which is very important when a failure is catastrophic in some applications (e.g., helicopter gearbox and nuclear power plant).

1.2 Fault Diagnosis Tasks and Methodologies

1.2.1 Fault Diagnosis Tasks

In general, there are three tasks for fault diagnosis, namely fault detection, fault isolation and fault identification [3].

1. Fault detection: it is the first step of fault diagnosis and tries to detect the presence of fault in the monitored system. Early detection of fault is very important before the fault possibly causes a catastrophic failure in the system.
2. Fault isolation: given that a fault has occurred and been detected, fault isolation aims to establish possible fault candidate that can explain the observed abnormal behavior. For single fault diagnosis, the objective is to obtain a unique single fault that can lead to the observations. It may not always be possible to determine a unique candidate given the sensors available to the monitored system. As for multiple fault diagnosis, the goal is to acquire sets of faults that, occurring together, are able to explain the observations.
3. Fault identification: this step is to determine the magnitude of the fault and its type. For abrupt fault, if multiple fault sets remain after fault isolation, then identification is required for each fault set and the fault set that matches the observations most closely is considered to be the true fault set. For incipient fault, the fault identification task is challenging since certain dynamic degradation behavior for this fault must be assumed in advance and sometimes this prior knowledge is not easy to obtain. If the severity of the identified fault is acceptable, this severity will be used in the reconfiguration design of the system's control law

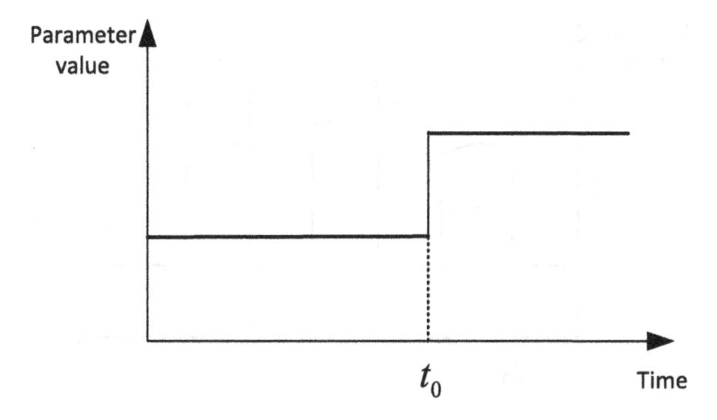

Fig. 1.2 Abrupt fault profile

to achieve fault tolerance. On the other hand, if the fault identification result indicates that the fault is too severe to be accommodated, then the corresponding faulty component has to be replaced.

The nature of possible faulty situations may be classified into three types as follows:

1. Abrupt fault: typically modeled as step-like deviation and is usually persistent as shown in Fig. 1.2, where t_0 is the time point at which the fault first starts. For abrupt fault, it is crucial that the fault diagnosis scheme is able to detect the sudden change in a timely manner to avoid catastrophic consequences. In such cases, early detection and accommodation are the key objectives of fault diagnosis.
2. Incipient fault: slow developing and are usually related to the wear and tear of the system components as shown in Fig. 1.3. It is relatively difficult to detect the

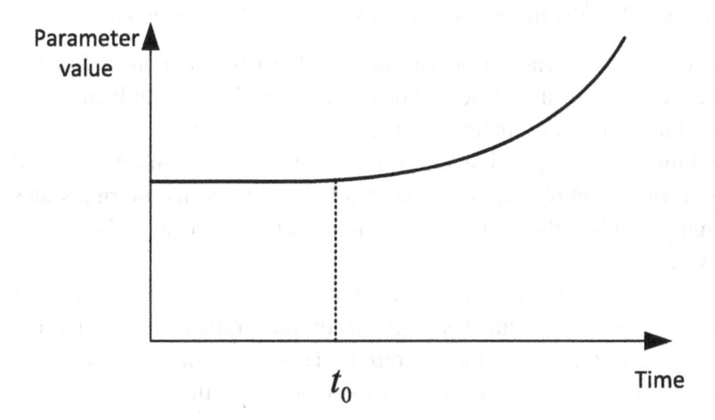

Fig. 1.3 Incipient fault profile

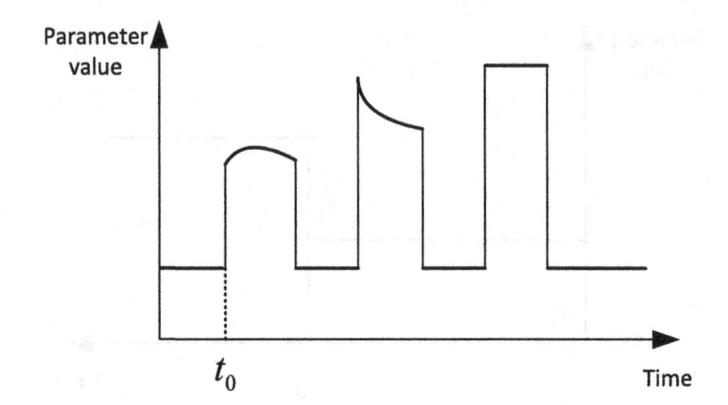

Fig. 1.4 Intermittent fault profile

incipient fault due to its slowly developing nature of the fault and the compensation effect of the system's feedback control.

3. Intermittent fault: usually manifests itself intermittently in an unpredictable manner as shown in Fig. 1.4. For example, a worn out roller in a printer may no longer be able to grip the paper consistently which in turn causes intermittent paper jams. A printer with a worn out roller usually operates correctly but will infrequently slip and cause a paper jam. Intermittent faults are hard to handle for several reasons as follows. First, if diagnosis process is not performed continuously, it might due to the intermittent faults that are not present when diagnosis is active. Second, fault signals are not persistent and detection is not consistent. It is difficult, in this case, to distinguish between intermittent faults and other types of faults like abrupt faults and incipient faults.

Basically, abrupt and incipient faults belong to persistent faults, which means that once they appear, do not disappear, while intermittent faults do. In the following, various sources of faults in the monitored system will be discussed.

1. Component fault: deviation of parameter value from its nominal one can cause condition change in the system. For example, a flat tire fault in a vehicle will increase the friction coefficient between ground and tire.
2. Sensor fault: sensors provide signal measurements of a monitored system, and convey information related to a system's behavior and its internal states. Sensor faults happen when there are discrepancies between measured signals and their actual values.
3. Actuator fault: actuators are the control effectors of a system. For most electromechanical systems, control signals from the controllers cannot be directly applied to the system. Actuators are required to transform control signals to proper actuation signals such as torques and forces to drive the system. Actuator faults occur when there are discrepancies between desired actuator output and actual actuator output to the system.

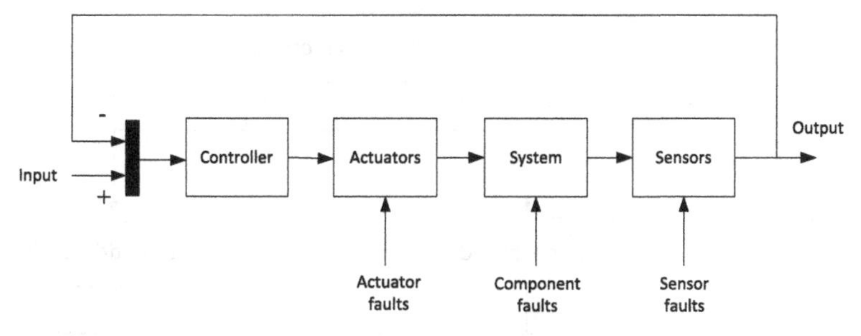

Fig. 1.5 Different faults in a monitored system

These different fault sources are shown in Fig. 1.5.

1. Uncertainties: uncertainties in modeling can be due to a bad estimation of dynamics in a system, a non-precise identification of the numerical values of the parameters or variation of their values because of heat, time or working conditions.
2. Disturbances: disturbances usually refer to noises in sensor measurements that are high frequency signals. Other factors like unmeasured friction, unknown inputs and backlash are also considered as disturbances.

A good diagnostic system should be robust to various disturbances and uncertainties but still maintain its fault sensitivity.

1.2.2 Fault Diagnosis Methodologies

Fault diagnosis methods can be broadly classified into two types (as shown in Fig. 1.6): model based method and data driven method. For model based methods, models serve as knowledge representation of a large amount of structural, functional and behavioral information and their relationship. This knowledge representation is capitalized to create complex cause-effect reasoning leading to construction of powerful and robust automatic diagnosis and isolation systems [4].

Qualitative model based approach provides an alternative when a numerical model of the system is unavailable. It utilizes qualitative abstractions to model complex systems while model structure is well defined. The models used in qualitative methods are relatively simple compared with numerical models. The sensitivity of fault diagnosis system to modeling errors and sensor noises may be alleviated [5]. Qualitative Simulation (QSIM) is a widely used modeling tool to describe continuous model qualitatively [6]. This approach is intended to simulate the behavior of physical systems using qualitative values, rather than providing explanations for behaviors of physical processes. In [7], several faulty models are built using QSIM. The observed faulty behavior is compared with that from the faulty models to choose the faults set which occur in the system. However, the faulty models determination needs

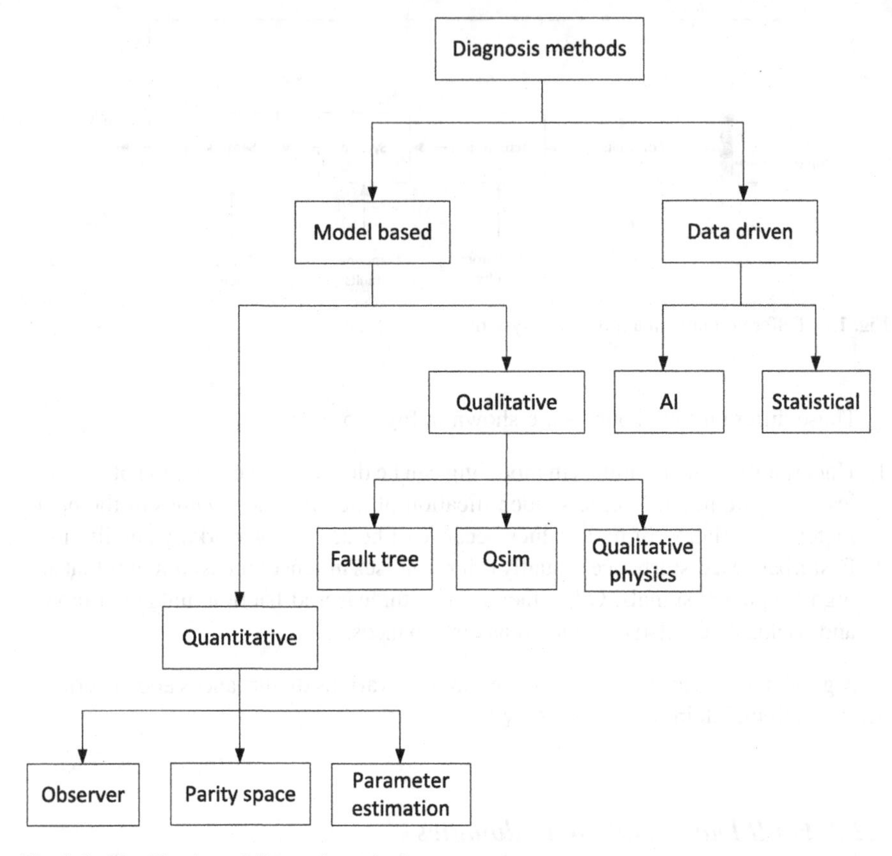

Fig. 1.6 Classification of diagnosis methods

prior knowledge. In [8, 9], QSIM based fault diagnosis is presented to handle multiple faults in continuous devices. The qualitative modeling framework quantizes the state space utilizing landmark values and specifies qualitative relations between the quantized states, which leads to a set of qualitative differential equations. A fuzzy qualitative simulation method is developed in [10]. The advantage of this method over QSIM based fault diagnosis is that if the observed behavior cannot match the predicted behavior of any faulty model, the candidate generator will check the modified models whose predicted behavior can match the observed one. Therefore, the modified models selected from the generator will be used to determine the fault candidates. The fuzzy qualitative simulation method can provide more precise information than QSIM based method because the utilization of fuzzy sets leads to a more accurate representation with respect to time. In continuous system diagnosis, time is an important factor to be considered during algorithm design. However, it is difficult to choose appropriate fuzzy sets number and membership function which is a common problem for fuzzy logic system design. In addition, there is no efficient method to determine the number of modified models to be searched by candidate generator.

Fault tree was originally introduced in 1961 at Bell Laboratories by H.A. Watson, under a U.S. Air Force Ballistics Systems Division project to evaluate the Minuteman I Intercontinental Ballistic Missile (ICBM) Launch Control System. The Boeing Company modified the concept for computer utilization later. Fault tree is now widely used in many fields [11, 12]. A fault tree is a model that graphically and logically represents the various combinations of possible events (faulty and normal), occurring in a system that lead to the top undesired event. It is a structured methodology to determine the potential causes of an undesired event, referred to as the top event. The top event usually represents a major accident causing safety hazards. While the top event is placed at the top of the tree, the tree is constructed downwards, dissecting the system with further detail until the primary events leading to the top event are known. The tree usually has layers of nodes. At each node, different logic operations like AND and OR are performed for propagation as shown in Fig. 1.7. Generally, a fault tree analysis includes the following four steps: (i) system definition, (ii) fault tree construction, (iii) qualitative evaluation and (iv) quantitative evaluation [13]. Before the building of the fault tree, a detailed understanding of the system is required. To carry out consistent diagnosis from fault trees, the trees should completely represent the system causal relationships, i.e., explain all fault scenarios. However, no formal methods can be used to verify the accuracy of the fault tree established.

Qualitative physics method usually derives qualitative equations from the differential equations. In [14], a Qualitative Bond Graph (QBG) method is developed based on the combination of qualitative reason and bond graph modeling theory in order to benefit from both methods. In QBG, qualitative equations form bond graph models instead of the differential equations are used for fault analysis. These equations represent the components' physical variables, locations, and their functional relations that can be stated directly from the model. This is particularly suitable for model based fault diagnosis because possible faults can be localized through analysis of relations

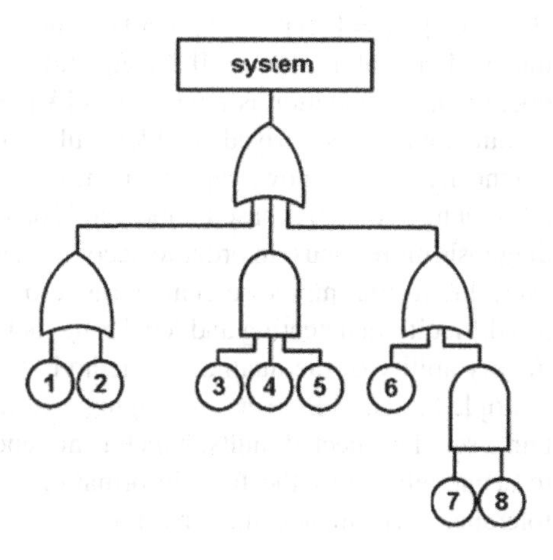

Fig. 1.7 A fault tree diagram

between the component states and observed abnormal behavior qualitatively. Such qualitative algorithm is done using available measurement, i.e., the history of past data. Therefore, the qualitative behavior equations are always written in differential causality [15].

For passive elements, qualitative equations are defined as:

$$\begin{aligned} \text{R} - \text{element: } & e(k) = Rf(k) \\ \text{C} - \text{element: } & f(k) = C[e(k) - e(k-1)] \\ \text{I} - \text{element: } & e(k) = I[f(k) - f(k-1)] \end{aligned}$$

where k denotes the current time and $k-1$ represents the time at the previous sample. Since only qualitative information is considered, the actual sampling time T, used for taking derivatives, does not appear in these equations.

A qualitative space $\{[+1], [+], [0], [-], [-1], [?]\}$ is defined to represent system behaviors and component states for qualitative fault diagnosis [16]. For system measurements, $[+]$ and $[-]$ stand for the positive and negative values in the measurements, respectively. $[0]$ represents the boundary between $[+]$ and $[-]$, while $[?]$ denotes an uncertain value. $[+1]$ and $[-1]$ represent very large and negative values, respectively. For power variables E and F, $[+1]$, $[-1]$ and $[0]$ represent different abnormal behaviors due to system faults, whereas $[+]$ and $[-]$ denote normal states. For system components R, C and I, $[+]$ represents a component blocked, e.g., valve blocked, which hinders power delivery, and $[0]$ denotes component leakage or short circuit. Since the operations on qualitative variables are different from numerical ones, a set of qualitative operators are defined as $\{+, -, \times, /, =\}$ and these qualitative operators have the same mathematical meanings as their counterparts in numerical domain. The inference mechanism of qualitative equations is different from normal mathematical equations according to the definitions of qualitative operation tables [17]. For example, consider qualitative equation $[+] = [+] + [X]$, a set of solutions $[X] = \{[+1], [+], [0]\}$ can be obtained. From the qualitative operation tables, $[+1] = [+1] + [0]$, $[+1] = [+1] + [+]$ and $[+1] = [+1] + [+1]$ are valid. However, $[+] = [+1] + [-1]$ is not true according to the qualitative operation tables, thus $[-1]$ is not a solution. If the equation is solved using a standard mathematical operation, the equation is rewritten as $[X] = [+1] - [+1] = [0]$, where only one unique solution is derived. Unlike fault-tree analysis and fuzzy-logic-based fault diagnosis, a priori knowledge about the cause-effect relationships of system faults is not required for QBG based approach [16, 18]. This feature makes the QBG based diagnosis more robust in order to accommodate different unforeseeable faults. However, due to the imprecise characteristic of qualitative representation, this method has difficulty in detecting and localizing incipient faults. Due to the limited discriminatory ability of the qualitative method, it is difficult to deal with the isolation of multiple faults. Moreover, the diagnosis results of QBG based method usually contain a set of suspected faults, which is not enough for some diagnosis tasks. In order to accurately know the fault information, i.e., fault severity, and to achieve fault tolerance performance, numerical models are preferred.

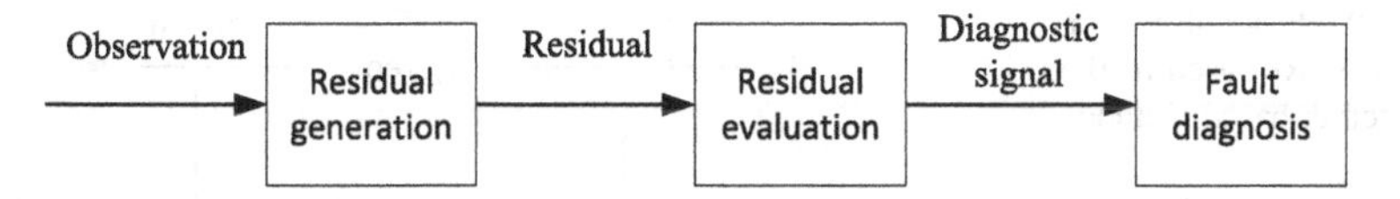

Fig. 1.8 General flowchart of a quantitative model based fault diagnosis method

Quantitative fault diagnosis method checks the consistency between actual system and its behavior model. Consistency checking is usually achieved through a comparison between the information obtained from the real system and information computed from a behavioral model. The resultant differences are called residuals. Each residual should be theoretically zero or near zero when the system is normal, but should distinguishably deviate from zero when a fault happens [19]. A fault is detected by monitoring the trend of the residuals, which usually involves setting a fixed threshold on a residual quantity. The models should be insensitive to modeling errors and at the same time sensitive to faults. In general, quantitative model based fault diagnosis method consists of two main stages: residual generation and residual evaluation as shown in Fig. 1.8.

The residual generation is essentially a procedure for extracting fault symptoms from the system measurements. In the residual evaluation, the trend of the generated residuals are inspected which usually involves setting a fixed threshold on a residual quantity. A well-designed residual makes residual evaluation process simple. Therefore, most of the published work is focused on the residual generation problem due to its higher importance. Residual generation can be developed using different methods, such as observer, parity relations (also called analytical redundancy relations (ARR)) and parameter estimation based methods. These two approaches of residual generation for FDI in real time are shown in Fig. 1.9. When there is a fault occurred in the system, some residuals will deviate sufficiently such that the deviation is distinguishable from the residual response during normal operation.

Observer based fault diagnosis is a well-known analytical model based FDI scheme, which compares the actual output from a system with reference output from an analytical model. An observer-based residual is simply the output estimation error itself or a combination of the output estimation errors. Various nonlinear observer design techniques have been used for residual generation, since no single, universal, optimal nonlinear observer exists for all nonlinear systems. The existing nonlinear observers have to be designed usually under certain assumptions on system structure, system inputs, and/or the degree of the system nonlinearity [20].

For deterministic framework, Hammouri et al. [21] utilized high-gain observers for fault detection of control affine nonlinear systems. Ding and Frank [22] developed adaptive nonlinear observers for fault detection. Sliding-mode observer is a useful tool for fault diagnosis. Edwards et al. [23] used a sliding mode observer to reconstruct faults, with no explicit consideration of the disturbances or uncertainty. Tan and Edwards [24] built on the work in [25] and presented a design algorithm for the observer, using Linear Matrix Inequalities (LMIs), such that the gain from the disturbances to the fault reconstruction is minimized. Ng et al. [26] extended the

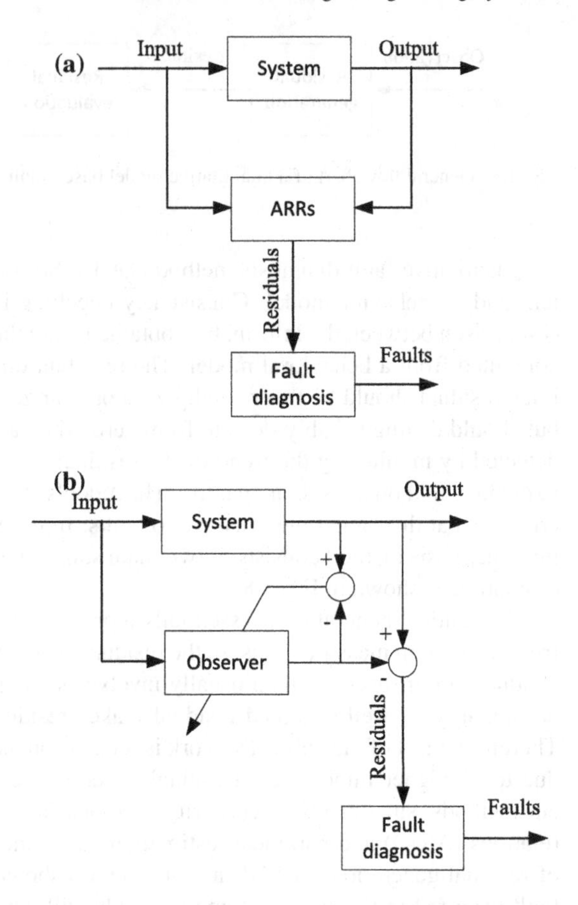

Fig. 1.9 Residual generators. **a** ARRs based method. **b** Observer based method

work of Tan and Edwards [25] to relax the requirement of a full rank first Markov parameter by exploiting two sliding mode observers in cascade; signals from the first observer were considered as outputs of a fictitious second system which has a first Markov parameter of full rank; then using the results in [25], a second sliding mode observer is designed based on the fictitious system to reconstruct the fault. Tan and Edwards [24] is based on the work of [26], i.e., using multiple observers in cascade. However the observer that is used exploits a super-twisting structure which will give a higher degree of accuracy for the fault estimation.

Since the disturbances of system under monitoring are random fluctuations with only their statistical parameters known, one solution to the fault diagnosis problem in such systems is to entail optimal state estimate with minimum estimation error. The Kalman filter in state space model is equivalent to an optimal predictor for a linear stochastic system using input-output model. It is well known that the Kalman filter is a recursive algorithm for state estimation and it has found wide applications in industrial applications. In [27], FDI schemes are constructed using banks of robust two-stage Kalman filters, which simultaneously estimate the state and the fault bias, and generate residual sets. In [28], an extended Kalman filter is proposed for the nonlinear dynamic estimation for an orbiting spacecraft. The developed methodology decides if a sensor fault has happened, locates the faulty sensor, and outputs the healthy sensor measurements. The approach is to design a fast convergence Kalman filter algorithm based on covariance matrix computation for rapid sensor fault detection. An interacting multiple model filters for both partial (soft) and total (hard) faults of reaction wheels in a spacecraft is developed in [29]. Different operating and faulty conditions due to changes and anomalies are considered in each reaction wheel related to the three axes of the satellite. Once a fault mode is detected and isolated, the recovery process can be carried out by invoking appropriate switching control strategies for the attitude control system. Alessandri et al. [30] used extended Kalman filter for detection of actuator faults in unmanned underwater vehicles.

The second method to residual generation is the parity space approach, which relies on analytical redundancy relations (ARR) to link time evolution of the known variables when a system operates according to its normal operation model. The ARRs can be automatically obtained from the model equations using various elimination algorithms [31]. In [32], a parity based fault estimation for nonlinear systems modelled by Takagi-Sugeno fuzzy models is proposed, the parity space approach for linear systems is generalized to Takagi-Sugeno fuzzy systems, and power spectra of the faults are incorporated into the design procedure. The design procedure is given in terms of a family of linear matrix inequalities (LMIs). Bond graph (BG) is a powerful modeling tool to model system in multiple domains. BG based fault diagnosis using ARRs has been successfully applied into many industrial areas [3]. In [33], a robust FDI technique is developed and applied to the traction system of an electric vehicle, in the presence of structured and unstructured uncertainties. Due to the structural and multi-domain properties of the bond graph, the generation of a nonlinear model and residuals for the studied system with adaptive thresholds is synthesized. The parameters and structured uncertainties are identified by using a least-square algorithm. A super-twisting observer is used to estimate both unstructured uncertainties and unknown inputs. A method to the generate the fault indicators and residual thresholds in the presence of parameter uncertainties by using a bond graph representation in linear fractional transformation form is developed in [34]. The residuals' sensitivity analysis, which is based on the fault detectability indexes, is used for residuals evaluation. The developed algorithms are applied on an electromechanical test bench system for on-line fault detection and isolation. A quantitative hybrid bond graph (QHBG) based FDI framework to address the FDI problems for hybrid systems is developed in [35, 36]. The framework efficiently monitors a hybrid system by utilizing the unified information of the HBG. This framework is composed of a GARR alarm generator, a fault detection module, a fault isolation module, a fault estimator, and a mode tracker. These modules are based on the HBG of the hybrid system. The mode tracker is utilized to determine the instantaneous operating mode according to sensor measurement and input information obtained from the hybrid system [37, 38]. This instantaneous mode information allows designers to evaluate the GARRs for residuals effectively at all operating modes. With these residuals, we can detect, isolate, and estimate the size of a fault. The fault detection module determines the occurrence of a fault when any of the residual signals is non-zero. For those isolable faulty components, the GARRs generate a set of unique residuals that allows the faulty parameters to be estimated. For those detectable but non-isolable parameters, the module will select a set of possible faults. Finally, the parameter estimation module estimates the selected possible faults to assess the health condition of the system [39]. This information can also be used to refine the non-isolable parameters. The major drawback of the parity space approach is that the residuals computed need the time derivatives of measured variables, which makes the approach sensitive to measurement noise compared to observer-based methods. Therefore, filtering and pre-processing are required to make it more practical in noisy environment.

The third approach to residual generation is parameter estimation. This approach assumes that the the faults of a dynamical system can be reflected by the physical parameters such as mass, friction, resistance, etc. Faults described as time dependent parameter drifts can be handled through parameter estimation [40]. The most important issue related to parameter estimation method for fault diagnosis is the complexity of the model. If the process model is a complex nonlinear model, then the parameter estimation problem is essentially a nonlinear optimization problem. Real-time application of complex nonlinear optimization problems is a serious bottleneck [13].

Another element of quantitative based method is residual evaluation, which determines whether any faults have occurred by checking the residuals and their trends. The decision making rules usually are designed specifically for different process [41].

Robust fault diagnosis tries to minimize misdetection and false alarms by considering the residual noises. Misdetection means missing to detect the presence of an actually occurred fault. On the contrary, false alarm refers to an indication of fault which in fact does not happen. There are various approaches to generate robust residuals, which are insensitive to modeling uncertainties and measurement noises. One of the robust methods, known as active approach, is based on generating residuals that are insensitive to modeling uncertainties, but sensitive to faults. Some techniques like unknown input observer and robust parity equations are proposed to achieve active robust performance [42–44]. An alternative approach to achieve robust, called as passive, attempts to accomplish robust in the decision making stage. In these methods, the effect of the parameter uncertainty is propagated to the residuals and then an adaptive threshold is used to envelop these residuals to achieve robust [34, 45]. The decision procedure for residual evaluation is improved using the detectability indexes, which represent the ability of the system to detect faults in presence of parameter uncertainties. It is concluded that the robustness in fault detection can be achieved in the residual generation or in the decision making stage. The residual preprocessing of active approach is analytical, while post processing in passive approach is usually numerical. Active method is easy to implement but it can only deal with limited types of system nonlinearity. Passive technique is not constrained by the forms of system nonlinearity; however, it is less robust compared with passive method due to the fact that there may exist some mis-detection of small magnitude faults. Some improvement can be made by using cumulative sums to detect small magnitude faults, but likelihood of false alarm increases. Consideration of tradeoff is then made depending on specific system.

It can be concluded that one of the major advantages of the quantitative model-based method is the ability to incorporate physical understanding of the underlying process into the monitoring scheme. However, several issues such as system nonlinearity, process complexity and lack of accurate data make it difficult, sometimes even impractical, to construct an accurate analytical model for the system. All these factors limit the usefulness of this approach in real industrial applications.

In contrast to the model based methods where a priori quantitative or qualitative knowledge about the system is required, only historical data is required by data driven based approaches [46]. There are different ways in which this data can be represented as a priori knowledge to a diagnostic system. This process is known as feature extraction. Data driven diagnostic approaches can be broadly classified as statistical methods and Artificial Intelligence (AI) methods [47].

Cluster analysis is a multivariate statistical classification method that groups signals into different fault categories based on the similarities found in the characteristics or features they possess. The classification method attempts to minimize within-group variance and maximize between-group variance. A number of heterogeneous groups with homogeneous contents are determined by the cluster analysis. There are substantial differences between the groups, but the signals within the same group are similar. A common way of signal grouping is based on certain distance measures or similarity measure between two signals. These measures are usually described by certain discriminant functions in statistical pattern recognition. Ding et al. [4] introduced a new distance metric called quotient distance for fault diagnosis of engine. Pan et al. [48] developed an extended symmetric Itakura distance for signals in time-frequency domains such as the Wigner-Ville distributions. Other than distance measures, correlation coefficient of feature vector is also a similarity measure commonly used for signal classification in machinery fault diagnosis [49].

A technique termed Support Vector Machine (SVM) is usually employed to optimize a boundary curve in which the distance of the closest point to the boundary curve is maximized. In [50], a novel frequency pattern and competent criterion are introduced for short-circuit-fault recognition in permanent-magnet synchronous motors (PMSMs). The frequency pattern is extracted from the monitored stator current analytically and the amplitude of sideband components at these frequencies is used as a proper criterion to determine the number of short-circuited turns. The occurrence and the number of short-circuited turns are predicted using SVM as a classifier. In view of the bad diagnosing capability of standard Support Vector Classifier (SVC) machine for fault diagnosis pattern series with Gaussian noises, Gaussian function is used as a loss function of SVC and a new SVC based on Gaussian loss function technique, by name g-SVC, is proposed in [51]. The results of its application to car assembly line diagnosis indicate that the diagnosing method is effective and feasible. A dual features functional support vector machine approach that uses both first and second derivatives of degradation profiles for early detection of faulty batteries with the reduced error rate is developed in [52]. The modified floating search algorithm for the repeated feature selection with newly added degradation path points is presented to find a few good features for the enhanced detection while reducing the computation time for online implementation. After that, an attribute sampling plan considering time-varying classification errors is presented to determine the optimal number of test cycles and sample sizes by minimizing the proposed cost function. The proposed method can be applied in a wide range of manufacturing processes to assess time-dependent quality characteristics.

Multivariate statistical approaches are powerful tools which are capable of compressing data and reducing its dimensionality so that essential information is retained and easier to analyze than the original huge data set. In addition, the statistical approaches can also deal with noise and correlation to extract true information effectively. The main objective of multivariate statistical techniques is to transform a number of related process variables to a smaller set of uncorrelated variables. Principal Component Analysis (PCA) is a standard multivariate technique. Generally, PCA is based on an orthogonal decomposition of the covariance matrix of the process variables along directions that represent the maximum variation of the data. The main purpose of the PCA method is to find factors that have a much lower dimension than the original data set which can properly describe the major trends of the original data set. A two-dimensional dynamic PCA is proposed to model and monitor such 2D dynamic batch processes, in which support region determination is a key step [53]. A properly computed support region ensures modeling accuracy, monitoring efficiency, and reasonable fault diagnosis. The support region determination method is implementable in many situations but still has certain limitations. To overcome these shortcomings, a two-dimensional dynamic PCA method with an improved support region determination procedure is developed by considering variable partial correlations and performing iterative stepwise regressions. Such a procedure expands support region batch by batch and is a generalization of the autoregressive model order selection to the 2D batch process cases. Some unique characteristics of semiconductor manufacturing processes have posed challenges for fault diagnosis applications, such as nonlinearity in most batch processes, and multimodal batch trajectories due to product mix. To explicitly handle these unique characteristics, a pattern recognition based fault detection method using the k-nearest-neighbor rule (FD-kNN) is proposed. In FD-kNN, historical data are used directly as the reference of normal process operation to determine whether a new measurement is a fault. For processes with a large number of variables, it can be computation and storage intensive, and hence may be difficult for online process monitoring. To address this difficulty, a fast pattern recognition based fault detection method, termed principal component-based kNN (PC-kNN) is proposed in [54], which takes advantages of both PCA for dimensionality reduction and FD-kNN for nonlinearity and multimode handling.

Artificial Intelligence (AI) makes a mathematical relation between the causes (faults) and the effects (sensor measurement and expert observations). This method is a process of mapping the information obtained in the measurement space and/or features in the feature space to faults in the fault space. AI techniques have been increasingly applied to fault diagnosis and have demonstrated better performance over conventional approaches. In practice, it is difficult to apply AI techniques due to the lack of efficient means to obtain training data and specific knowledge, which are required to train the models. In the literature, two popular AI techniques for fault diagnosis are artificial neural networks and Expert Systems (ESs). Other AI techniques include fuzzy-neural networks, fuzzy logic systems and Evolutionary Algorithms (EAs).

An artificial neural network is a computational model which is able to mimic the human brain structure. It consists of several elements connected in a complex layer structure which enables the network to approximate a complex nonlinear function. The artificial neural network approximates the unknown function by tuning its weights with input and output observations. This process is usually known as training of an artificial neural network. There are different neural network models and feed-forward neural network structure is the most commonly used structure in system fault diagnosis [55, 56]. The advantage of a cascade correlation neural network is that the initial network structure and the number of nodes are not determined. Thus, this method can be applied in condition where online training is preferable. Spoerre [57] applied cascade correlation neural network for bearing fault classification and showed that cascade correlation neural network can result in utilizing the minimum network structure for fault recognition with satisfied accuracy. Wang and Too [58] utilized the unsupervised neural networks, self-organizing map and learning vector quantization for rotating machine fault diagnosis. Tallam et al. [59] developed some self-commissioning and on-line training algorithms for feed-forward neural network with particular application to electric machine fault diagnostics. Two main limitations of neural networks are the lack of physical explanations of the trained model and the difficulty in the training process.

Compared with neural networks, which learn knowledge based on training on data, ESs use domain expert knowledge with an automated inference engine to carry out reasoning to solve problem. Three widely used reasoning methods for ES in fault diagnosis field are (1) rule-based reasoning, (2) case-based reasoning, and (3) model-based reasoning [60].

ESs and neural networks have their own limitations. One main limitation of rule-based ESs is the combinatorial explosion, which leads to the computation problem. This problem exists when the number of rules increases exponentially as the number of variables increases. Another major limitation is the consistency maintenance, which means that the process by which the system decides when some of the variables need to be recomputed in other values. Obviously, combination of both techniques would significantly improve the performance. Yang et al. [61] developed an approach to combine case-based reasoning ES with ART-Kohonen neural network to realize fault diagnosis. It was demonstrated that the proposed method outperforms the self-organizing feature map-based system in terms of classification rate.

EAs, which uses mechanisms inspired by biological evolution, have also been shown to have merits in applications to machine diagnosis. Genetic algorithms (GAs) are the most widely used form of EA. Sampath et al. [62] developed a GA-based optimization method for gas turbine diagnosis. Several examples of other EA algorithms for fault classification and diagnosis can be found in [63, 64].

Data driven methods have been shown to perform well in terms of robustness to system noise. The merit of data driven method lies in its ability to transform high-dimensional data into a low dimension where important information is obtained.

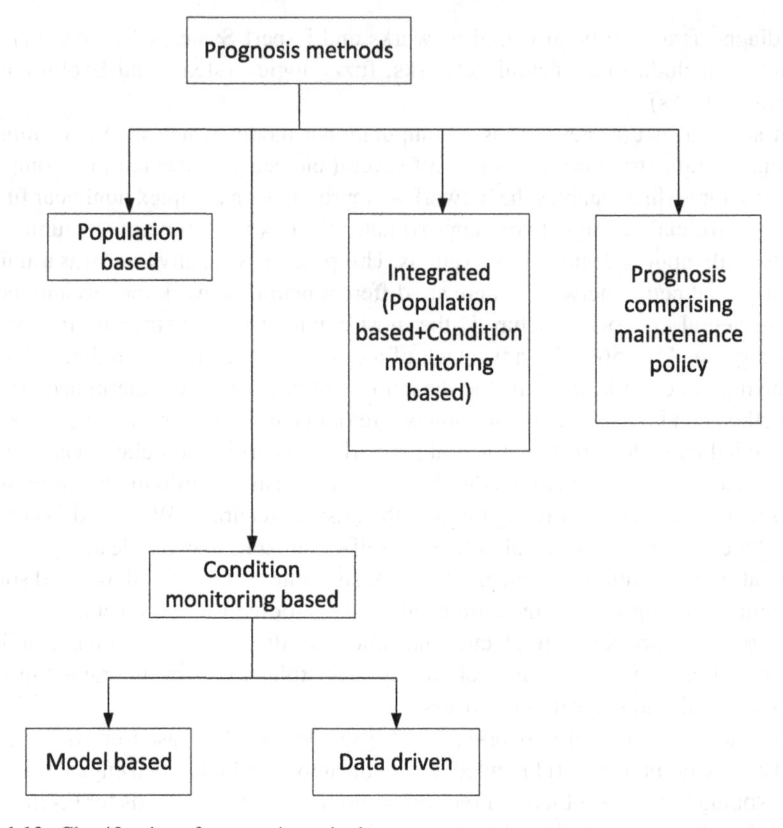

Fig. 1.10 Classification of prognosis methods

There are some limitations to those methods which are based solely on historic process data. One limitation is their generalization capability outside of the training data. Besides its lack of generalization ability, neural networks also have a difficulty in dealing with multiple faults. This limitation leads to an outstanding distinction between model-based approaches and data driven methods.

1.3 Failure Prognosis Tasks and Methodologies

1.3.1 Failure Prognosis Tasks

The term prognosis has been used widely in medical field to describe the likely outcome of an illness. In the industrial arenas, prognosis is interpreted to answer the question, "What is the remaining useful lifetime of a machine or a component once an impending failure condition is detected, isolated, and identified [65]?"

Failure prognosis is to determine how soon and likely a failure will occur. Prognosis could significantly reduce expensive downtime and maintenance costs. However, prognosis is a relatively new research field which has yet to receive its prominence compared to the other aspects of CBM [66]. Recent reviews on prognosis have been reported in the literature [47, 67, 68]. The most widely used prognosis is to predict how much time is left before a failure happens given the current system profile and past operation condition. The time left before a failure occurs is called Remaining Useful Life (RUL).

1.3.2 Failure Prognosis Methodologies

Generally, failure prognosis methodologies can be classified into four categories (as shown in Fig. 1.10):

1. Population based.
2. Condition monitoring based.
3. Integrated approaches, utilizing both population based methods and condition monitoring based methods.
4. Prognosis comprising maintenance policies.

For population based methods, a family of identical components operated under the same conditions is analyzed and the mean-time-to-failure of the whole population of such components is estimated. These approaches basically utilize historical time-to-failure data to estimate the population characteristics. The accuracy of the population based estimation becomes better with increasing amounts of available statistical data. Several failure models, such as Weibull, Poisson and exponential distributions have been used to model failure behaviors of different systems. Weibull analysis is one of the most popular methods utilized in maintenance procedures due to its ability to accommodate various types of failure behavior. All these methods have been extensively investigated over the past few decades and have applications in a wide range of industries such as electronics, automotive, aerospace, power system [69–71]. However, these approaches only provide general overall estimates for the entire population of identical units. This type of estimations is useful to manufacturers that produce units in high volumes but are of little value to end users [66]. For instance, a maintenance engineer would be more interested in the reliability information of a particular component currently running in the machine, rather than in the mean-time-to-failure of the whole population of such a component. To estimate the current condition of an operating component, a more "engineering" approach based on the actual change in component health condition is required. Recent advances in condition monitoring technologies have enabled the collection of non-intrusive degradation measurements of a component in operation. These condition monitoring data contain useful information related to individual component. With these data, prognosis models that estimate the future health of a monitored component are feasible. It is worth to note that the condition monitoring data do not replace the data that reflect population characteristics, because they are corroborative data that reflect the current health of a component in operation.

The second category is condition based technique which can be divided into two types: model-based and data-driven methods. The model-based prognostic method utilizes an analytical model (dynamical model) to describe the system degradation for predicting the remaining time to reach a failure state. Compared with the data driven approach which requires costly data to train the prognosis system, the model-based technique may provide an economical mean to predict RUL. These models usually involve building technically comprehensive mathematical models to describe the physics of the system and failure modes, such as crack propagation in suspension, blockage of pipe in the hydraulic actuator and air leakage of tire in vehicle. These models attempt to combine system-specific mechanical knowledge, defect growth formulas and CBM data to provide "knowledge-rich" prognosis output.

A common model-based approach is crack growth modelling. Li et al. [72] proposed a recursive least squares based prognosis method for rolling element bearing using Paris-Erdogan law. The parameters in the model are tuned by an adaptation method based on features extracted from vibration signal. Least-square scheme enables adaptation of model parameters to changes in condition. Ray and Tangirala [73] adopted a nonlinear stochastic model to describe the fatigue crack dynamics, and the time-dependent damage rate and accumulation in mechanical structures are calculated in a real time manner. An extended Kalman filter was used to predict the remaining service life. However, the state equations for the dynamic system are not considered. Adams and Nataraju [74] modeled damage accumulation in a structural dynamic system as first/second order nonlinear differential equations. However, their damage state equations are functions only of the damage variables. The state variables (reflecting different usage) certainly influence the damage evolution. Chelidze et al. [75] modeled degradation as a "slow-time" process, which is coupled with a "fast-time" observable subsystem. The model was utilized to describe the battery degradation of an electro-mechanical system. Qiu et al. [76] considered the stiffness-based prognostic method for bearing systems which is based on vibration response analysis and damage mechanics. Luo et al. [77] adopted an Interacting Multiple Model (IMM) to track the hidden damage. RUL prediction is carried out by mixing mode-based life predictions according to time-averaged mode probabilities. The failure mode for this example is defined as a crack in the suspension spring caused by fatigue. This failure mode is linked to a suspension parameter (stiffness) that can be estimated (tracked) over time to assess the remaining life of the suspension subsystem. In this example, the post-action is based on prognosis to replace the suspension spring when the crack length reaches a predetermined critical length. Therefore, there are no interim maintenance actions.

A method to predict the RUL of a battery by using model-based prediction approach is developed in [78]. The health of a battery can be indicated by a term named State-of-charge (SOC), which is defined as the ratio between the remaining capacity and the initial or rated capacity. To compute the RUL of a battery, an ARMA model is used to describe the behavior of the battery's SOC. Note that SOC is a slow varying term related to system damage. To compute the parameters of the ARMA model, a randles circuit is used to model the battery so that the necessary parameters that are required for identifying the ARMA model parameters can be computed. Some failure modes of the SOC are passivation (battery does not accept charge), separation, bridging (battery is short-circuited), dry-out, sulfation (the acid density is decreased), softening, that indicates a high discharge rate, a thermal run-away (increase in battery temperature), and grid corrosion. A preliminary SOC prediction method has been proposed based on the ARMA model with the parameters trained for a certain failure mode. The battery is modelled based on its operating principles to make sure the estimations are as close to the actual states as possible. Expert knowledge of the system is critical for model-based prognosis. Some other similar works on battery's SOC estimation can also be found in [79, 80]. In [81, 82], spur gear crack growth is modeled by Paris' law. A 2D Finite Element Analysis (FEA) model was integrated to calculate stress and strain fields based on gear tooth load, geometry and material properties. A stochastic version of the Yu-Harris bearing life equation is used to predict spall initiation and the Kotzalas-Harris progression model is adopted to estimate the time to failure [83]. In order to enhance the above method, a framework for physics-based prognostics which integrates material-level models, system-level data fusion algorithms and parameter tuning techniques is developed in [84].

A single fault growth model might not sufficiently capture a sequence of fault behaviors. Consider, for example, a rolling element bearing as a critical component of rotating machinery. The bearing may begin to corrode under certain operating conditions and, in parallel or sequentially, may be spalling and eventually, cracking. For accurate model-based fault prognosis, it is essential for fault progression models to be developed to capture these evolving behaviors. An approach to multi-fault modeling of prognosis of a rolling element bearing of a helicopter's oil cooler is developed in [85]. A simple and cost-effective on-line parameter adaptation solution is introduced to improve the performance of modeling.

A different way of model-based prognostic methods is to establish an explicit relationship between the condition variables and the lifetimes (current lifetime and failure lifetime) via mechanistic modelling. Two representative examples, [86] for machines considered as energy processors subject to vibration monitoring, and [76] for bearings with vibration monitoring. Lesieutre et al. [87] proposed a hierarchical modelling method to assess RUL. Engel et al. [88] discussed some practical issues related to accuracy, precision and confidence of the RUL prediction.

The main advantage of the model-based approach is its incorporation of physical understanding of the system for health monitoring. Another advantage is that, in some cases, the changes in the feature vector are closely related to model parameters. Thus, the model-based approach can also establish a relationship between the drifting parameters and the features. In addition, as the understanding of the system degradation improves, the model can be modified to increase its accuracy and to handle subtle performance problems. However, since prognosis involves projecting current abnormal condition into the future in the absence of future observations, assumptions and simplifications are often inevitable in prognosis modelling. Nevertheless, care must be taken to minimize these assumptions and simplifications.

Currently, most physical model-based prognosis methods focus on the prediction of crack propagation. However, in many cases other failure modes tend to dominate and the maintenance engineer needs to correctly identify the fault type in question. Crack growth models are also difficult to apply in practice because they require knowledge of the exact geometry or/and orientation of the crack, which are usually very irregular and cannot be identified without disassembling the machine component [66].

Data-driven prognostic methods try to establish models directly from routinely collected data instead of establishing mathematical models based on first principle and human expertise. They are built based on historical data and output predictions directly from data. Related review on statistical data driven based prognosis has been reported in the literature [89]. The conventional data-driven methods include simple projection models, such as exponential smoothing [90] and auto-regressive model [91]. One major advantage of these techniques is the simplicity of calculations. However, most of these trend prediction techniques assume that there is some underlying stability in the monitored system. They also rely on past observations of degradation to project future degradation. ANN is the most widely used data-driven approach in the prognosis literature. Wang et al. [92] utilized recurrent neural networks and neural-fuzzy inference systems to predict the RUL in rotary machinery. The neural-fuzzy inference systems performed better than the recurrent neural networks when the training data is sufficient. However, it could not provide reliable prediction if there was no sufficient training data or there were fast dynamic fluctuations. In order to overcome these problems, an adaptive training technique was proposed by Wang [93] to improve the neural-fuzzy model. Gebraeel et al. [94] developed a method to predict the actual bearing failure time instead of the future condition index considered in [92, 93].

In [95], a prognostic framework based upon the concept of dynamic wavelet neural networks is developed. The dynamic wavelet neural network incorporates temporal information and storage capacity into their functionality so they can carry out prognostic tasks for future. Both constructs are based on a dynamic wavelet neural networks model which functions as the mapping tool. A dynamic wavelet neural networks based degradation model that utilizes condition-based sensory signals to calculate and continuously update residual life distributions of partially degraded components is developed in [96]. Initial predicted failure times are estimated through "supervised" trained dynamic wavelet neural networks using real-time sensory signals. These estimates are used to derive a prior failure time distribution for the component that is being monitored and update the prior distributions by using a Bayesian approach. The novelty of this methodology lies in the ability to update a component's remaining life distribution using condition-based sensory signals. The real-time sensory signals capture the latest degradation state of the component, and the resulting updated distributions are directly linked to the physical degradation state of the component.

Sheppard and Kaufman [97] developed a diagnostic approach based on Bayesian networks that incorporates information on failure probability, instrument uncertainty, and the predictions for false indication. They also used a dynamic Bayesian network, which is an extension of the Bayesian network, to perform prognosis by modeling changes over time. Bunks et al. first pointed out that Hidden Markov Models (HMM) could be applied in the area of failure prognosis in machining processes [98]. HMM is a parametric model where its parameters can be estimated by the experimental data using statistical techniques. HMMs have some distinct features that are not possessed by most traditional methods. They are not only able to reflect the randomness of machine behaviors but also reveal their hidden states and changing processes. Furthermore, HMMs have a well-established theoretical basis and easy to realize in software. However, HMMs have some inherent limitations. One is the assumption that successive system behavior observations are independent. The other limitation is the Markov assumption where the probability in a given state at time t only depends on the state at time $t - 1$ is clearly untenable in most practical applications. Baruah and Chinnam presented a novel method to employ HMM for performing both diagnosis and prognosis [99]. This method applies HMM for modeling sensor signals and, in turn, identifies the health states as well as estimates RUL.

In [100], an on-line particle filtering based framework for fault diagnosis and failure prognosis is developed. This framework consists of two autonomous modules, a FDI module and a failure prognostic module. The FDI module utilizes a particle filtering to estimate the probability density function of the system state and calculates the probability of a fault condition in a real time manner. Once the abnormal condition is detected, the estimated state probability density function are used as initial conditions in the prognosis module. The failure prognostic module predicts the evolution of the fault indicator and calculates the RUL of the faulty subsystem, using a non-linear state-space model with unknown time-varying parameters and a particle filtering which updates the current state estimate. The proposed method is validated through the data from a seeded fault test for a UH-60 planetary gear plate.

PCA is usually used to extract the significant characteristics of structural failures out of the sensors. Its main function is to retain the most important characteristics of its inputs by using a small amount of data. The advantage of PCA is that it significantly reduces the dimension of input data to enhance training speed and recognition speed. Furthermore, PCA has the self-learning capability. PCA methods have been applied to many applications in image processing, signal processing, and pattern recognition. Recently, it has also been introduced into the prognosis for the data preprocess purpose. Zhang et al. utilized PCA for the principal signal features extraction work that could help to generate a component's health/degradation index as the input of an on-line RUL prediction [101]. Kwan and Zhang employed PCA for the data extraction process in their fault diagnostics and prognostic model [102].

Data-driven methods have some advantages, such as capturing complicated phenomenon without a priori knowledge, and performing faster than traditional system identification techniques in multivariate prognosis for artificial neural network. However, the linkage between fault degradation and changes in physical parameters is generally lost with data-driven models. Moreover, it is a tough and high cost task to collect abundant data. Many algorithms and data training models require a large amount of historic data, including normal state data and failure (even fatal failure) data that need to destroy the components/systems artificially.

The third category for failure prediction is the integration methods which utilize both information from population based approaches and condition monitoring based approaches. Since condition monitoring information usually reflects the state of individual operating units, it does not replace the reliability data that reflect population characteristics. Condition monitoring data mainly provide information for short-term condition prediction only. These techniques based on condition monitoring information require further research because prognosis with such short a prediction horizon is not useful for optimal maintenance scheduling. Sufficiently long lead time is often required for effective and economical preparation of spare units and human resources. A longer prediction horizon is also necessary for deciding whether a unit will last until the next maintenance action. On the other hand, these approaches based on population data provide general estimations for the whole population of identical components to facilitate time-based maintenance. Maintenance or repair at pre-established intervals tends to incur even higher scheduled downtime. Besides, failure behavior of each component is a function of changes in work schedule, operating environment and other duty parameters, and of failure interaction between components. Therefore, current condition of an operating component needs to be monitored online [66]. The integration methods generally produce longer-range failure forecasts than methods that only use individual component condition information.

Mazhar et al. [103] developed a two-step approach. The first stage applies the Weibull analysis to the time-to-failure data to assess the mean life of components. In the second stage, an artificial neural network model is used to analyze the degradation and condition monitoring data. Finally, the Weibull analysis and the developed artificial neural network model are integrated to assess the remaining useful life of components for reuse. Goode et al. [104] introduced a statistical method to predict the RUL of pumps in a hot strip steel mill. Alarm limits were first determined using the statistical process control theory, with the assumption that healthy state data follow a normal distribution. The "installation to potential failure" and the "potential failure to functional failure" intervals were represented using Weibull distribution. Time to failure was calculated by taking into vibration data account. This work presented a relatively simple approach to forecasting failure with utilization of both population and condition monitoring data.

Heng et al. [105] proposed an intelligent reliability model named the intelligent product limit estimator, which was able to include suspended condition monitoring data in machinery fault prognosis. The accurate modeling of suspended data was found to be of great importance, since in practice machines are rarely allowed to run to failure and data are commonly suspended. The model includes a neural network whose training targets are asset survival probabilities estimated using a variation of the Kaplan-Meier estimator and the true survival status of historical units. This work presented a concept of utilizing available information more fully and accurately, and of providing longer-range prediction in a probabilistic sense with minimal assumptions.

The last category for failure prediction is the prognosis comprising maintenance policies [47]. The purpose of prognosis is to provide decision support for maintenance actions. As a result, it is natural to include maintenance policies when considering the machine prognosis process. This makes the situation more complex since extra effort is required to describe the nature of maintenance policies. Maintenance in this situation is called CBM. The main idea behind prognosis incorporating maintenance policies is to optimize the maintenance policies based on certain criteria such as cost, risk, availability and reliability.

Wang [106] proposed a CBM model using a random coefficient growth model where the coefficients of the regression growth model are assumed to follow known distribution functions. The model was utilized to determine the optimal critical level and inspection interval in CBM according to a criterion of interest, which can be downtime, reliability, or cost. Dieulle et al. [107] considered the maintenance of a technical device subject to a continuous-time random deterioration, and monitored through perfect inspections. Once the monitored condition exceeds a pre-set critical threshold, the device is considered as too much wear and tear and a preventive replacement is triggered. When the device condition reaches a failure level, a breakdown occurs with all the associated negative effects such as an expensive unplanned replacement has to be carried out, the unexpected unavailability of the device can cause production losses or delay. Thus the objective of a maintenance policy is to avoid failure occurrence at the lowest cost. This work assumed a one-level replacement policy, and then acquired the optimal threshold and inspection scheduling by minimizing the global cost per unit time.

1.4 Organization of the Book

This book is devoted to introduce and discuss various issues related to hybrid bond graph based hybrid system health monitoring. To treat these advanced methods in hybrid system health monitoring, the text of the book is divided into seven chapters.

Chapter 1 presents an introduction to the basic concepts of fault diagnosis and failure prognosis together with their classifications. The state of the art techniques developed in the published works are also discussed.

Chapter 2 describes the concept of hybrid system and different modeling methods for hybrid system are discussed as well. The hybrid bond graph modeling technique is introduced in this chapter.

Chapter 3 considers the hybrid bond graph based FDI. In this chapter, first, the bond graph based fault diagnosis for continuous system using ARRs is introduced. Two existing Bond Graph (BG)-based methods to generate symbolic ARRs, i.e., covering path method and causality inversion method, are introduced. An integrated strategy which combines the gists of both symbolic ARR generation methods is developed. The ARR method has been extended to hybrid systems, based on a new concept of Global Analytical Redundancy Relations (GARRs). In order to avoid causality reassignment due to changes in configuration through the various operating modes, two causality assignment methods from fault diagnosis perspective are introduced to achieve a HBG with a desirable causality assignment that leads to a unified description of system's behavior (i.e., GARRs). These results are discussed in this chapter.

Chapter 4 discusses the problem of fault parameter estimation for hybrid system. The estimation with and without mode change information are considered. Several examples, including automotive suspension and electrical circuit, are studied using the proposed methods.

Chapter 5 presents mode tracking techniques for hybrid system based on hybrid bond graph. In this chapter, two mode tracking techniques are introduced. The first method is based on mode-change signatures, it demands less computing resources but it can be used only until fault detection. The second method which is based on ARRs requires more computing resources but it can be used after the fault detection and isolation.

Chapter 6 considers a real world example, i.e., a mobile robot test-bed. The methodologies developed in the previous chapters are applied to monitor the steering system of an electric vehicle. Fault detection, isolation, and estimation are applied, experimental setup is described, and real time supervision results are discussed

Chapter 7, the final chapter, illustrates some results about failure prognosis for hybrid system. First, failure prognosis based on augmented global analytical redundancy relations is developed and verified both in simulation and experiment. Multiple faults are considered in this chapter and these faults may be developed when the system is at a non-detectable mode. In addition, failure prognosis based on dynamic fault isolation is introduced. In the dynamic fault isolation scheme, a waiting time is introduced during decision making process to handle the situation in which not all the faults are developed when the system is at a detectable mode.

References

1. J. Lee, J. Ni, D. Djurdjanovic, H. Qiu, Intelligent prognostics tools and e-maintenance. Comput. Ind. **57**(6), 476–489 (2006)
2. J.H. Luo, M. Namburu, K.R. Pattipati, L. Qiao, S. Chigusa, Integrated model-based and data-driven diagnosis of automotive antilock braking systems. IEEE Trans. Syst. Man Cybern. A Syst. Hum. 40(2), 321–336 (2010)
3. A.K. Samantaray, B. Ould Bouamama, *Model-Based Process Supervision: A Bond Graph Approach* (Springer, London, 2008)
4. S.X. Ding, *Model-Based Fault Diagnosis Techniques: Design Schemes, Algorithms, and Tools* (Springer, Berlin, 2008)
5. C. Ghiaus, Fault diagnosis of air conditioning systems based on qualitative bond graph. Energy Build. **30**(3), 221–232 (1999)
6. B. Kuipers, Qualitative simulation. Artif. Intell. **29**(3), 289–338 (1986)
7. B. Kuipers, Qualitative simulation as causal explanation. IEEE Trans. Syst. Man Cybern. **17**(3), 432–445 (1987)
8. H.T. Ng, Model-based multiple fault diagnosis of time-varying, continuous physical devices. In *the Sixth Conference on Artificial Intelligence Applications*, vol. 1, May 1990, pp. 9–15
9. S. Subramanian, R. J. Mooney, Qualitative multiple-fault diagnosis of continuous dynamic systems using behavioral modes. In *the 13th National Conference on, Artificial Intelligence*, August 1996, pp. 965–970

10. Q. Shen, R.R. Leitch. Qualitative model based of continuous dynamic systems. In *Proceedings of the First International Conference on Intelligent Systems Engineering*, Heriot-Watt University, Edinburgh, 1992, pp. 147–152
11. E.E. Hurdle, L.M. Bartlett, J.D. Andrews, System fault diagnostics using fault tree analysis. J. Risk Reliab., Part O (Proc. IMechE) **221**(1), 43–55 (2007)
12. M.G.M. Madden, P.J. Nolan, Monitoring and diagnosis of multiple incipient faults using fault tree induction. IEE Proc.-Control Theory Appl. 146(2), 204–212 (1999)
13. V. Venkatasubramanian, R. Rengaswamy, K. Yin, S.N. Kavuri, A review of process fault detection and diagnosis: part I: quantitative model-based methods. Comput. Chem. Eng. **27**(3), 293–311 (2003)
14. S. Xia, D.A. Linkens, S. Bennette, Automatic modeling and analysis of dynamical physical systems using qualitative reasioning and bond graphs. Intell. Syst. Eng. **2**(4), 201–212 (1993)
15. D. A. Linkens, H. Wang, Fault diagnosis based on a qualitative bond graph model, with emphasis on fault localization. In *IEE Colloquium on Qualitative and Quantitative Modeling Methods for Fault Diagnosis*, pp. 1–6, 1995
16. C.H. Lo, Y.K. Wong, A.B. Rad, Model-based fault diagnosis in continuous dynamic systems. ISA Trans. **43**(3), 459–475 (2004)
17. C.H. Lo, Y.K. Wong, A.B. Rad, Intelligent system for process supervision and fault diagnosis in dynamic physical systems. IEEE Trans. Ind. Electron. **53**(2), 581–592 (2006)
18. C.H. Lo, Y.K. Wong, A.B. Rad, Fusion of qualitative bond graph and genetic algorithms: a fault diagnosis application. ISA Trans. **41**(4), 445–456 (2002)
19. J. Chen, R. Patton, *Robust Model-Based Fault Diagnosis for Dynamic Systems* (Kluwer Academic Publishers, Boston, 1999)
20. S. Ehsan, K. Khashayar, *Fault Diagnosis of Nonlinear Systems Using a Hybrid Approach* (Springer, New York, 2009)
21. H. Hammouri, M. Kinnaert, E.H. El Yaagoubi, Observer-based approach to fault detection and isolation for nonlinear systems. IEEE Trans. Autom. Control **44**(10), 1879–1884 (1999)
22. X. Ding, P.M. Frank, An adaptive observer-based fault detection scheme for nonlinear systems. In *Proceedings of the 12th IFAC World Congress*, Sydney, Australia, pp. 63–68, 1993
23. C. Edwards, S.K. Spurgeon, R.J. Patton, Sliding mode observers for fault detection and isolation. Automatica **36**(4), 541–553 (2000)
24. C.P. Tan, C. Edwards, Robust fault reconstruction in uncertain linear systems using multiple sliding mode observers in cascade. IEEE Trans. Autom. Control **55**(4), 855–867 (2010)
25. C.P. Tan, C. Edwards, Sliding mode observers for robust detection and reconstruction of actuator and sensor faults. Int. J. Robust Nonlinear Control **13**(5), 443–463 (2003)
26. K.Y. Ng, C.P. Tan, C. Edwards, Y.C. Kuang, New results in robust actuator fault reconstruction in linear uncertain systems using sliding mode observers. Int. J. Robust Nonlinear Control **17**(14), 1294–1319 (2007)
27. J.H. Lee, J. Lyou, Fault diagnosis and fault tolerant control of linear stochastic systems with unknown inputs. Syst. Sci. **27**(3), 59–76 (2001)
28. J. Sanjay, A new fast converging Kalman filter for sensor fault detection and isolation. Sens. Rev. **30**(3), 219–224 (2010)
29. N. Tudoroiu, K. Khorasani, Satellite fault diagnosis using a bank of interacting Kalman filters. IEEE Trans. Aerosp. Electron. Syst. **43**(4), 1334–1350 (2007)
30. A. Alessandri, M. Caccia, G. Veruggio, Fault detection of actuator faults in unmanned underwater vehicles. Control Eng. Pract. **7**(3), 357–368 (1999)
31. M. Staroswiecki, G. Comtet-Varga, Analytical redundancy relations for fault detection and isolation in algebraic dynamic systems. Automatica **37**(5), 687–699 (2001)
32. S.K. Nguang, P. Zhang, S. Ding, Parity based fault estimation for nonlinear systems: An LMI approach. In *Proceedings of American Control Conference*, Minneapolis, USA, 2006, pp. 5141–5146
33. M.A. Djeziri, R. Merzouki, B. Ould Bouamama, Robust monitoring of an electric vehicle with structured and unstructured uncertainties. IEEE Trans. Veh. Technol. **58**(9), 4710–4719 (2009)

34. M.A. Djeziri, R. Merzouki, B. Ould-Bouamama, G. Dauphin-Tanguy, Robust fault diagnosis by using bond graph approach. IEEE/ASME Trans. Mechatron. **12**(6), 599–611 (2007)
35. C.B. Low, D. Wang, S. Arogeti, M. Luo, Quantitative hybrid bond graph-based fault detection and isolation. IEEE Trans. Autom. Sci. Eng. **7**(3), 558–569 (2010)
36. C.B. Low, D. Wang, S. Arogeti, J.B. Zhang, Causality assignment and model approximation for hybrid bond graph: fault diagnosis perspectives. IEEE Trans. Autom. Sci. Eng. **7**(3), 570–580 (2010)
37. S. Arogeti, D. Wang, C. B. Low, D. H. Zhang, J. Zhou. Mode tracking of hybrid systems in FDI framework. In *the 3rd International Conference on Robotics Automation and, Mechatronics (RAM2008)*, 2008, pp. 841–846
38. S. Arogeti, D. Wang, C.B. Low. Mode tracking and FDI of hybrid systems. In *the 10th International Conference on Control, Automation*, 2008, pp. 112–118
39. C.B. Low, D. Wang, S. Arogeti, M. Luo. Fault parameter estimation for hybrid systems using hybrid bond graph. In *the 3rd IEEE Multi-conference on Systems and Control (MSC 2009)*, 2009, pp. 1338–1343
40. R. Isermann, Fault diagnosis via parameter estimation and knowledge processing. Automatica **29**(4), 815–835 (1994)
41. I. Hwang, S. Kim, Y. Kim, C.E. Seah, A survey of fault detection, isolation, and reconfiguration methods. IEEE Trans. Control Syst. Technol. **18**(3), 636–653 (2010)
42. J. Chen, R. Pattern, H. Zhang, Design of unknown input observers and robust fault detection filters. Int. J. Control **63**(1), 85–105 (1996)
43. S.F. Lin, A.P. Wang, Design of observers with unknown inputs using eignstructure assignment. Int. J. Syst. Sci. **31**(6), 705–711 (2000)
44. D.T. Horak, Failure detection in dynamic systems with modelling errors. J. Guidance Conrol Dyn. **131**(6), 508–516 (1998)
45. Z. Shi, F. Gu, B. Lennox, A.D. Ball, The development of an adaptive threshold for model-based fault detection of a nonlinear electro-hydraulic system. Control Eng. Pract. **13**(11), 1357–1367 (2005)
46. V. Venkatasubramanian, R. Rengaswamy, S.N. Kavuri, K. Yin, A review of process fault detection and diagnosis Part III: process history based methods. Comput. Chem. Eng. **27**(3), 327–346 (2003)
47. A.K.S. Jardine, D. Lin, D. Banjevic, A review on machinery diagnostics and prognostics implementing condition-based maintenance. Mech. Syst. Signal Process. **20**(7), 1483–1510 (2006)
48. M.C. Pan, P. Sas, H. Van Brussel, Machine condition monitoring using signal classification techniques. J. Vib. Control **9**(10), 1103–1120 (2003)
49. X. Lou, K.A. Loparo, Bearing fault diagnosis based on wavelet transform and fuzzy inference. Mech. Syst. Signal Process. **18**(5), 1077–1095 (2004)
50. B.M. Ebrahimi, J. Faiz, Feature extraction for short-circuit fault detection in permanent-magnet synchronous motors using stator-current monitoring. IEEE Trans. Power Electron. **25**(10), 2673–2682 (2010)
51. Q. Wu, Fault diagnosis model based on Gaussian support vector classifier machine. Expert Syst. Appl. **37**(9), 6251–6256 (2010)
52. J.I. Park, S.H. Baek, M.K. Jeong, S.J. Bae, Dual features functional support vector machines for fault detection of rechargeable batteries. IEEE Trans. Syst. Man Cybern. Part C Appl. Rev. 39(4), 480–485 (2009)
53. Y. Yao, F.R. Gao, Statistical monitoring and fault diagnosis of batch processes using two-dimensional dynamic information. Ind. Eng. Chem. Res. **49**(20), 9961–9969 (2010)
54. Q.P. He, J. Wang, Large-scale semiconductor process fault detection using a fast pattern recognition-based method. IEEE Trans. Semicond. Manuf. **23**(2), 194–200 (2010)
55. B. Li, M.Y. Chow, Y. Tipsuwan, J.C. Hung, Neural-network-based motor rolling bearing fault diagnosis. IEEE Trans. Ind. Electron. **47**(5), 1060–1069 (2000)
56. Y. Fan, C.J. Li, Diagnostic rule extraction from trained feedforward neural networks. Mech. Syst. Signal Process. **16**(6), 1073–1081 (2002)

57. J.K. Spoerre, Application of the cascade correlation algorithm (CCA) to bearing fault classification problems. Comput. Ind. **32**(3), 295–304 (1997)
58. C.C. Wang, G.P.J. Too, Rotating machine fault detection based on HOS and artificial neural networks. J. Intell. Manuf. **13**(4), 283–293 (2002)
59. R.M. Tallam, T.G. Habetler, R.G. Harley, Self-commissioning training algorithms for neural networks with applications to electric machine fault diagnostics. IEEE Trans. Power Electron. **17**(6), 1089–1095 (2002)
60. M. Stanek, M. Morari, K. Frohlich, Model-aided diagnosis: An inexpensive combination of model-based and case-based condition assessment. IEEE Trans. Syst. Man Cybern. Part C Appl. Rev. **31**(2), 137–145 (2001)
61. B.S. Yang, T. Han, Y.S. Kim, Integration of ART-Kohonen neural network and case-based reasoning for intelligent fault diagnosis. Expert Syst. Appl. **26**(3), 387–395 (2004)
62. S. Sampath, S. Ogaji, R. Singh, D. Probert, Engine-fault diagnostics: an optimisation procedure. Appl. Energy **73**(1), 47–70 (2002)
63. B. Samanta, C. Nataraj, Use of particle swarm optimization for machinery fault detection. Eng. Appl. Artif. Intell. **22**(2), 308–316 (2009)
64. Q. Wu, Car assembly line fault diagnosis based on robust wavelet SVC and PSO. Expert Syst. Appl. **37**(7), 5423–5429 (2010)
65. G. Vachtsevanos, F. Lewis, M. Roemer, A. Hess, B. Wu, *Intelligent Fault Diagnosis and Prognosis for Engineering Systems* (John Wiley & Sons Inc., New Jersey, 2006)
66. A. Heng, S. Zhang, A.C.C. Tan, J. Mathew, Rotating machinery prognostics: state of the art, challenges and opportunities. Mech. Syst. Signal Process. **23**(3), 724–739 (2009)
67. V. Venkatasubramanian, Prognostic and diagnostic monitoring of complex systems for product lifecycle management: challenges and opportunities. Comput. Chem. Eng. **29**(6), 1253–1263 (2005)
68. M. Luo, D. Wang, M. Pham. Model-based fault diagnosis/prognosis for wheeled mobile robots: a review. In *the 31st Annual Conference of IEEE*, 2005, pp. 2267–2272
69. A. Schomig, O. Rose, On the suitability of the Weibull distribution for the approximation of machine failures. In *Proceedings of the Industrial Engineering Research Conference*, Portland, USA, 2003
70. P.G. Groer, Analysis of time-to-failure with a Weibull model. In *Proceedings of the Maintenance and Reliability Conference*, Knoxville, USA, 2000, pp. 5901–5904
71. W. Li, Evaluating mean life of power system equipment with limited end-of-life failure data. IEEE Trans. Power Syst. **19**(1), 236–242 (2004)
72. Y. Li, T.R. Kurfess, S.Y. Liang, Stochastic prognostics for rolling element bearings. Mech. Syst. Signal Process. **14**(5), 747–762 (2000)
73. A. Ray, S. Tangirala, Stochastic modeling of fatigue crack dynamics for on-line failure prognostics. IEEE Trans. Control Syst. Technol. **4**(4), 443–451 (1996)
74. D.E. Adams, M. Nataraju, A nonlinear dynamical systems framework for structural diagnosis and prognosis. Int. J. Eng. Sci. **40**(17), 1919–1941 (2002)
75. D. Chelidze, J.P. Cusumano, A. Chatterjee, Dynamical systems approach to damage evolution tracking-Part I: The experimental method. J. Vibr. Acoust. Trans. ASME **124**(2), 250–257 (2002)
76. J. Qiu, B.B. Set, S.Y. Liang, C. Zhang, Damage mechanics approach for bearing lifetime prognostics. Mech. Syst. Signal Process. **16**(5), 817–829 (2002)
77. J.H. Luo, K.R. Pattipati, L. Qiao, S. Chigusa, Model-based prognostic techniques applied to a suspension system. IEEE Trans. Syst. Man Cybern. Part A Syst. Hum. 38(5), 1156–1168 (2008)
78. J.D. Kozlowski, C.S. Byington, A.K. Garga, M.J. Watson, T.A. Hay, Model-based predictive diagnostics for electrochemical energy sources. In *Proceedings of the IEEE Aerospace Conference*, vol. 6, 2001, pp. 3149–3164
79. J.H. Aylor, A. Thieme, B.W. Johnson, A battery state-ofcharge indicator for electric wheelchairs. IEEE Trans. Ind. Electron. **39**(5), 398–409 (1992)

80. O. Caumont, P.L. Moigne, C. Rombaut, X. Muneret, P. Lenain, Energy gauge for lead-acid batteries in electric vehicles. IEEE Trans. Energy Convers. **15**(3), 354–360 (2000)
81. C.J. Li, S. Choi, Spur gear root fatigue crack prognosis via crack diagnosis and fracture mechanics. In *Proceedings of the 56th Meeting of the Society of Mechanical Failures Prevention Technology*, Virginia Beach, 2002, pp. 311–320
82. C.J. Li, H. Lee, Gear fatigue crack prognosis using embedded model, gear dynamic model and fracture mechanics. Mech.l Syst. Signal Process. **19**(3), 836–846 (2005)
83. R.F. Orsagh, J. Sheldon, C.J. Klenke, Prognostics/diagnostics for gas turbine engine bearings. In *Proceedings of the IEEE Aerospace Conference*, vol. 7, pp. 3095–3103
84. G.J. Kacprzynski, A. Sarlashkar, M.J. Roemer, A. Hess, B. Hardman, Predicting remaining life by fusing the physics of failure modeling with diagnostics. JOM J. Miner. Met. Mater. Soc. **56**(2), 29–35 (2004)
85. B. Zhang, C. Sconyers, R. Patrick, G. Vachtsevanos, A multi-fault modeling approach for fault diagnosis and failure prognosis of engineering systems. In*Annual Conference of the Prognostics and Health Management Society*, 2009
86. C. Cempel, H.G. Natke, M. Tabaszewski, A passive diagnostic experiment with ergodic properties. Mechan. Syst. Signal Process. **11**(1), 107–117 (1997)
87. G.A. Lesieutre, L. Fang, U. Lee, Hierarchical failure simulation for machinery prognostics. In *Joint Conference of the 51st Meeting of the SMFPT / 12th Biennial Conference on Reliability, Stress Analysis and Failure Prevention (RSAFP Committe of ASME)*, 1997, pp. 103–110
88. S.J. Engel, B.J. Gilmartin, K. Bongort, A. Hess, Prognostics, the real issues involved with predicting life remaining. In *Proceedings of IEEE Aerospace Conference*, vol. 6, 2000, pp. 457–469
89. X.S. Si, W.B. Wang, C.H. Hua, D.H. Zhou, Remaining useful life estimation–a review on the statistical data driven approaches. Eur. J. Oper. Res. **213**(1), 1–14 (2011)
90. W. Batko. Prediction Method in Technical Diagnostics. PhD Thesis, Cracov Mining Academy, 1984
91. K. Kazmierczak. Application of autoregressive prognostic techniques in diagnostics. In *Proceedings of the Vehicle Diagnostics Conference*, 1983, pp. 128–134
92. W. Wang, F. Golnaraghi, F. Ismail, Prognosis of machine health condition using neuro-fuzzy systems. Mech. Syst. Signal Process. **18**(4), 813–831 (2004)
93. W. Wang, An adaptive predictor for dynamic system forecasting. Mech. Syst. Signal Process. **21**(2), 809–823 (2007)
94. N. Gebraeel, M. Lawley, R. Liu, V. Parmeshwaran, Residual life predictions from vibration-based degradation signals: a neural network approach. IEEE Trans. Ind. Electron. **51**(3), 694–700 (2004)
95. P. Wang, G. Vachtsevanos, Fault prognostics using dynamic wavelet neural networks. Artif. Intell. Eng. Des. Anal. Manuf. **15**(4), 349–365 (2001)
96. N. Gebraeel, M. Lawley, A neural network degradation model for computing and updating residual life distributions. IEEE Trans. Autom. Sci. Eng. **5**(1), 154–163 (2008)
97. J.W. Sheppard, M.A. Kaufman, Bayesian diagnosis and prognosis using instrument uncertainty. In *Proceedings of, AUTOTESTCON*, 2005, pp. 417–423
98. C. Bunks, D. McCarthy, T. Al-Ani, Condition-based maintenance of machines using hidden Markov models. Mech. Syst. Signal Process. **14**(4), 597–612 (2000)
99. P. Baruah, R.B. Chinnam, HMMs for diagnostics and prognostics in machining processes. Int. J. Prod. Res. **43**(6), 1275–1293 (2005)
100. M. Orchard, G. Vachtsevanos, A particle filtering approach for on-line fault diagnosis and failure prognosis. Trans. Inst. Meas. Control **31**(3), 221–246 (2009)
101. X.D. Zhang, R. Xu, K. Chiman, S.Y. Liang, Q.L. Xie, L. Haynes, An integrated approach to bearing fault diagnostics and prognostics. In *Proceedings of the American Control Conference*, vol. 4, 2005, pp. 2750–2755
102. C. Kwan, X. Zhang, R. Xu, L. Haynes, A novel approach to fault diagnostics and prognostics. In *Proceedings of the IEEE International Conference on Robotics and Automation*, vol. 1, 2003, pp. 604–609

103. M.I. Mazhar, S. Kara, H. Kaebernick, Remaining life estimation of used components in consumer products: life cycle data analysis by Weibull and artificial neural networks. J. Oper. Manage. **25**(6), 1184–1193 (2007)
104. K.B. Goode, J. Moore, B.J. Roylance, Plant machinery working life prediction method utilizing reliability and condition-monitoring data. In *Proceedings of Institution of Mechanical Engineers*, vol. 214, 2000, pp. 109–122
105. A. Heng, A. Tan, J. Mathew, B.S. Yang, Machine prognosis with full utilization of truncated lifetime data. In *Proceedings of the Second World Congress on Engineering Asset Management*, Harrogate, UK, 2007, pp. 775–784
106. W. Wang, A model to determine the optimal critical level and the monitoring intervals in condition-based maintenance. Int. J. Prod. Res. **38**(6), 1425–1436 (2000)
107. L. Dieulle, C. Berenguer, A. Grall, M. Roussignol, Sequential condition-based maintenance scheduling for a deteriorating system. Eur. J. Oper. Res. **150**(2), 451–461 (2003)

Chapter 2
Hybrid Systems and Hybrid Bond Graph Models

2.1 Hybrid Systems

In general, the term "hybrid" refers to heterogeneous in nature and "hybrid system" means a system which consists of interacting continuous and discrete dynamics [1, 2]. The discrete states are usually represented by modes. Figure 2.1 shows a diagram of hybrid systems. Discrete events (transitions) trigger the system to undergo from one mode to another. At each mode, the system is governed by continuous dynamics, and different modes correspond to different continuous dynamic models. For continuous dynamic systems, the state variables continuously change through time and hence are considered as time-driven. The possible values of these continuous state variables are real number, and differential equations are the general tools to model continuous systems. In contrast to continuous dynamic systems, discrete event dynamic systems are discrete-state, event-driven systems where the state evolution depends entirely on the occurrence of instantaneous discrete events. These systems contain solely of discrete state spaces and event-driven state transition mechanisms [3, 4]. The state variables of discrete event dynamic systems remain constant between events and are evolved in a discontinuous way by the occurrence of discrete events. Examples of instantaneous events like switching on and off a valve or acquisition roll contacting the paper of a printer. The possible values of discrete state variables take from an enumerable set of values. Hybrid dynamical systems possess state variables of both continuous and discrete values, i.e., state changes either continuously or discretely. In other words, hybrid dynamical systems can be treated as both time-driven and event-driven [5, 6]. The transition in hybrid dynamical systems could happen autonomously as a result of the continuous evolution of system variables or due to a discrete event such as a control command (a jump from one mode to another).

In real world applications, there are many examples of hybrid systems such as high speed printer, automotive, switched mode power converter and so forth [7, 8]. For example, the feeding motor of a printer can be in a ramp-up mode, a rotational constant velocity mode, a ramp-down mode, or an idle mode [9, 10]. Each of these modes is governed by a different time-driven continuous model. On the other hand,

D. Wang et al., *Model-based Health Monitoring of Hybrid Systems*,

DOI: 10.1007/978-1-4614-7369-5_2,

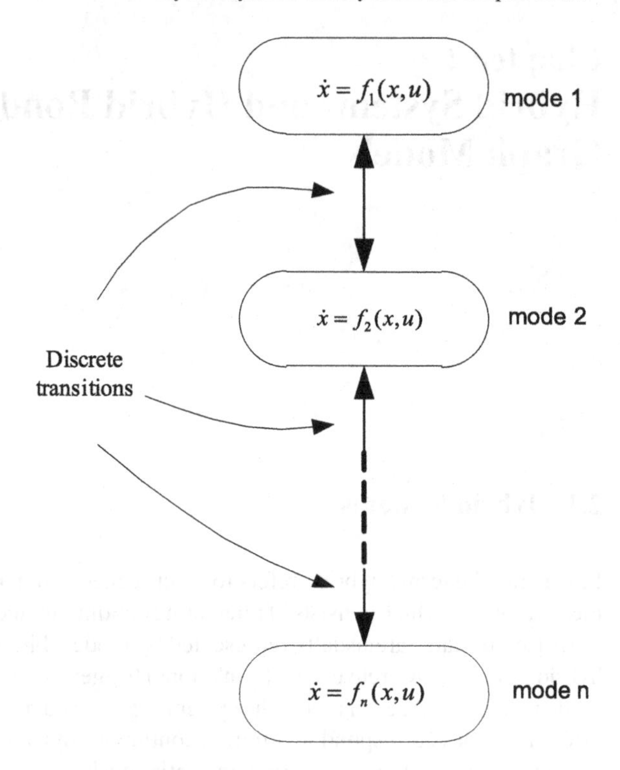

Fig. 2.1 Diagram of hybrid systems

the system is also characterized by discrete events, some of which can be attributed to control events and others are caused by continuous state variables crossing threshold values. For instance, the transition from idle mode to ramp-up mode for the motor is caused by a "turn_motor_on" control event. However, a transition that represents the acquisition roll contacting the paper is autonomous, and must be estimated using model and sensor data [11].

A simple example of hybrid system in process control is illustrated in Fig. 2.2. In the figure, $h(t)$ represents the height of the liquid in the tank, A denotes the cross section area of the tank, q_i is the constant flow in and $q_0(t) = Rh(t)$ is the flow out where R is a positive real number. Assuming that $q_i(t) > Rh(t)$, the tank will fill if both valves are open. The height of the liquid in the tank can be measured and both the empty and fill valves can be controlled to be in either open or closed state. Let $u_d(t)$ represent the command input signal which controls the state of the empty and fill valves. The possible values of the discrete variable $u_d(t)$ could be taken as {0,1,2, 3} where 0 means that both valves are open, 1 denotes that the fill valve open and the empty valve is closed, 2 represents the fill valve is closed and the empty valve is open and 3 denotes both valves are closed. The dynamic of the tank is changed with different discrete command inputs and could be represented as [12]

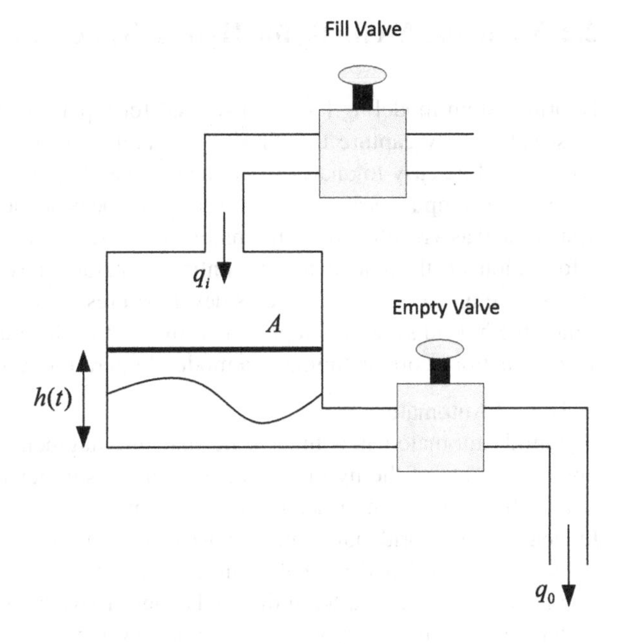

Fig. 2.2 An example of hybrid system

$$\frac{dh(t)}{dt} = \begin{cases} \frac{q_i - Rh(t)}{A} & \text{if } u_d(t) = 0 \\ \frac{q_i}{A} & \text{if } u_d(t) = 1 \\ \frac{Rh(t)}{A} & \text{if } u_d(t) = 2 \\ 0 & \text{if } u_d(t) = 3 \end{cases} \tag{2.1}$$

It is worth to note that the choice of modeling a real world system as continuous, discrete or hybrid is problem specific. In general, the classification of a system as continuous system, discrete system and hybrid system may be considered as an abstraction of the real world system. The same system could be modeled and classified as more than one type. For example, a printer can be modeled as a set of cause-effect rules, e.g., "when the printer is turned on, the motor begins to ramp up", "when the motor is turned off, the motor ramps down and returns to the idle position" and "when the acquisition roll contacting the paper, the feed roll starts to feed paper", and in this case it is classified as discrete event system. The printer can also be modeled as continuous system with a set of differential equations that describe the dynamic behavior of the feed motor. The same printer may also consist of sets of differential equations that describe the feed motor behavior and several discrete commands such as "turn_motor_on" and "turn_motor_off". As a result, such system can be modeled as a hybrid system.

2.2 Modeling Methods for Hybrid Systems

Hybrid system modeling has been studied for a period of time, and a good model must sufficiently capture the physical phenomenon of a process and at the same time should be easy to analyze. There are several approaches for modeling hybrid systems. In computer science area, researchers focus on the discrete part of the hybrid systems and as a result hybrid automaton and Petri nets are developed to provide such information. On the other hand, the continuous nature of hybrid systems are of interest to system engineers and the models developed resemble switched system models in which the hybrid system is represented by a set of differential equations with each set of equations corresponding to a mode of operation as shown in Fig. 2.1.

A. Hybrid Automaton
A hybrid automaton is a finite state machine augmented with differential equations. The state of the hybrid automaton changes either instantaneously according to a discrete transition or according to a continuous evolution of continuous state. In general, a hybrid automaton consists of a finite set of real value variables $X = \{x_1, \ldots, x_n\}$ and a labeled directed graph (V, E) [13], where V is a finite set of vertices and E is a set of directed edges or arcs between vertices. The vertices V denote the continuous part of the hybrid system and they are labeled with constraints on the derivatives of the variables in X. The labeled directed graph is used to model the discrete part of the hybrid system in which the each vertex is represented by a circle in the graph and the directed arc (edge) is represented by an arrow. For instance, the directed arc (v_i, v_j) is represented by an arrow starting at vertex v_i and terminating at vertex v_j.

In order to introduce the modeling principle of hybrid automaton, a simple example of thermostat system that is shown in Fig. 2.3 is analyzed. For this example, the temperature of the room evolves according to laws of thermodynamics and the state of the heater (on/off). The thermostat senses the temperature and performs certain computations and turns the heater on and off. The real-valued variable x represents the temperature of the room. The thermostat system includes two control modes, namely on and off. The initial temperature of the room is $x = 22\,°\mathrm{C}$ and the thermostat is in off mode where the temperature of the room evolves according to the differential equation $\dot{x} = -kx$ (k is a constant). The heater switches to on once the falling temperature reaches $x = 20\,°\mathrm{C}$ and the system switches to on mode. The arrow with condition $x = 20$ is an edge that represents a control switch where the

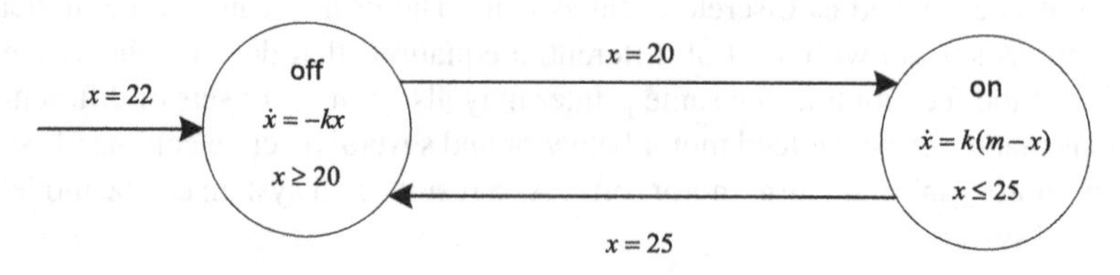

Fig. 2.3 Hybrid automaton of a thermostat

transition is enabled when the condition is satisfied. When the system is in on mode, the continuous dynamics follows the equation $\dot{x} = k(m - x)$ (m is a constant) until the temperature reaches $x = 25$ °C. The heater will go off and hence the temperature will drop. For each control mode, for example mode off, there is an invariant condition $x \geq 20$ which means the system will stay in this mode only when the invariant condition $x \geq 20$ is satisfied. In other words, the invariant condition $x \geq 20$ ensures that the heater goes to on when the temperature reaches $x = 20$ °C.

B. Petri Nets

Petri nets were invented in August 1939 by Carl Adam Petri with the focus of describing chemical processes [14]. It is a modeling language for modeling distributed systems. A Petri net is a directed bipartite graph, in which the nodes represent transitions (signified by bars) and places (signified by circles). Petri nets offer a graphical notation for stepwise processes that include choice, iteration, and concurrent execution and also have an exact mathematical definition of their execution semantics, with a well-developed mathematical theory for process analysis.

In general, a Petri net consists of places, arcs, and transitions. Arcs run from a place to a transition or vice versa, but can not run between transitions or between places. The places from which an arc runs to a transition are input places of the transition; the places to which arcs run from a transition are the output places of the transition. Generally, places may contain a discrete number of marks called tokens in a Petri net. The distribution of tokens over the places denotes a configuration (or state) of the net. This distribution of tokens is called marking. The marking indicates the state of the system and it changes when transitions occur. In a Petri net diagram, a transition of a Petri net may fire whenever there are sufficient tokens at the start of all input arcs. Once it fires, it consumes tokens, and then places tokens at the end of all output arcs. Execution of Petri nets is nondeterministic, which means that when multiple transitions are triggered at the same time, any one of them may fire. If a transition is activated, the Petri net may fire, but it doesn't have to. Since firing is nondeterministic, and multiple tokens may be present anywhere in the net, sometimes even in the same place. Petri nets are well suited for modeling and describing the concurrent behavior of distributed systems.

An example of a Petri net is demonstrated in Fig. 2.4. The Petri net models the behavior of a manufacturing process. The system is composed of two buffers (Buffer 1 and Buffer 2) and one machine. The tokens in place p_1 represent the parts in Buffer 1 and the tokens in p_3 represent the parts in Buffer 2. The marking illustrated in Fig. 2.4 indicates that the system is currently with two parts in Buffer 1 and one being processed by the machine. Each transition of a Petri net is associated with an

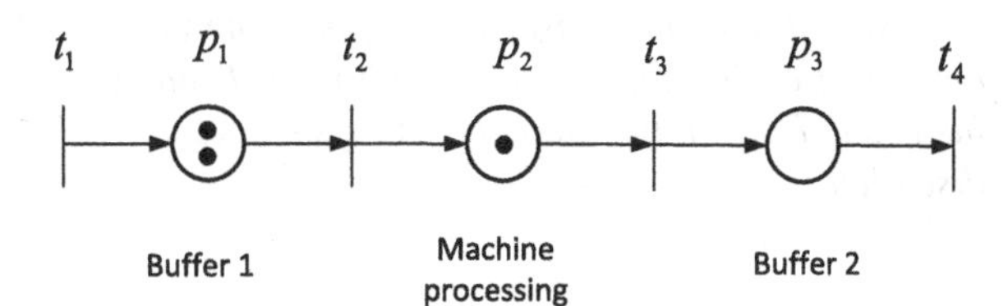

Fig. 2.4 Petri net of a manufacturing system

event of the system. For the Petri net in Fig. 2.4, t_1 denotes the arrival of a new part in Buffer 1. Transition t_2 represents the event of removing a part from Buffer 1 and beginning the process in the machine. Similarly, t_3 describes the moving one part from the machine to Buffer 2. Transition t_4 dispatches a batch of one part from the manufacturing system. The system behavior is simulated by the firing of transitions. A transition firing represents the occurrence of an event. When a transition fires, the system state is modified and then the Petri net marking is changed. A transition fires only if it is enabled. A transition is enabled if each input place has at least one token. If the transition is enabled, it may fire. When it fires, it removes one token from the input places, and adds one token to the output places. Figure 2.5 illustrates Petri net evolutions where Fig. 2.5a presents the initial state (marking). If transition t_1 is fired, the new marking presented in Fig. 2.5b is reached. Similarly, if transition t_2 and t_3 are fired, the corresponding marking is shown in Fig. 2.5c.

Another modeling tool for hybrid system is the Hybrid Bong Graph (HBG) developed by Mosterman and Biswas [15]. Like bond graph (BG), HBG supports component-level modeling practices. System components are defined as HBG fragments that interface through energy ports (for bonds) and signal ports (for signals). The HBG extends the benefits of the BG to hybrid systems and utilizes controlled junctions to model discrete mode changes. For a 1-type controlled junction, flows to all connected bonds are zero when it is turned off; similarly, a 0-type controlled junction forces the effort value to zero at all connected bonds during the OFF state. Because HBG defines switching locally, they provide a concise representation of the hybrid model. This contrasts to hybrid automata models, where system modes are pre-enumerated, which, in practical systems, is not often feasible. Unlike other approaches that use switching bonds, switch elements, or modulated transformers and gyrators to model switching in bond graph, HBG introduces controlled junctions. Controlled junctions act as ideal switches, enabling a junction to be either in the on or the off state. In the remaining parts of this section, the principles of BG modeling and HBG modeling will be introduced in details.

2.3 Basics of Bond Graph

Bond graph was invented by Henry Paynter at the Massachusetts Institute of Technology (MIT) in April 1959 [16] and subsequently developed into a systematic methodology by Rosenberg, Margolis and Karnopp, all of whom are former Ph.D. students of Henry Paynter [17]. As wrote by Paynter [18]:

So it was that on April 24, 1959, as the writer was about to give a seminar lecture at Case Institute (now Case-Western) on "Interconnected Engineering Systems", he awakened earlier that morning with the 0,1-junctions somehow finally planted in his head! Thus on that date the BG system was complete and constituted a formal discipline.

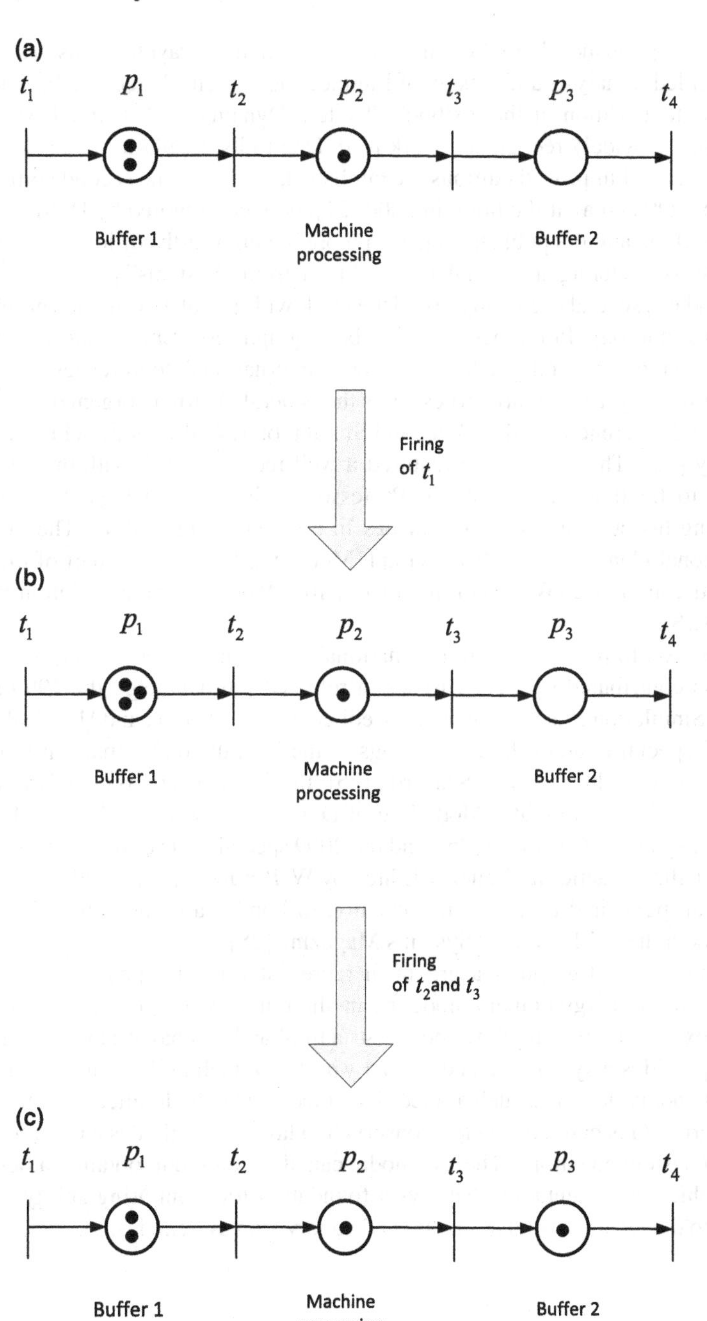

Fig. 2.5 Petri net evolutions

The first published books of these pioneers include Paynter's historic lecture notes entitled "Analysis and Design of Engineering Systems", tracking back to 1961 [19]. The first edition of the textbook "System Dynamics—A Unified Approach" has become a widely recognized work [20]. Due to its popularity in academia and industry, several improved editions are published, including the second edition published in 1990 and a third edition in 2000 [21], both co-authored by D. Margolis. In 2006, the three authors published an even more mature fourth edition titled "System Dynamics—Modeling and Simulation of Mechatronic Systems" [22].

Nowadays, several events suggested the worldwide popularity of bond graph modelling methodology. In the past decades, bond graph researchers contributed many special sessions on bond graph modelling to international conferences. For example, W. Borutzky, a Germany professor, is the general chairman organizing the 2006 European Conference on Modelling and Simulation (ECMS 2006) held near Bonn, Germany [23]. This conference featured a well received track with three sessions devoted to bond graph modelling. Professor Cellier gave a keynote speech by addressing his current research activities in bond graph modelling. The bi-annual International Conference on Bond Graph Modelling (ICBGM) is part of the Western Multi-conference (WMC) of the Society for Modelling and Simulation International (SCS).

In addition to publications in international conferences, bond graph researchers have also contributed to special issues in refereed journals, e.g., the 1999 special issue of Simulation Practice and Theory edited by J. U. Thoma and H. J. Halin [24], the 2002 special issue of the Proceedings of the Institute of Mechanical Engineers edited by P. Gawthrop and S. Scavarda [25], the 2006 special issue of the journal Mathematical and Computer Modelling of Dynamical Systems edited by I. Troch, W. Borutzky and P. Gawthrop [26], and the 2009 special issue of the journal Simulation Modelling Practice and Theory Edited by W. Borutzky [27]. In 2007, Gawthrop and Bevan published a tutorial introduction to bond graph modelling for control engineers in the IEEE Control Systems Magazine [28].

In brief, a bond graph is a graphical representation of a physical system. As a unified multi-energy domain modeling method, it provides an approach to model a complex system, which allows both a structural and a behavioral system analysis [22]. It provides a systematic and formal way for modeling dynamic systems with different energy domains, such as electrical, mechanical, hydraulic, etc., in a unified framework. BG is based on energy conservation law that facilitates automatic generations of system equations. The BG model can also be used to obtain mathematical and graphical representation that lays a foundation for monitoring ability analysis (ability to detect and to isolate faults) and supervision system design.

2.3.1 Bonds, Power and Causality

In BG theory, multiport is an important modeling concept that is used to model physical subsystems. It can be used to model components such as DC motor, pump,

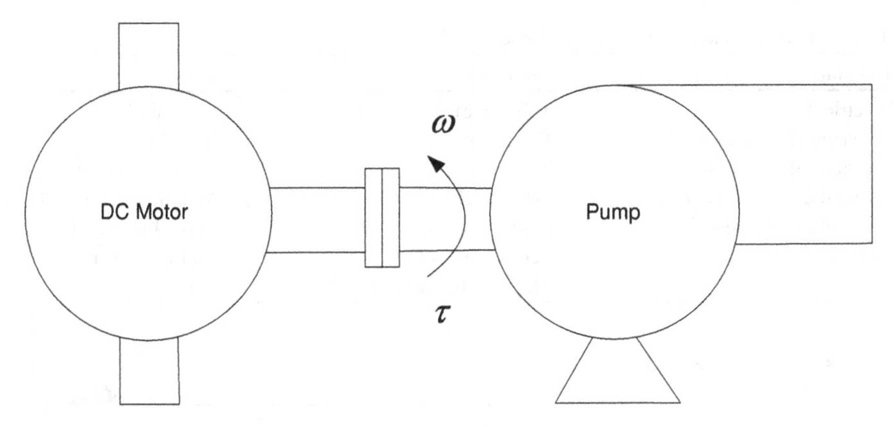

Fig. 2.6 Engineering multiport connection

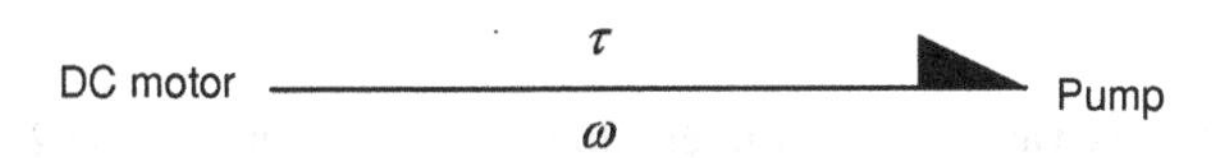

Fig. 2.7 Word bond graph for system of Fig. 2.6

gearbox, etc. Places at which subsystems can be interconnected are places at which power can flow between subsystems. Such places are called ports, and subsystem with one or more ports is call multiport. A system with single port is called a 1-port, and a system with two ports is called a 2-port, and so on.

The device sketched in Fig. 2.6 can be treated as two multiport elements (DC motor and pump) with ports that can be connected to each other to form a system. Figure 2.7 represents a word bond graph of the system. A word bond graph is a first step towards a bond graph, in which words define the multiport components. Figure 2.7 implies that a port of the motor, indicated by a single line emanating from the word representing motor, and a port of the pump that been connected. The single line represents a bond is formed between the two subsystems. When two subsystems are physically connected together, a pair of complementary variables (called power variables) is constrained to be equal for the two subsystems. For the example in Fig. 2.7, the motor and the pump have a common angular speed ω and torque τ at the coupling. The half arrow on the bond indicates the direction of power flow which is given by the product of the angular speed and the torque. When the product of the angular speed and the torque ($\omega\tau$) is positive, power is flowing from the motor to the pump. Similarly, power can be expressed as the product of a force and a velocity in mechanical domain, where the product of voltage and current represents the power in electrical domain. Since power interactions are always present when two multiports are connected, two power variables, effort (indicated by symbol e) and flow (indicated by symbol f) represent various power variables in different energy domains in a universal way. The product of these two power variables is the power $P = e \cdot f$.

Table 2.1 Effort and flow in different domains [22]

Domain	Effort (e)	Flow (f)
Electrical	Voltage (e)	Current (i)
Mechanical (translational)	Force (F)	Velocity (v)
Mechanical (rotational)	Torque (τ)	Angular velocity (ω)
Hydraulic (incompressive)	Pressure (P)	Volume flow rate (Q)
Hydraulic (compressive)	Enthalpy (h)	Mass flow rate ($\dot{m}$)
Thermal	Temperature (T)	Entropy flow rate ($\dot{s}$)
Magnetic	Magneto-motive force (F)	Flux rate ($\dot{\phi}$)

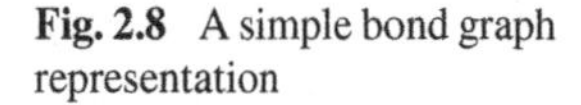

Fig. 2.8 A simple bond graph representation

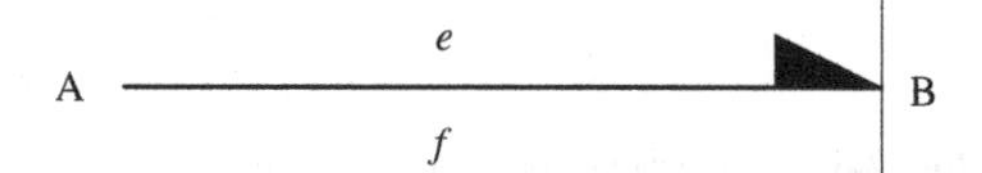

These two power variables in different domains are given in Table 2.1. The flow represents either: current, flow rate or velocity, effort represents either force, voltage or pressure. The effort and flow of the bond are numbered according to its bond number. These analogies (which exist in different physical domains) enable the description of complex physical phenomena that integrate different physical domains and transformations of energy between different physical domains with a very small set of basic elements.

In bond graph theory, the way in which inputs and outputs of the subsystem or component are specified by means of the causal stroke. The causal stroke is a short, perpendicular line added at one end of a bond. The casual stroke location indicates the direction of the effort, while the flow points in the opposite direction. The causality (cause and effect relationship) is a unique feature of bond graph theory. By using components A and B to represent multiport subsystems in Fig. 2.8, a concise bond graph notation consisting of four pieces of information is given as follows: (1) the existence of link between two systems is defined by the bond, (2) the type of power (electrical, hydraulic, and so on) by the power variables e and f, (3) the positive power direction defined by the half arrow and (4) the causality defined by the stroke [29]. From this figure, e is the input to B and f is the output from B, while f is the input to A and e is the output from A. The half-arrow sign convention for power flow and the causal stroke are independent. For the example shown in Fig. 2.8, the causal stroke may be put at the end near A, which implies that e is the input to A and f is the output from A.

2.3.2 Bond Graph Elements

By using bond graphs and the classification of power variables, a few basic types of multiport elements are utilized to represent models in a variety of energy domains. By using BG modeling, any physical system or subsystem can be described by generic BG components which include source elements *Se* and *Sf*, dissipative element *R*, storage elements *C* and *I*, two junctions 0, 1, and two transducers *TF* and *GY*.

2.3.2.1 1-Port Elements

A 1-port element is addressed through a single port component where a single pair of effort and flow variables exists. This type of element can be connected to another system through its one port. 1-port bond graph elements include passive 1-port elements and active 1-port elements. Passive 1-port elements (1) transform received power into dissipated power (element *R*), (2) store received power as kinetic energy (element *I*), and (3) as potential energy (element *C*). For these passive 1-port elements, the power convention is established by means of a half arrow pointing toward the element as shown in Fig. 2.9.

The 1-port resistor (or resistance) is an R-element where its effort and flow variables are related by a static function. Figure 2.10 shows a bond graph of an electrical resistor. In the figure, u is the supply voltage (effort variable) and i is the current (flow variable). The electrical resistor is an *R*-element with linear voltage-current constitutive relation according to Ohm's law $u = Ri$. Similarly in mechanical domain, the damper is considered as an *R*-element which is characterized by linear force-velocity relation.

In general, the constitutive equation of *R*-element can be represented as

$$\phi_R(e, f) = 0 \tag{2.2}$$

In bond graph convention, the physical parameter of an element can be attached to the symbol of the element by a colon. For the example shown in Fig. 2.10, R_e is the parameter of electric resistor which is attached to the 1-port resistor symbol *R*.

Next, consider a 1-port device in which a static constitutive relation exists between an effort and a displacement (time integral of flow). This element is called a *C*-element (capacitor) which represents any system transforming the received power into potential energy without loss. Figure 2.11 shows a bond graph of spring in mechanical domain. In linear case, the constitutive law of the spring with stiffness k is

$$F = \frac{1}{C}\int \dot{x}dt = kx \tag{2.3}$$

where x is the relative displacement across the spring, $\dot{x}$ is the spring velocity, and F is the force. In this example, the compliance parameter $C = \frac{1}{k}$.

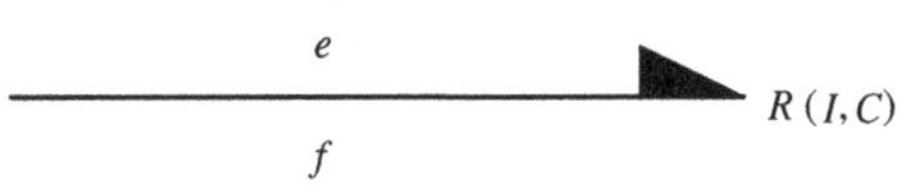

Fig. 2.9 Representation of 1-port passive elements

e — f — $R\,(I, C)$

Similarly, the electrical capacitor in electrical domain and the storage tank in hydraulic domain can also be considered as 1-port capacitors. The constitutive law for C-element which relates the effort to the time integral of flow is

$$\phi_C(e, \int f dt) = \phi_C(e, q) = 0 \tag{2.4}$$

where $q = \int f dt$. The linear form of such constitutive law is $e = \frac{1}{C}\int f dt = \frac{q}{C}$.

A second energy-storing 1-port element arises if the momentum p ($p = \int e dt$) is related by a static constitutive law to the flow f. Such an element is called inertia (I-element) in bond graph terminology. This element describes any system that transforms the received power into kinetic energy without loss. The I-element is used to model inductance of electrical systems and mass or inertia effects in mechanical or hydraulic process. The 1-port inertia is characterized by the following constitutive law

$$\phi_I(f, \int e dt) = \phi_I(f, p) = 0 \tag{2.5}$$

If the relation is linear, the constitutive relation will have the form of $f = \frac{1}{I}\int e dt = \frac{p}{I}$, where I is the inertia parameter. An example of 1-port inertia in mechanical domain is depicted in Fig. 2.12, the mass is ideally represented by the constitutive law

$$v = \frac{p}{M} = \frac{\int F dt}{M} \tag{2.6}$$

where v is the velocity, F is the force and M is mass of the block. The inertia parameter is M for this example.

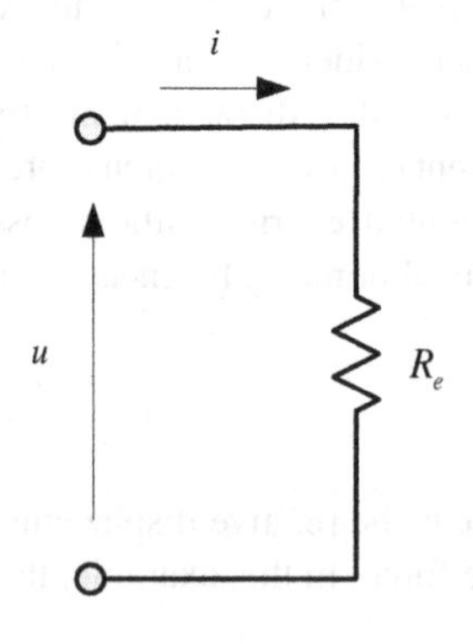

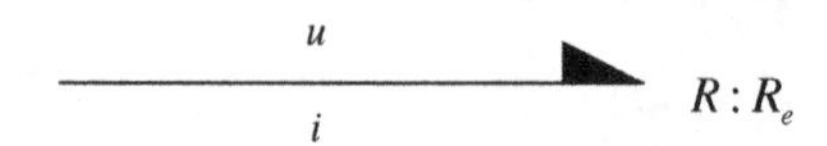

Fig. 2.10 1-port resistor in electrical domain

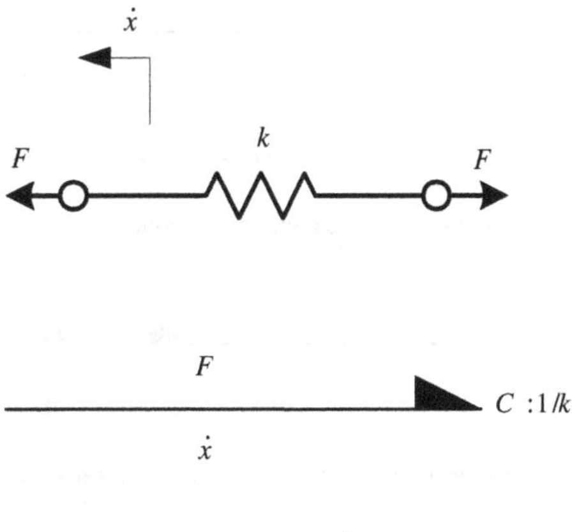

Fig. 2.11 1-port capacitor in mechanical domain

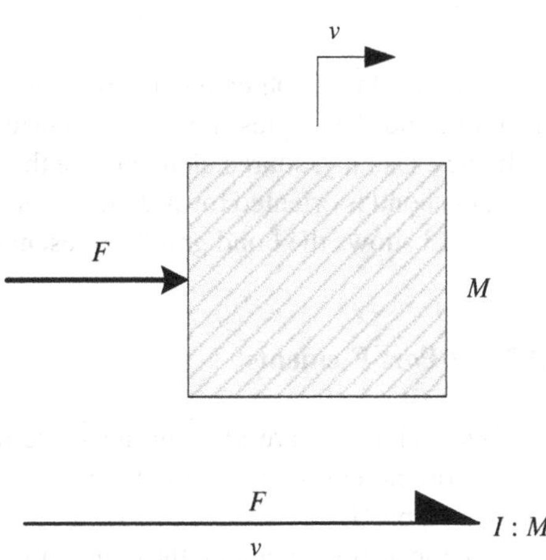

Fig. 2.12 1-port inertia in mechanical domain

It is worth to note that passive 1-port elements can be modulated by external signal. For example, consider a continuous controlled valve. The volume flow rate across the valve will be zero when the valve is closed. Under this case, the flow resistance is not a constant and is determined by the valve state, this 1-port element is called modulated resistor. The same idea also apply to I-element and C-element, and if these elements are not constant, they are called modulated inertia and modulated capacitor.

Next, the effort and flow source are introduced. These active 1-port elements are called active elements in bond graph terminology because they supply power to the process. Depending on the type of power variable the source provides, there are two 1-port source elements: effort source (Se) and flow source (Sf). Effort source imposes an effort on a system, which can be constant or a function of time, but independent of the flow variable. Effort sources include examples like voltage supplies, pressure pump, gravity, and so forth. As an example of a constant effort source, gravity force on a mass is essentially independent of the velocity of the mass near the surface

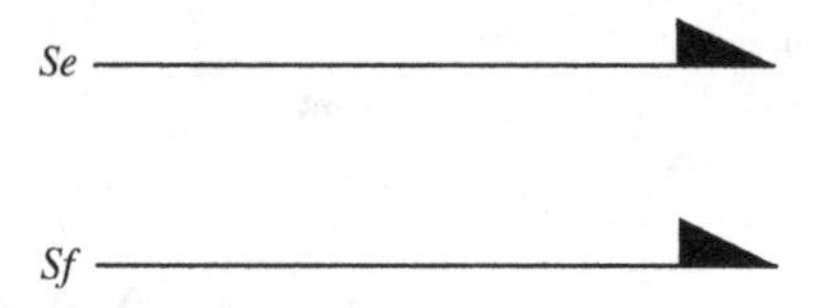

Fig. 2.13 1-port active elements in bond graph

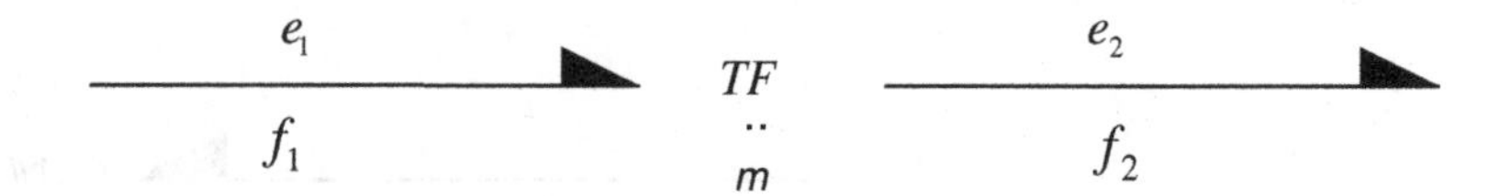

Fig. 2.14 Bond graph representation of a transformer element

of the earth. On the other hand, flow source provides a flow independent of the effort variable. Examples are electric current generator, and imposed velocity on mechanics. Usually, source elements are thought of as supplying power to a system, hence, the bond is oriented toward the system connected to the source by convention. Figure 2.13 shows the bond graph representation of source elements.

2.3.2.2 2-Port Elements

The 2-port elements are ideal in the sense that power is conserved. There are two basic 2-port elements: transformer element is denoted by *TF*, and gyrator element is represented by *GY*.

For a transformer element, the constitutive equations algebraically relate the inlet and outlet efforts, and also relate the inlet and outlet flows. The input effort is proportional to the output effort and the output flow is proportional to the input flow. The bond graph representation of a transformer element is shown in Fig. 2.14. In the figure, the subscripts 1 and 2 correspond to the two bonds. The constitutive law for a transformer element is given as

$$e_1 = me_2, \;\; f_2 = mf_1 \tag{2.7}$$

where parameter m is called the modulus of the transformer. It is not difficult to find that $e_1 f_1 = me_2 \times f_2/m = e_2 f_2$. The sign convention of a transformer is represented by one bond points toward the element and the other one is oriented away from the element.

Figure 2.15 depicts an ideal rigid lever which represents a transformer element in mechanical domain. In the figure, F_1 and F_2 are the forces imposed on the ends of the lever and v_1 and v_2 are the velocities of the ends. a and b are the length parameters where the modulus of the transformer of this example is b/a. The lever is an ideal transformer because kinetics indicates that $(b/a)v_1 = v_2$ and moment equilibrium

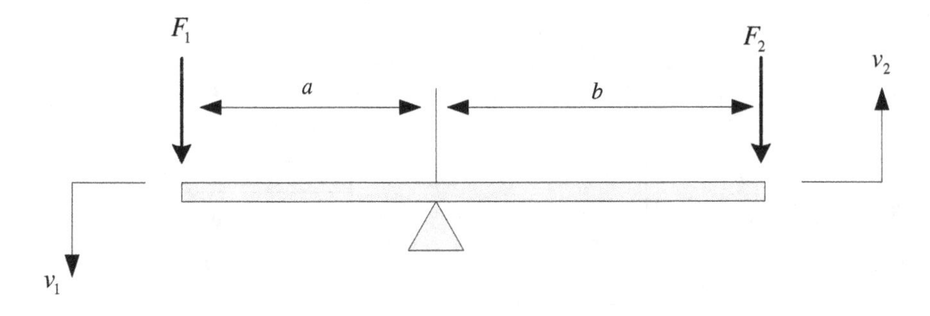

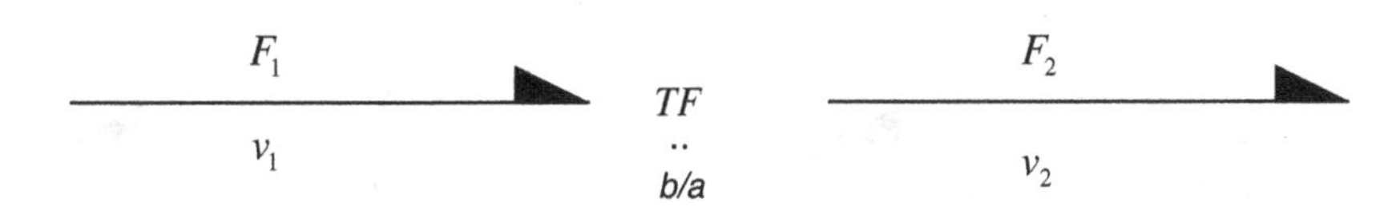

Fig. 2.15 Transformer element in mechanical domain

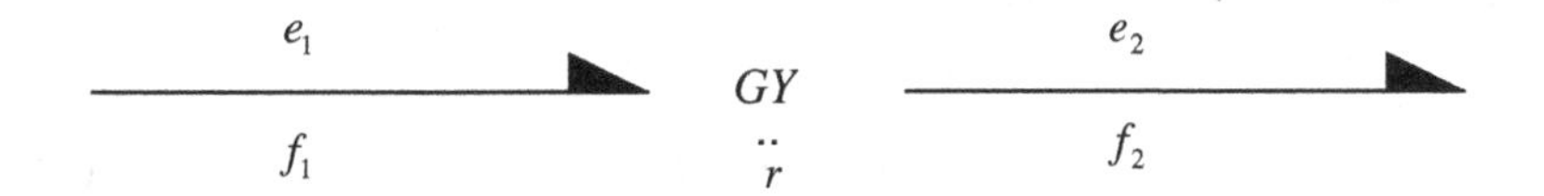

Fig. 2.16 Bond graph representation of a gyrator element

requires $F_1 = (b/a)F_2$. This is exactly the definition of transformer element in (2.7). Other examples of transformers (TF) are pumps and cylinders in hydraulic domain, gear boxes in mechanical domain and electrical transformers in electrical domain.

Another energy conservative 2-ports bond graph element is gyrator (GY), generally transforming energy from one domain into another. For a gyrator, the input effort is proportional to the output flow and the output effort is proportional to the input flow. The constitutive law of the gyrator is

$$e_1 = rf_2, \; e_2 = rf_1 \tag{2.8}$$

where r is the gyrator modulus. The general bond graph representation of a gyrator is depicted in Fig. 2.16 where the subscripts 1 and 2 correspond to the two bonds. The power sign convention of a gyrator is the same as the sign convention of transformer.

An example of electrical gyrator is depicted in Fig. 2.17, where u and i are back EMF voltage and current of the DC motor, ω and τ are output angular speed and torque, respectively. k_e is the current to torque ratio of the motor. The motor transforms electric power to mechanical rotary power, where the corresponding constitutive law of this device is

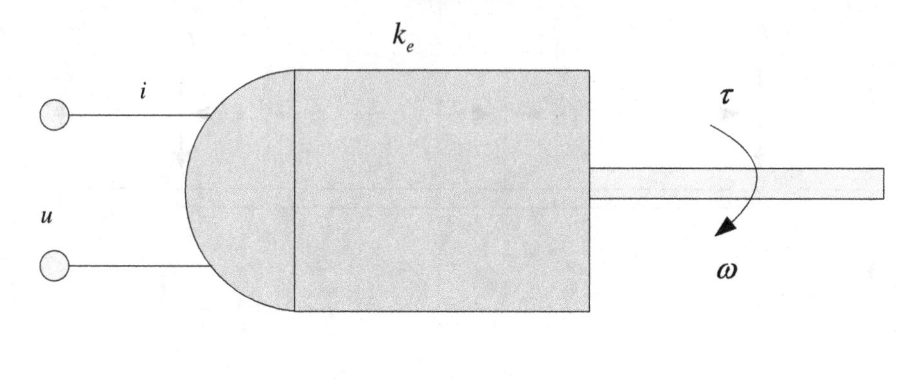

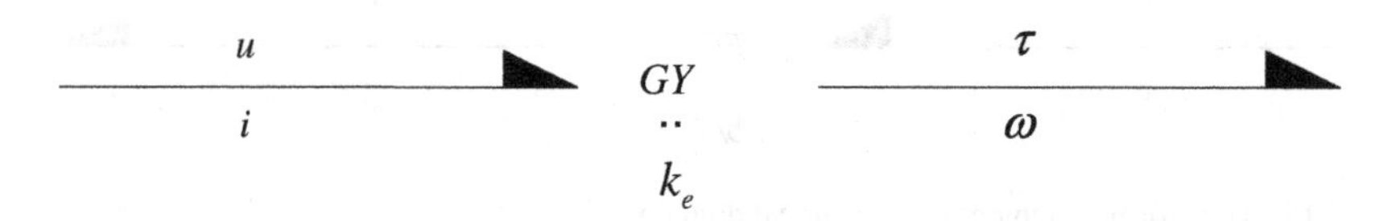

Fig. 2.17 Electrical gyrator element

$$u = k_e\omega,\ \tau = k_e i \tag{2.9}$$

The modulus of the gyrator of this example is k_e. Other examples of gyrators include voice coil transducer and gyroscope. If the transformer modulus m and the gyrator modulus r are not constant, but depend on time or any other parameter,

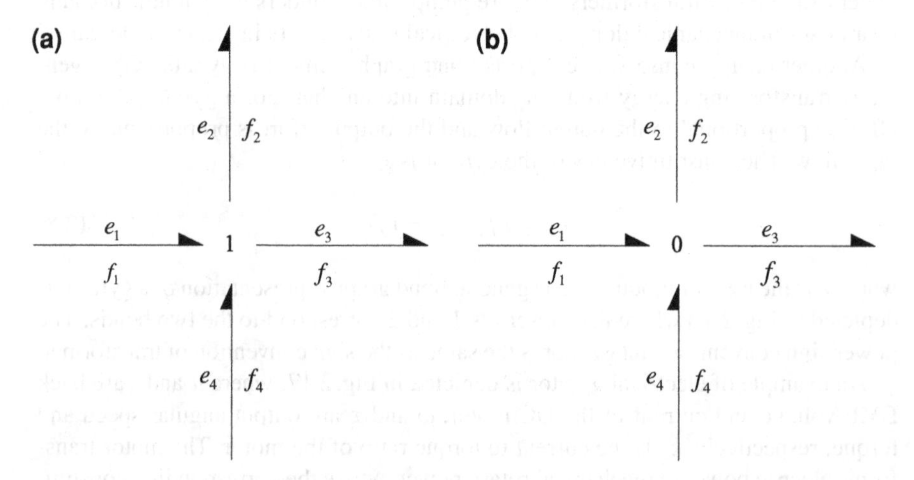

Fig. 2.18 Bond graph representations of 1-juntion and 0-junction

then these 2-port elements are called Modulated Transformer (MTF) element and Modulated Gyrator (MGY) element.

2.3.2.3 Junctions

There are two types of connections that connect between BG elements. They are the series and parallel connections. In bond graph terminology, these connections are called 1-junction (common flow junction) and 0-junction (common effort junction). Figure 2.18 demonstrates the graph representations of 1-juntion and 0-junction in bond graph theory. These two kinds of junctions are ideal in the sense that power is neither dissipated nor stored. The constitutive law of 1-junction and 0-junction can be expressed as

$$\sum_{i=1}^{n} e_i f_i = 0 \tag{2.10}$$

where n is the total number of bonds connected to the junction and Σ is the algebraic sum. The sign is taken (+) when the power is oriented towards the junction and (-) when it is outwards from the junction. Consider the model in Fig. 2.18a, bond 1 and bond 4 have power directions towards the junction and the power directions of bond 2 and bond 3 are out of the junction, thus the constitutive law is represented as

$$e_1 f_1 - e_2 f_2 - e_3 f_3 + e_4 f_4 = 0 \tag{2.11}$$

The 1-junction is defined such that all flows are the same, thus

$$f_1 = f_2 = f_3 = f_4 \tag{2.12}$$

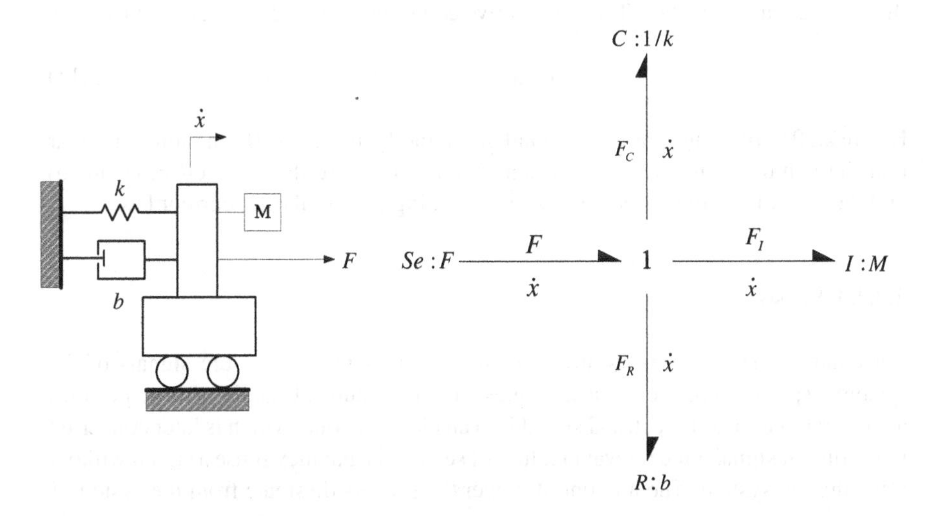

Fig. 2.19 An example of bond graph representation involving 1-juntion

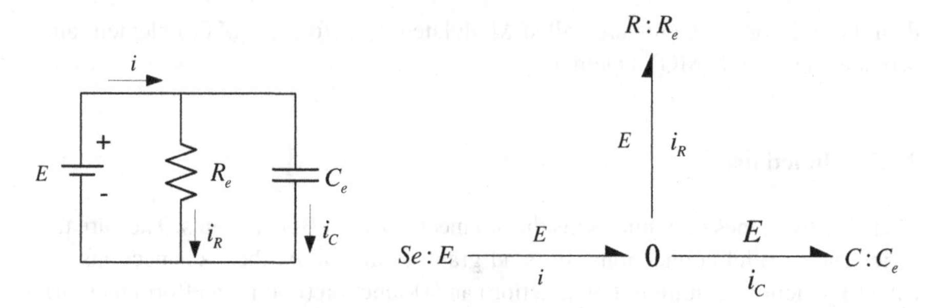

Fig. 2.20 An example of bond graph representation involving 0-juntion

Combining (2.11) and (2.12) yields

$$e_1 - e_2 - e_3 + e_4 = 0 \tag{2.13}$$

In summary, the flows on all bonds of 1-junction are always identical, and the algebraic sum of the efforts is equal to zero.

An example of bond graph model involving a 1-junction is depicted in Fig. 2.19. In the figure, k is the spring stiffness, b is the damping coefficient of the damper and M is the mass of the cart. F is the force imposed on the cart and $\dot{x}$ is the velocity. In the corresponding bond graph model, the input force is modeled by the effort source Se, the spring is modeled by C-element with $C = 1/k$, the damper is modeled by R-element with $R = b$, and the cart is modeled by I-element with $I = M$. F_C, F_R and F_I are forces imposed on the spring, damper and mass, respectively. A common velocity exists among these elements and a dynamic equilibrium of forces indicates that the algebraic sum of the efforts is equal to zero.

For 0-junction, the associated bonds share the same effort, and the algebraic sum of the flows always vanishes. The constitutive equations in Fig. 2.18b can be written as

$$f_1 - f_2 - f_3 + f_4 = 0, \; e_1 = e_2 = e_3 = e_4 \tag{2.14}$$

Figure 2.20 depicts an example of bond graph model involving 0-junction. It is clear that the voltage source, R-element and C-element share the same effort (voltage) and the sum of currents is equal to zero according to Kirchhoff's current law.

2.3.2.4 Sensors

For control systems, sensors are necessary components to measure signals of the system responses, e.g., velocities, torques, or temperatures. Usually, sensors perform a conversion of a non-electrical signal into an electrical one, which is later converted into a digital signal. One relevant feature of sensors is that they sense a signal without affecting the system. The amount of power the sensors dissipate from the system is

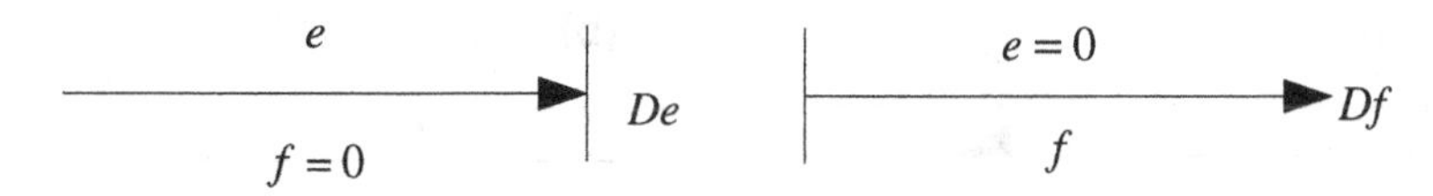

Fig. 2.21 Bond graph representations of sensors

very small and can be neglected. For example, an ideal ammeter indicates current but introduces no voltage drop, and an ideal voltmeter reads a voltage while passing no current. In bond graph model, the signal measured by a sensor is represented by a full arrow bond.

Figure 2.21 depicts the bond graph representations of effort sensor (De) and flow sensor (Df). The causal stroke location decides the type of the sensor, where the causal stroke is located near the sensor element for effort sensor and the causal stroke is put away from the sensor element for flow sensor. The full arrow bond is also called information bond where one of its conjugate power variables is set to zero. For example, the effort sensor provides only information about the effort measurement, such as pressure, force, and voltage. There is no power interaction from the sensor to the system. Thus the flow of the effort sensor is zero.

2.3.3 Causality of Basic Bond Graph Elements

Bond graph has the notion of causality, indicating which side of a bond defines the instantaneous effort and flow. This section presents some of the specific constraints of causalities of the basic bond graph elements.

2.3.3.1 Causality for 1-Port Elements

For bond graph source elements, the causalities are depicted in Fig. 2.22. In the figure, the location of causal stroke indicates the direction of the effort variable.

Fig. 2.22 Bond graph causality assignment of sources

Fig. 2.23 Bond graph causality assignment of resistor. **a** resistive causality, **b** conductive causality

Fig. 2.24 Bond graph causality assignment of capacitor. **a** integral causality, **b** derivative causality

In bond graph theory, the 1-port resistor is normally indifferent to the causality imposed on it. This implies that the constitutive equation remains the same for the two different causalities. Figure 2.23 shows the two possibilities of causalities assigned to resistor. Figure 2.23a represents the resistive causality of resistor where the input is flow and output is effort, while Fig. 2.23b denotes the conductive causality of resistor where input is effort and output is flow.

These two possibilities can be represented in equation form as follows

$$e = \phi_R(f) \tag{2.15}$$

$$f = \phi_R^{-1}(e) \tag{2.16}$$

where (2.15) is the constitutive law of the resistive causality of resistor, (2.16) is the constitutive law of the conductive causality of resistor.

As for energy-storing 1-port element C, there are two types of causality, namely integral causality and derivative causality as shown in Fig. 2.24. For integral causality of C-element, f is the input and e is given by a static function of time integral of f

$$e = \phi_C^{-1}\left(\int f dt\right) \tag{2.17}$$

and for derivative causality, e is the input, f is the time derivative of a static function of e

$$f = \frac{d}{dt}\phi_C(e) \tag{2.18}$$

Another energy-storing 1-port element is I-element. Similarly, it has integral causality and derivative causality as depicted in Fig. 2.25.

In contrast to C-element, the input is e for integral causality of I-element and derivative causality exists when f is the input variable. These two causality representations can be written as

$$f = \phi_I^{-1}\left(\int edt\right) \tag{2.19}$$

$$e = \frac{d}{dt}\phi_I(f) \tag{2.20}$$

2.3.3.2 Causality for 2-Port Elements

For transformer elements, denoted by *TF*, there are two possible causality assignments as shown in Fig. 2.26. The constitutive law of the transformer in Fig. 2.26a can be represented by

$$e_1 = me_2, \ f_2 = me_1 \tag{2.21}$$

The constitutive law of the transformer shown in Fig. 2.26b is represented by

$$f_1 = f_2/m, \ e_2 = e_1/m \tag{2.22}$$

For a gyrator (*GY*), the causality assignments are depicted in Fig. 2.27.
For Fig. 2.27a, the constitutive law is

$$e_1 = rf_2, \ e_2 = rf_1 \tag{2.23}$$

The constitutive law of the gyrator in Fig. 2.27b is

$$f_1 = e_2/r, \ f_2 = e_1/r \tag{2.24}$$

2.3.3.3 Causality for Junctions

For 1-junction, all the flows of all the bonds are equal and the efforts of all the bonds must sum to zero. If the flow on any single bond is an input to the junction, the flows on all other bonds can be determined. The bond that determines the flow is called the strong bond of 1-junction and this bond has causal stroke away from the junction. All other bonds are called weak bonds. The output of the 1-junction is the effort variable of the strong bond and the effort of the strong bond can be expressed as

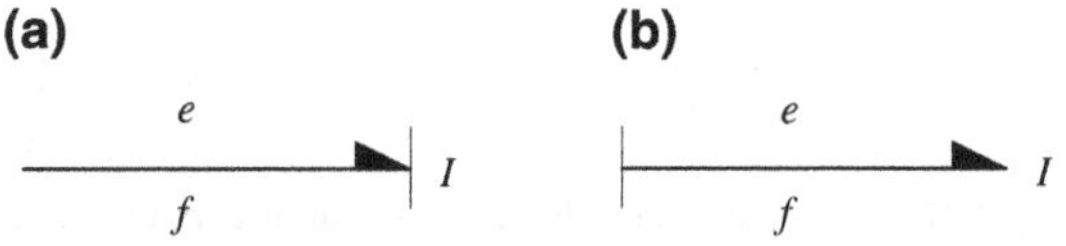

(a) e f I (b) e f I

Fig. 2.25 Bond graph causality assignment of inertia. **a** integral casuality, **b** derivative causality

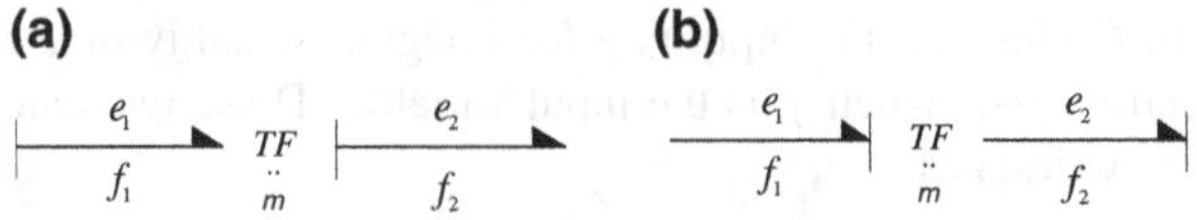

Fig. 2.26 Bond graph causality assignment of transformer. **a** one possibility, **b** another possibility

signed sum of efforts of other bonds. The causality of 1-junction is shown in Fig. 2.28. In the figure, bond 1 is the strong bond and all other bonds are weak bonds. For 1-junction, there is only one bond which has the causal stroke at the end assigned away from the junction and other bonds have causal strokes at the end near the junction.

For 0-junction, there is only one bond which has causal stroke at the end near the junction and other bonds must have causal strokes at the end away from the junction. The bond that decides the effort of the junction is called the strong bond of 0-junction and it is causalled near the junction. All other bonds are weak bonds of 0-junction. The output of the 0-junction is the flow variable of the strong bond and the flow variable of the strong bond can be expressed as algebraic sum of flows of other bonds. Figure 2.29 depicts the causality for 0-junction where bond 1 is the strong bond and other bonds are weak bonds.

2.3.4 Sequential Causality Assignment Procedure

Besides the generalized modeling capability, BG modeling also utilizes the concept of the causality to derive equations from the graph. Every bond of the graph has a stroke marked at one end of the bond. For a bond in BG, the causal stroke location indicates the direction of the effort, while the flow points in the opposite direction. This indication allows users to derive relations between system's variables systematically. Causal path is a graphical representation that explicitly shows the relation between system variables based on the concept of causality. In bond graph theory, when a signal (effort or flow) reaches a junction, it is either distributed (flows any one of the weak bonds if the signal is the variable of the strong bond of the junction) or goes to the strong bond (if it is a variable of the weak bond). When the signal reaches a passive element (*R*, *I* or *C*) it returns back through it but undergoes a qualitative change (flow variable becomes effort variable and vice verse). When a signal reaches a source or sensor, it is terminated there [29].

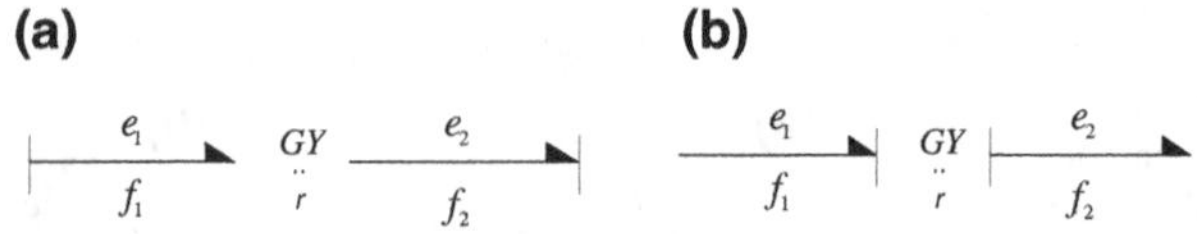

Fig. 2.27 Bond graph causality assignment of gyrator. **a** one possibility, **b** another possibility

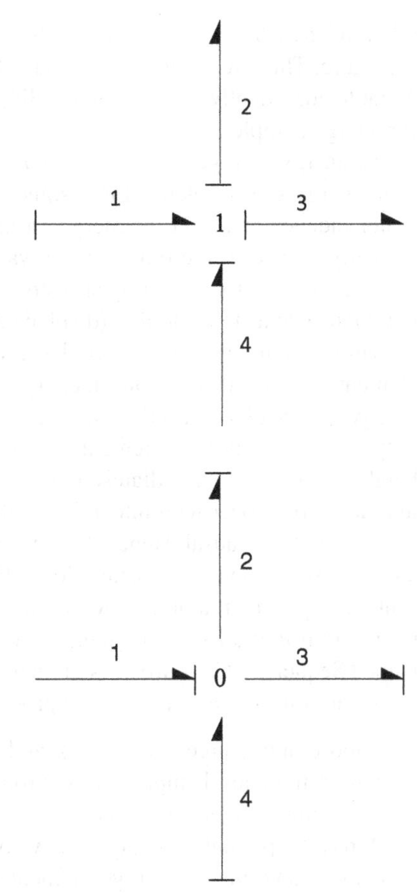

Fig. 2.28 Bond graph causality assignment of 1-junction

Fig. 2.29 Bond graph causality assignment of 0-junction

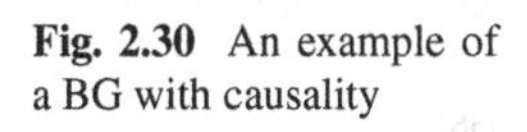

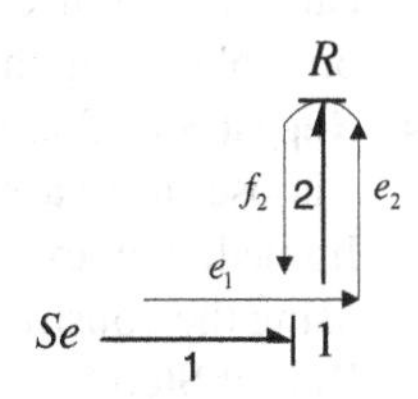

Fig. 2.30 An example of a BG with causality

Figure 2.30 illustrates a simple application of using causality to derive an analytical expression of a variable systematically. In the figure, there are three components: *Se*, *R* and 1-junction. The causality assigned to the graph is represented by the strokes at the end of bond 1 and 2. Consider the case where it is desirable to find the analytical equation of variable f_2. For the R component in the figure, e_2 is the input and f_2 is the output. This implies that the constitutive relation (ϕ_R^{-1}) of R is $f_2 = \phi_R^{-1}(e_2)$. Similarly, the causal stroke of bond 1 reveals that $e_2 = e_1 = Se$. Hence, the analytical connection leads to $f_2 = \phi_R^{-1}(Se)$. Note that the analytical relation between f_2 and Se can be shown explicitly using causal paths as shown in the figure. This illustration shows that BG provides a method to derive relation of a variable analytically and systematically. This advantage would be more significant for a large complex system.

An energy storage element (C or I) is independent if it uses integral causality. If an energy storage element is assigned with derivative causality, then the element is not independent, and its energy variable (displacement or momentum) is algebraically related to the other energy variables in the system. Such energy storage element in derivative causality still stores energy, however, such representation does not contribute a state variable (displacement or momentum) to the system. Thus, in causality assignment, it is preferred to assign integral causality to each energy storage element [22]. Another reason choosing integral causality over derivative causality for energy storage element is the numerical problem in computer simulation because the output of an energy store element with derivative causality exhibits a pulse of infinite height if the input immediately jumps to another value at certain time. This problem does not appear with integral causality. The derivative causality is not physical because it is not causal. Hence, the use of derivative causality in a model is artificial and represents modeling assumptions. Therefore, in a behavioral model that represents the systems relation between inputs (sources) and outputs (state variables) it is preferred that all storage elements are assigned integral causality. A procedure, named Sequential Causality Assignment Procedure (SCAP) was developed to assign these causalities to a BG [22]. The procedure can be described as follows:

1. Choose any source (Se or Sf), and assign its required causality. Immediately extend the causal implications through the graph as far as possible, using the constraint elements (0, 1, *TF*, *GY*).
2. Repeat Step 1 until all sources have been assigned.
3. Choose any energy storage element (C or I), and assign its preferred integral causality. Immediately extend the causal implications through the graph as far as possible, using the constraint elements (0, 1, *TF*, *GY*).
4. Repeat Step 3 until all energy storage elements have been assigned.
5. Choose any unassigned R-element and assign a causality to it (basically arbitrary). Immediately extend the causal implications through the graph as far as possible, using the constraint elements (0, 1, *TF*, *GY*).
6. Repeat Step 5 until all remaining bonds have been assigned.

An example in electric domain is shown in Fig. 2.31. Causality assignment is carried out step by step in the example using SCAP. In Fig. 2.32a, bond 1 is directed according to the source element Se. Since the strong bond for the 1-junction has not decided, other bonds cannot be assigned a causality. There are no more source elements, and then the process proceeds to Step 3. In Fig. 2.32b, bond 2 and bond 5 are causally directed to produce integral causalities for I-element and C-element. Bond 3 is directed using the constraint of 1-junction where bond 4 is causally according to the constraint of 0-junction. The complete bond graph model is depicted in Fig. 2.32b. From Fig. 2.32b with proper causality assignment, we know that the components C and R have the same effort, i.e., they are connected in parallel using 0-junction. L and

V_{in} share the same flow, since they are connected in series. It can also be observed from the diagram that how the power of the source V_{in} is distributed between the different components.

2.3.5 Example of a Quarter Car System Modeling

Let us look at a mechanical example with the focus on the systematic approach to generate a bond graph model in simulation environment (SIMULINK® in Matlab®). The objective of this part is to introduce readers a systematic modeling process by using the bond graph method. The steps to take on this approach are as follows:

1. Identify the kind of BG elements that make up the system.
2. Lay out a bond graph showing how the elements are connected, then enumerate all the bonds and BG junctions.
3. Apply physics principles to represent the constitutive relations of all basic BG components and complete the BG model in the Matlab® SIMULINK® environment.
4. Set the physical parameter values of the 1-port and 2-port BG components, and assign the input values for source components.

The quarter car model is shown in Fig. 2.33. The model consists of the sprung mass M_s (car body, engine, etc.) and the unsprung mass M_u that accounts for the wheel and axle masses supported by the tire. The suspension is modeled as a spring K_s and a damper C_s in parallel, which connects the unsprung mass to the sprung mass. The tire is modeled as a spring K_t representing the transfer of the road force to the unsprung mass through the tire's elastic property. Note that gravity is not considered in the model since only motion over and above the static deflections are considered [29].

The detailed procedures of modeling is described as follows:

Step 1. Identify the elements that make up the system (Fig. 2.33a). Here the masses M_s and M_u are elements that store kinetic energy, hence they can be modeled as

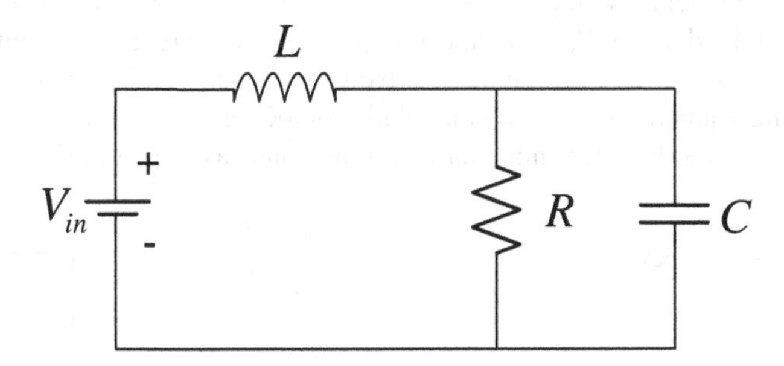

Fig. 2.31 A RLC electric circuit

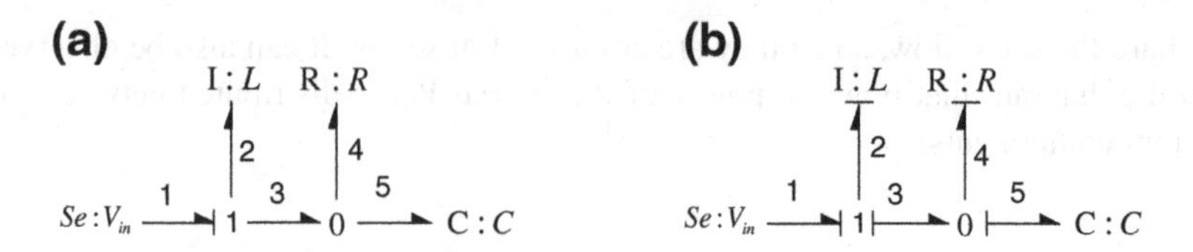

Fig. 2.32 Causality assignment process of the electric circuit

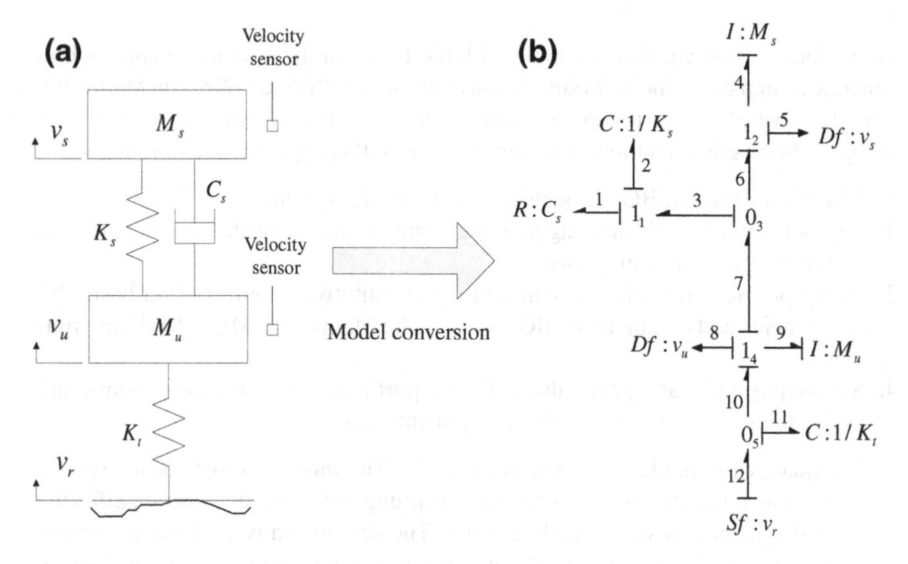

Fig. 2.33 Quarter car model representations. **a** physical model, **b** bond graph model

I-elements in BG terminology. The springs K_s and K_t are elements for storing potential energy, and they can be modeled as C-elements. As for the damper C_s, it dissipates energy and can be modeled as R-element.

Step 2. In order to develop the whole BG model of the system, first, two 1-junctions (1_2, and 1_4) are used to represent distinct velocities (v_s and v_u) of the system. The junctions are attached the physical elements (M_s and M_u) that move with the distinct velocities. 0-junctions (0_3, and 0_5) are used to establish the relative velocities across the remaining elements, so the relative velocity across the suspension (modeled as a spring K_s and a damper C_s in parallel) is $v_u - v_s$ and the relative velocity across the tire spring is $v_r - v_u$. Note that the relative velocity across the suspension is also modeled using 1 junctions (1_1) with the relative velocities (K_s, K_t and C_s). Finally, complete the BG model after enumerating all the bonds as shown in Fig. 2.33b.

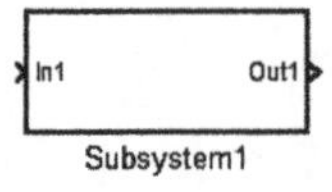

Fig. 2.34 Subsystem block in the Simulink® model

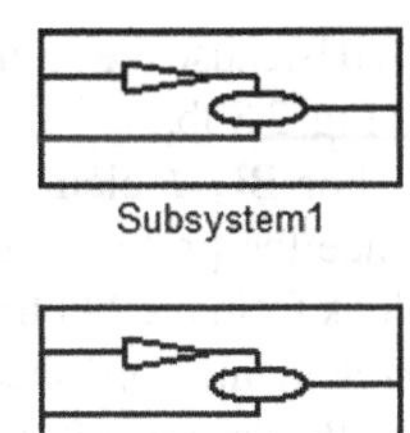

Fig. 2.35 Subsystem block after deleting all ports inside the block

Fig. 2.36 Subsystem block with name hidden

Step 3. All BG basic elements and bonds are built by using the "subsystem" block in the Simulink® library.

A. Built bonds (half arrow bond and information bond)
Copy the "subsystem" block from the Ports and Subsystems library into the model as shown in Fig. 2.34. Open the Subsystem block by double-clicking it and delete all ports inside the block as shown in Fig. 2.35. Right click the Subsystem block and set Format as Hide Name as shown in Fig. 2.36. Right click the Subsystem block and select the Mask Subsystem. Then put "plot ([0.9 0.1 0.4], [0.5 0.5 0.7], [0.9 0.9], [0.3 0.7], [0.92 0.92], [0.3 0.7]) text (0.45, 0.2, num2str (BondNo))" in the mask editor and choose invisible for block frame as shown in Fig. 2.37. The purpose of the plot command is to draw the half arrow figure in the subsystem through defining the coordinates of different points. Then choose Parameters in mask editor and add a new parameter named BondNo in the editor as shown in Fig. 2.38. Click apply in the mask editor and press ok as shown in Fig. 2.39. It is obvious that different directions of half arrow can be drawn by setting different plot command. Finally, double-click the bond and set the required bond number as shown in Fig. 2.40. As for full arrow bond (information bond), designer can use the similar steps to establish it.

B. Build 1-port components (R, I, C)
For 1-port R-element, repeat the steps as shown from Figs. 2.34–2.36. Then right click the Subsystem block and select Mask Subsystem. Put disp (['R:Cs']) in the mask editor and choose invisible for block frame as shown in Fig. 2.41. In the figure, symbol R is the 1-port resistor and C_s is the physical parameter which is attached to this symbol by bond graph convention. Next, define the variable for R-element (C_s under this case) as shown in Fig. 2.42.

The next step is to build the constitutive relation of R-element under the mask of the subsystem block. Right click the subsystem block and choose "Look Under Mask". Copy the "From" block and "Goto" block from the Signal Routing library into the model. Drag the "Product" block from the Math Operations library into the model. Copy the "Constant" block from the Sources library into the model. Next, define "From" block as f_1 by double-clicking it. In a similar way, define "Goto" block as e_1 and "Constant" block as C_s. Finally connect block f_1 and block C_s to the two input ports of "Product" block and connect block e_1 to the output port of "Product" block as shown in Fig. 2.43. Note that subscribe 1 corresponds to the number of bond connected to the R-element (Cs) as shown in Fig. 2.33b. Here resistive causality is used because the input of the R-element is flow (f_1) and output is effort (e_1) according to Fig. 2.33b.

For BG I-element, we can define the variable Ms as the physical parameter. Since the preferred integral causality is required for I-element, drag the "Divide" block from the Math Operations library into the model instead of "Product" block for R-element. Copy the "Integrator" block from the Continuous library into the model. For the case of M_s connected to bond 4 in Fig. 2.33b, define "From" block as e_4 and "Goto" block as f_4. The constitutive relation model of M_s is shown in Fig. 2.44.

Finally, consider the C-element with physical variable $1/K_s$, define "From" block as f_2 and "Goto" block as e_2. The constitutive relation model is shown in Fig. 2.45.

C. Junctions

For 0-junction, consider the modeling of 0_3 in Fig. 2.33b, the constitutive equations can be written as

$$f_7 - f_6 = f_3, \; e_6 = e_6 = e_3 \tag{2.25}$$

The constitutive relation model of junction 0_3 is shown in Fig. 2.46.

By taking junction 1_1 as an example of modeling of BG 1-junction, the constitutive laws can be represented as

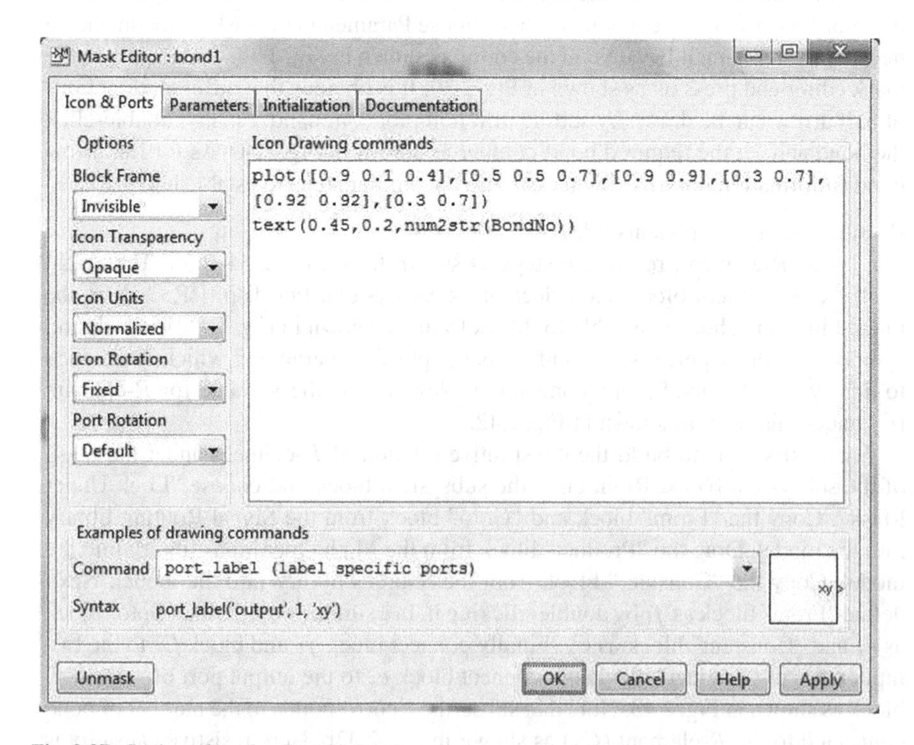

Fig. 2.37 Setting of mask editor for bond

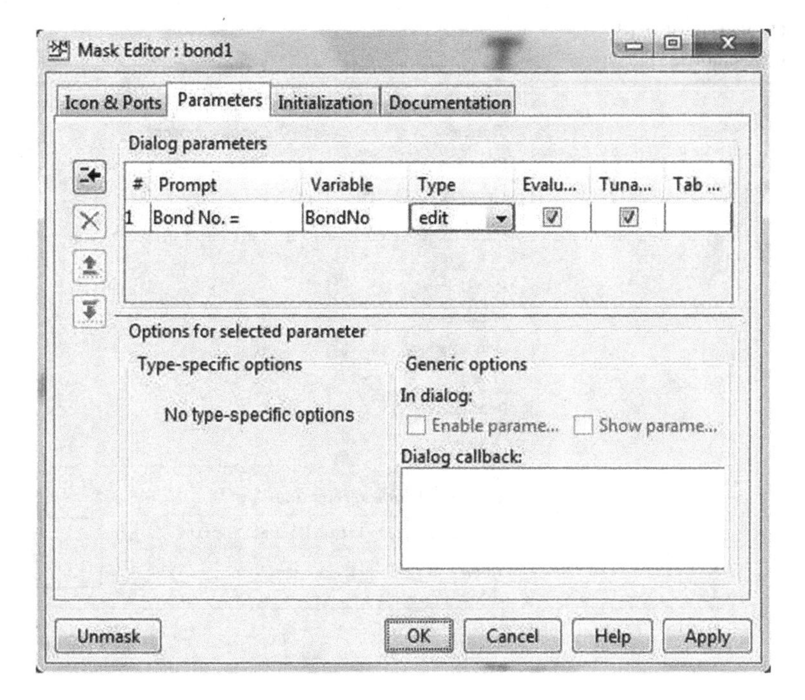

Fig. 2.38 Setting of mask editor with adding new variable

Fig. 2.39 A half arrow bond in Simulink®

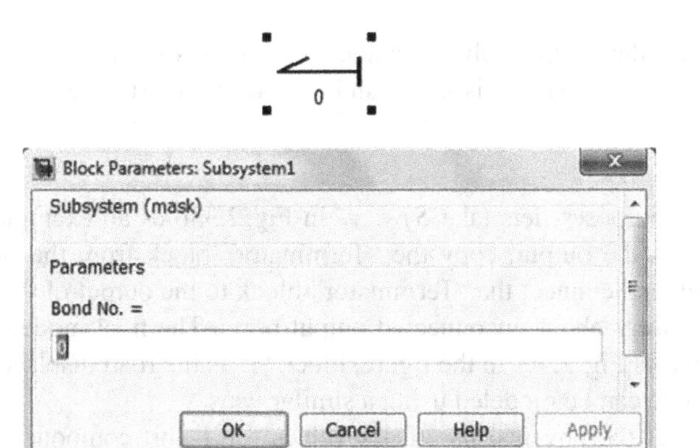

Fig. 2.40 Setting the bond number in Simulink®

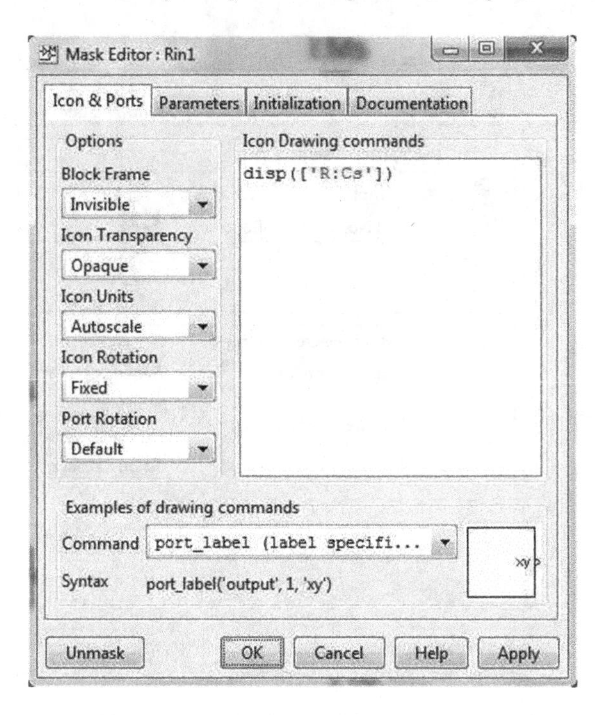

Fig. 2.41 Setting of mask editor for R-element

$$e_1 + e_2 = e_3, \; f_1 = f_2 = f_3 \tag{2.26}$$

The constitutive relation model of junction 1_1 is shown in Fig. 2.47.

D. Sensors

For flow sensor, the effort is always equal to zero, thus the model of constitutive relation of flow sensor $Df : v_s$ is shown in Fig. 2.48. Similarly, the flow of the effort sensor is zero.

E. Sources

For illustrative purposes, lets take $Sf : v_r$ in Fig. 2.33b as an example. Since the effort variable is the output, copy the "Terminator" block from the Sinks library into the model, and connect the "Terminator" block to the output of the effort (e_{12}) to prevent warnings about unconnected output ports. The final model of this flow sensor is shown in Fig. 2.49. In the figure, block "r" is the road displacement input. The effort source can be modeled using a similar way.

At Step 4, set the physical parameter values for 1-port components. Consider example of R-element (C_s) as shown in Fig. 2.33b, key in 1,000 after double-clicking the $R : C_s$ block as shown in Fig. 2.50. Other physical parameters can be defined in a similar way. Table 2.2 lists all the physical parameters used in the mechanical example.

Then assign the input value for source flow. For simplicity, the road input is written as

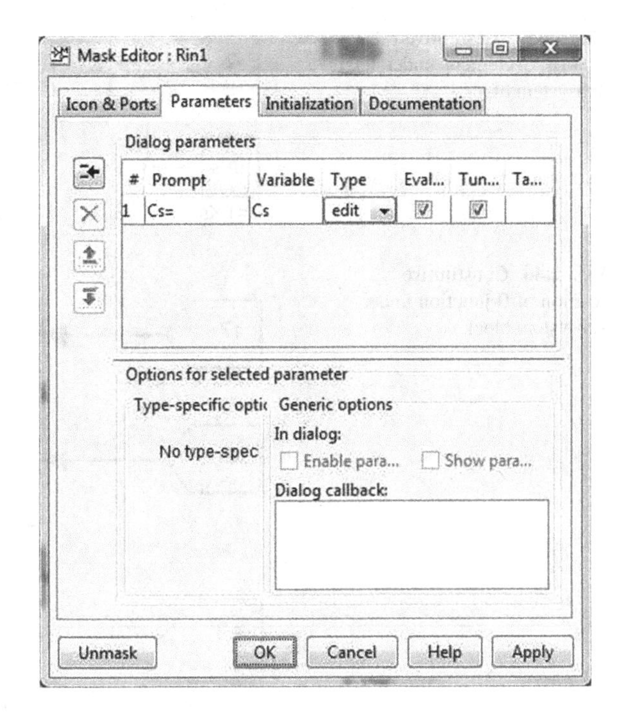

Fig. 2.42 Define variable for R-element

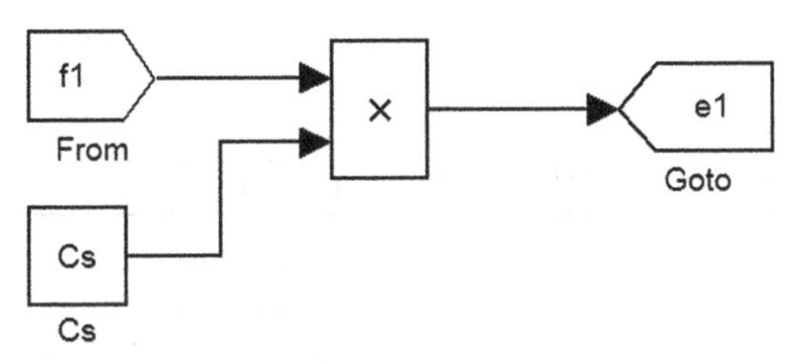

Fig. 2.43 Constitutive relation of R-element under subsystem block

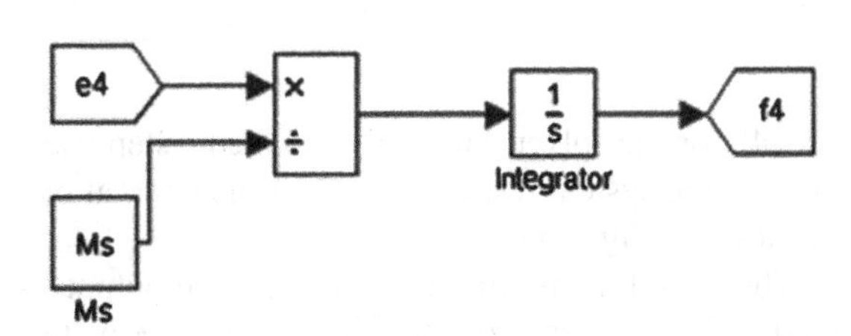

Fig. 2.44 Constitutive relation of I-element under subsystem block

$$v_r = \frac{d}{dt}r = \frac{d}{dt}[0.04sin(2\pi t)] = 0.08\pi cos(2\pi t) \tag{2.27}$$

The whole BG model of the quarter car system is shown in Fig. 2.51. In order to run the system and observe the system responses, select "Simulation" in the tool bar and click "Configuration Parameters". Then key in 10 for "Stop time" and choose

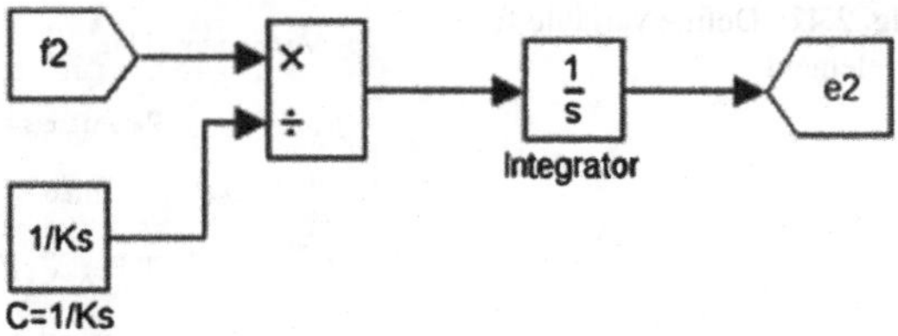

Fig. 2.45 Constitutive relation of C-element under subsystem block

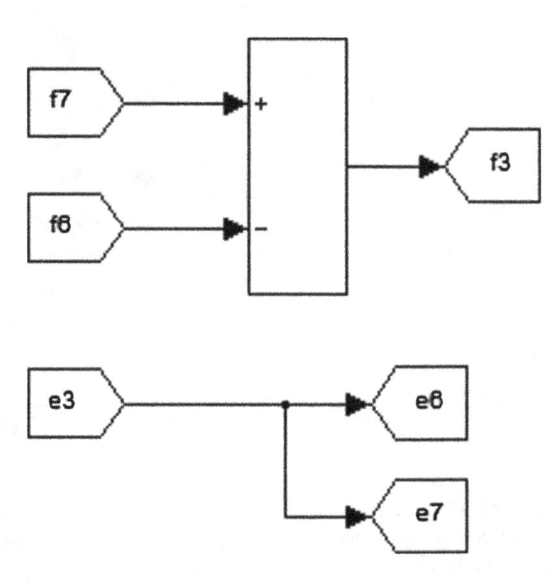

Fig. 2.46 Constitutive relation of 0-junction under subsystem block

Table 2.2 Nominal physical parameters of the quarter car system

Parameter	Value	Unit
M_s	290	kg
M_u	59	kg
K_s	16800	N/m
K_t	190000	N/m
C_s	1000	N· s/m

"ode4" for the solver. Put 0.05 for Fixed—step size as shown in Fig. 2.52. We can observe the system responses at the time interval of 10 s with a sample time 0.05 s as shown in Fig. 2.53.

The causal paths from the source to the outputs, i.e., v_s and v_u, can be decided for Fig. 2.33b. Start from the flow signal f_{12} of the source input, this signal will go to the strong bond of the connected junction 0_5, denoted as $f_{12} \rightarrow f_{11}$. Then f_{11} encounters a 1-port element $C : 1/K_t$ through which the causal path returns with a qualitative change. The causal path under this condition can be represented as $f_{12} \rightarrow f_{11} \rightarrow C : 1/K_t \rightarrow e_{11}$. The effort signal e_{11} will be distributed by junction 0_5: one goes back to the source where the causal path terminates, denoted as $f_{12} \rightarrow f_{11} \rightarrow C : 1/K_t \rightarrow e_{11} \rightarrow e_{12}$ and the other path is represented as $f_{12} \rightarrow f_{11} \rightarrow C : 1/K_t \rightarrow e_{11} \rightarrow e_{10}$. Since the effort signal e_{10} is the

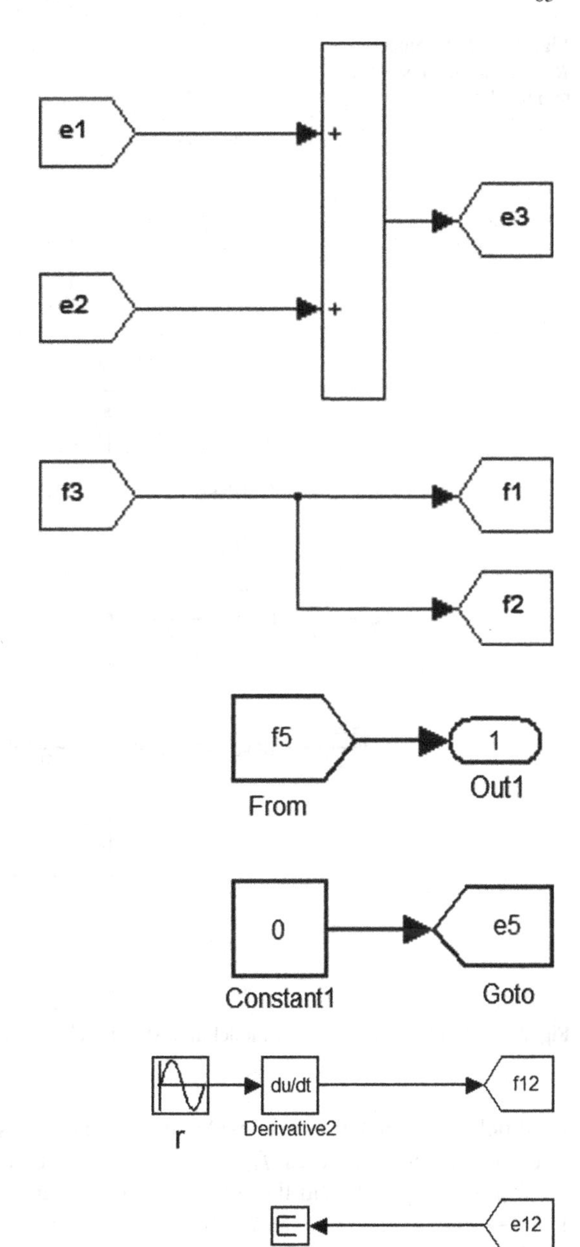

Fig. 2.47 Constitutive relation of 1-junction under subsystem block

Fig. 2.48 Constitutive relation of flow sensor under subsystem block

Fig. 2.49 Constitutive relation of flow source under subsystem block

weak variable of the next junction 1_4, the path will go to strong bond 9, denoted as $f_{12} \rightarrow f_{11} \rightarrow C : 1/K_t \rightarrow e_{11} \rightarrow e_{10} \rightarrow e_9$. The signal of strong bond 9 will undergo a qualitative change through $I : M_u$ denoted as $f_{12} \rightarrow f_{11} \rightarrow C : 1/K_t \rightarrow e_{11} \rightarrow e_{10} \rightarrow e_9 \rightarrow I : M_u \rightarrow f_9$. Signal f_9 will follow two paths from junction

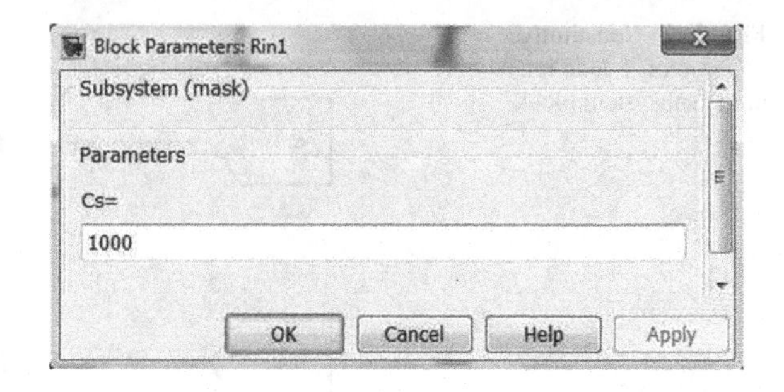

Fig. 2.50 Set value for R-element under subsystem block

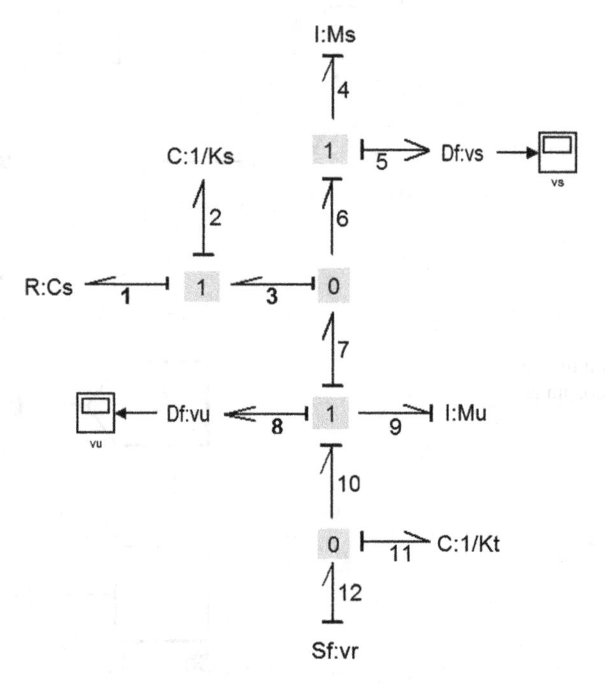

Fig. 2.51 Bond graph quarter car model in SIMULINK® environment

1_4 (if only forward paths are considered): the first one is the causal path from source to output v_u represented as $f_{12} \rightarrow f_{11} \rightarrow C : 1/K_t \rightarrow e_{11} \rightarrow e_{10} \rightarrow e_9 \rightarrow I : M_u \rightarrow f_9 \rightarrow f_8$ and the other one can be denoted as as $f_{12} \rightarrow f_{11} \rightarrow C : 1/K_t \rightarrow e_{11} \rightarrow e_{10} \rightarrow e_9 \rightarrow I : M_u \rightarrow f_9 \rightarrow f_7$. Since signal f_7 is not the strong bond of the next junction 0_3, this signal goes to the strong bond (bond 3) denoted as $f_{12} \rightarrow f_{11} \rightarrow C : 1/K_t \rightarrow e_{11} \rightarrow e_{10} \rightarrow e_9 \rightarrow I : M_u \rightarrow f_9 \rightarrow f_7 \rightarrow f_3$. Bond 3 is the strong bond and f_3 will be divided into two paths: one is represented as $f_{12} \rightarrow f_{11} \rightarrow C : 1/K_t \rightarrow e_{11} \rightarrow e_{10} \rightarrow e_9 \rightarrow I : M_u \rightarrow f_9 \rightarrow f_7 \rightarrow f_3 \rightarrow f_1$ and the other is denoted as $f_{12} \rightarrow f_{11} \rightarrow C : 1/K_t \rightarrow e_{11} \rightarrow e_{10} \rightarrow e_9 \rightarrow I : M_u \rightarrow f_9 \rightarrow f_7 \rightarrow f_3 \rightarrow f_2$. Two causal paths will encounter 1-port elements ($C : 1/K_s$ and $R : C_s$ respectively) and then qualitative changes will

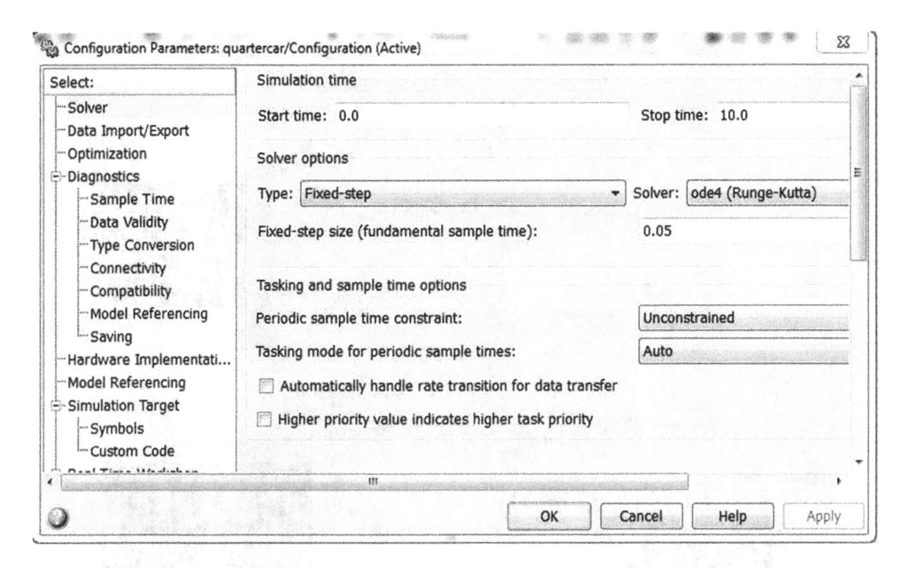

Fig. 2.52 Configuration Parameters in SIMULINK® environment

happen. Thus these two causal paths can be written as $f_{12} \rightarrow f_{11} \rightarrow C : 1/K_t \rightarrow e_{11} \rightarrow e_{10} \rightarrow e_9 \rightarrow I : M_u \rightarrow f_9 \rightarrow f_7 \rightarrow f_3 \rightarrow f_1 \rightarrow R : C_s \rightarrow e_1$ and $f_{12} \rightarrow f_{11} \rightarrow C : 1/K_t \rightarrow e_{11} \rightarrow e_{10} \rightarrow e_9 \rightarrow I : M_u \rightarrow f_9 \rightarrow f_7 \rightarrow f_3 \rightarrow f_2 \rightarrow C : 1/K_s \rightarrow e_2$. These two weak variables (i.e., e_1 and e_2) will go to the strong bond 3 denoted as $f_{12} \rightarrow f_{11} \rightarrow C : 1/K_t \rightarrow e_{11} \rightarrow e_{10} \rightarrow e_9 \rightarrow I : M_u \rightarrow f_9 \rightarrow f_7 \rightarrow f_3 \rightarrow f_1 \rightarrow R : C_s \rightarrow e_1 \rightarrow e_3$ and $f_{12} \rightarrow f_{11} \rightarrow C : 1/K_t \rightarrow e_{11} \rightarrow e_{10} \rightarrow e_9 \rightarrow I : M_u \rightarrow f_9 \rightarrow f_7 \rightarrow f_3 \rightarrow f_2 \rightarrow C : 1/K_s \rightarrow e_2 \rightarrow e_3$. Finally, these two causal paths will reach f_5 as follows: $f_{12} \rightarrow f_{11} \rightarrow C : 1/K_t \rightarrow e_{11} \rightarrow e_{10} \rightarrow e_9 \rightarrow I : M_u \rightarrow f_9 \rightarrow f_7 \rightarrow f_3 \rightarrow f_1 \rightarrow R : C_s \rightarrow e_1 \rightarrow e_3 \rightarrow e_6 \rightarrow e_4 \rightarrow I : M_s \rightarrow f_4 \rightarrow f_5$ and $f_{12} \rightarrow f_{11} \rightarrow C : 1/K_t \rightarrow e_{11} \rightarrow e_{10} \rightarrow e_9 \rightarrow I : M_u \rightarrow f_9 \rightarrow f_7 \rightarrow f_3 \rightarrow f_2 \rightarrow C : 1/K_s \rightarrow e_2 \rightarrow e_3 \rightarrow e_6 \rightarrow e_4 \rightarrow I : M_s \rightarrow f_4 \rightarrow f_5$.

There are other BG modeling softwares like 20-sim®, SYMBOLS (SYstem Modeling by BOnd graph Language and Simulation) Shakti and CAMPG (Computer Aided Modeling Program with Graphical Input). CAMPG is a model generator, which takes the topological description of a physical system model described by a Bond Graph, and transforms it into a dynamic simulation model in source code form [30]. The CAMPG output files contain the differential equations of the system, input, output and initial conditions variables automatically generated by the computer. CAMPG is a preprocessor that generates computer models for linear or non-linear systems in multi energy domains using a unified approach. It will interface with MATLAB®/SIMULINK®, FORTRAN, C or the user's own program. In other words, CAMPG provides a graphical user interface for the creation of bond graphs and can generate files that are accepted as input files by MATLAB® and Simulink® for further model processing. The 20-sim® software developed by the

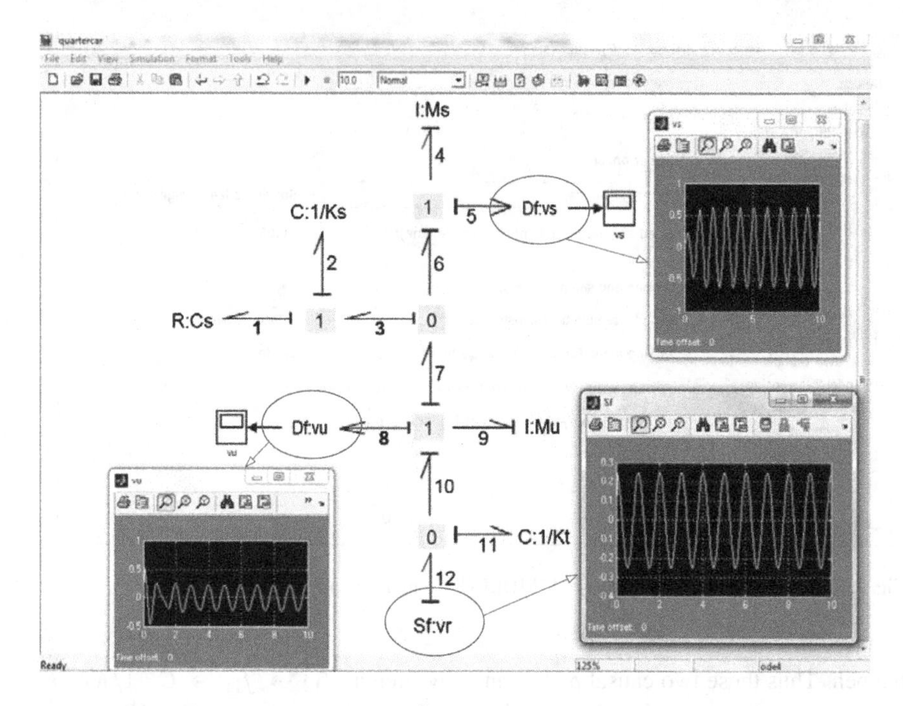

Fig. 2.53 System responses in SIMULINK® environment

University of Twente is based on the well-known block oriented TUTSIM simulation program [31]. 20-sim® is a modern modeling and simulation program that runs under Microsoft Windows operating system. With 20-sim® you can simulate the behavior of dynamic systems, such as electrical, mechanical and hydraulic systems or any combination of these. SYMBOLS Shakti is an objective oriented hierarchical modeling, simulation and control analysis software. It allows users to create models using bond graph, block diagram and equation models [32].

2.4 Hybrid Bond Graph

Hybrid systems consist of continuous dynamics and discrete behavior represented by modes. At each mode, the system is governed by continuous dynamics, and each different mode is modeled by different continuous models. There are various efforts to model the discontinuous behavior in hybrid systems. In [33], a modulated transformer is developed to model the discrete behavior in hybrid systems. When the switch is closed, the modulation parameter m of the modulated transformer is 1, while $m = 0$ when the switch is open. However, when the switch opens the transformer connected junction, incorrect flow values will present. Sometimes this junction fails to disconnect the subsystems due to the existence of freewheeling flows.

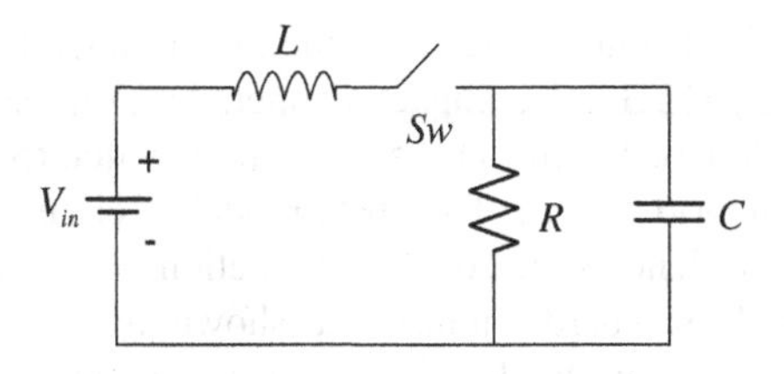

Fig. 2.54 A simple hybrid system: electrical circuit

Moreover, causality reassignment is required when the switching occurs. In order to overcome this problem, a combination of modulated transformer and resistance is proposed in [34], where a fixed causality BG is achieved even if the physical configuration changes. However, the presence of resistance gives rise to dissipation, which indicates a non-ideal switching. The resistance also results in numerically stiff systems that will increase the simulation time. An ideal switch is proposed in [35] as a new BG element. The element enforces zero effort on the junction when the switch is ON and zero flow when the switch is OFF. Since effort or flow is always zero, the power (effort × flow) associated with the switch element is always zero. The disadvantage of this approach is that one needs to reassign the causality if the switching happens. To handle this problem, a concept of switched power junction is developed to represent switching phenomena in systems [36, 37]. The switched power junction (SPJ) models ideal switching, so that the problem of stiff systems and associated numerical stability problems while simulating the system are eliminated. The SPJ are a generalization of the standard 0 and 1 junctions. Thus, the 0 and 1 junctions are special cases of the more general switched power 0 and switched power 1 junctions. BG is geared towards modeling of continuous systems. To model a hybrid system using BG language, additional features are necessary to capture the discrete mode changes of the hybrid system.

HBG extends BG modeling by incorporating the controlled junctions to enable the hybrid system to be modeled using the BG components [38–40]. For 1-type controlled junction, it enforces zero flow to all connected bonds when it is turned off. Similarly, a 0-type controlled junction forces the effort value to zero at all connected

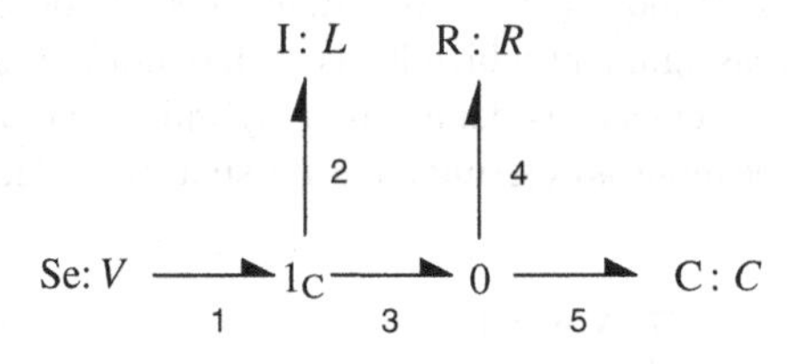

Fig. 2.55 HBG model of the circuit when the switch is ON

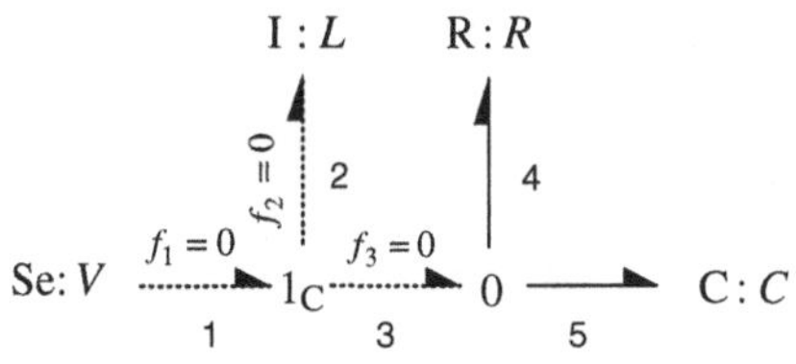

Fig. 2.56 HBG model of the circuit when the switch is OFF

bonds during the OFF state. Consequently, the hybrid systems can be represented by HBG in a compact manner. An example of hybrid system, i.e., electrical circuit, is depicted in Fig. 2.54, where a switch (Sw) is used to control the ON/OFF of the hybrid system. The HBG model of circuit is presented in Fig. 2.55, where junction 1_C denotes the controlled junction. When the controlled junction is ON, it functions like standard 1-junction as shown in Fig. 2.55; when the junction is OFF, it enforces zero flows to all adjacent connected bonds. This deactivation of bond 1, 2 and 3 are represented in Fig. 2.56 by dash lines.

2.4.1 Causality Properties and Causality Assignment for HBG

HBG is a BG-based modeling approach which provides an avenue to model complex hybrid systems. In this section, a study on the HBG is presented, and some properties pertaining to the HBG are introduced. Based on these findings, a causality assignment procedure technique is presented to achieve a HBG with a desirable causality assignment that leads a unified description of system's behavior.

2.4.1.1 Causality Properties of HBG

HBG extends the ability of BG to model hybrid systems using controlled junctions. However, changes in configuration through the various operating modes of the system can result in a need to reassign the causality of the HBG. In [41], the properties of HBG are studied so that it can be efficiently utilized for simulations of hybrid systems. A hybrid SCAP algorithm is developed to reassign causality incrementally after mode changes, starting from the junction directly affected by the switching, and then propagating the changes only to those junctions whose causal assignments are affected. In [42], this procedure is improved by using the concept of junctions in fixed causality and bonds with persistent causality. To exploit the rich information of a HBG, it is desirable to have a consistent causality description for the hybrid system at all modes. This consistent causality description eliminates the need for causality reassignment which leads to derivation of unified constraint relations.

Let us consider a simple hybrid system in electrical domain with two modes, and the mode is determined by the state of an electrical switch as depicted in Fig. 2.57. Two

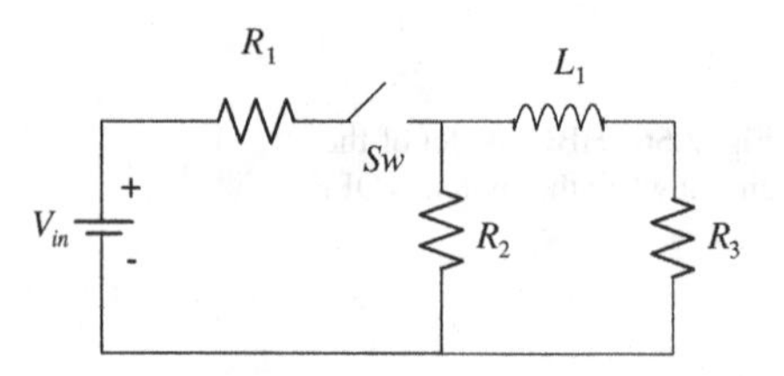

Fig. 2.57 A hybrid system of electrical circuit

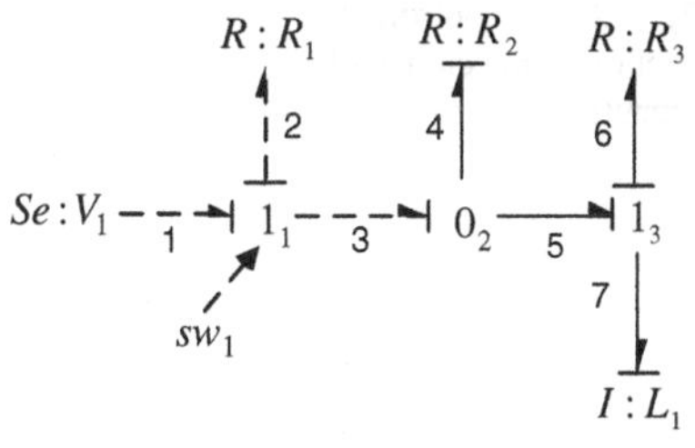

Fig. 2.58 HBG with a causality assignment when the switch is OFF

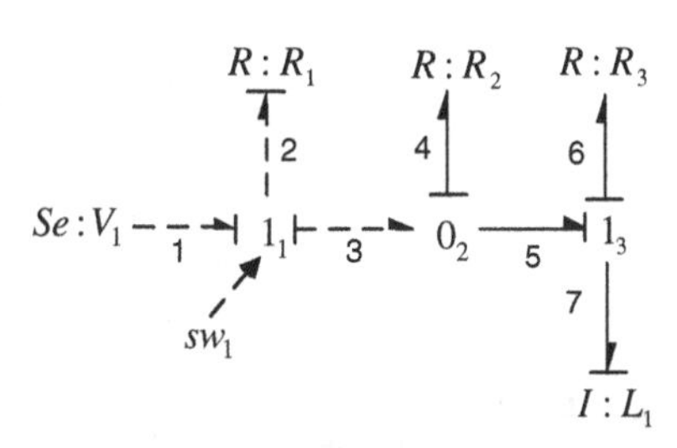

Fig. 2.59 HBG with an alternate causality assignment when the switch is OFF

HBG models with different causality assignments are depicted in Figs. 2.58 and 2.59. For reference purposes, all the bonds and BG components of the HBG are enumerated. The full arrow bond of switch sw_1 means that a signal flow occurs at essentially zero power flow. In the figures, the active bonds are represented by solid lines while inactive bonds are represented by dash lines. Here active bonds refer to bonds which are not deactivated by an OFF state of the controlled junction.

Figure 2.58 describes three inactive bonds during the OFF state of the controlled junction 1_1. Notice that the inactive bonds cause invalid causality to junction 0_2 which leads to the invalid causality of the HBG during the OFF state. This invalid causality obstructs the generation of the constraint equations at the OFF mode and the problem is significant for complex systems with many controlled junctions. Next consider the same HBG with an alternate causality assignment as shown in Fig. 2.59. In the figure, the inactive bonds during the OFF state of the controlled junction 1_1 don't pose any invalid causality to the active BG components. For example, the junction 0_2 still has a valid causality when the bond 3 is deactivated. This observation suggests that it is possible to describe the behavior of a hybrid system based on one unified causality assignment where all active bonds' causalities are consistent, and the causality is valid at all modes.

Without loss of generality, the general case where a controlled junction is connected to a junction 0, 1, a 1-port component, or a 2-port component is considered. First, the case of BG junctions is considered.

Figure 2.60 depicts a 1-controlled junction (1_c) and a BG junction with the given causality assignment shown in the figure. The causal stroke of bond 3 indicates that the effort (e_3) is the output variable of 1_c, which is also an input variable of the 0-junction; where f_3 is the output variable of 0-junction and an input variable of 1_c. By inspection, if the controlled junction is OFF, the causality of the 0-junction

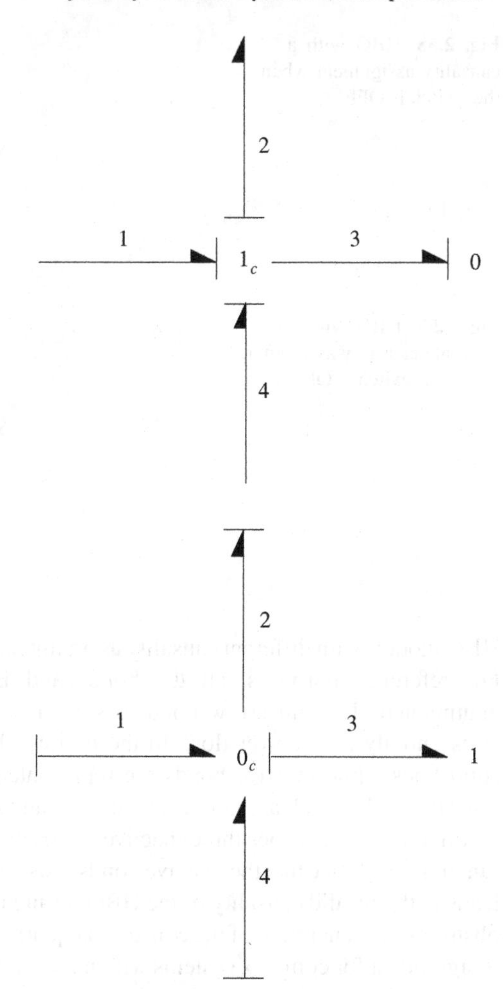

Fig. 2.60 1-controlled junction connected with 0-junction

Fig. 2.61 0-controlled junction connected with 1-junction

becomes invalid. This behavior is due to the OFF state of 1_c, which causes the output variable of the 0-junction to be undefined. Similarly, this behavior also applies to a 0-controlled junction, as shown in Fig. 2.61. This property is stated as follows.

Property 2.1 Suppose a controlled junction is adjacently connected to a normal junction 0, 1. If the output variable of the controlled junction is an input variable of the normal junction, then the OFF state of the controlled junction poses an invalid causal form to the normal junction.

In HBG language, all 1-port components that are adjacently connected to a controlled junction are disconnected from the remaining graph when the junction is OFF (i.e., their single bond is deactivated). Therefore, consideration of causal validity of any 1-port component is not required.

Property 2.1 provides a causal condition on the controlled-junctions which is needed to be avoided when assigning the causality of the HBG. Besides property 2.1, a condition that the OFF state of the controlled junction poses no invalid causal form to the HBG can be deduced. One such condition can be stated as follows.

Property 2.2 If the output variable of a controlled junction is an input variable of a 1-port component, then the OFF state of the controlled junction poses no invalid causal form to the HBG.

To show this property, let us consider the following cases: (1) a controlled junction that is adjacently connected to a normal junction, and (2) a controlled junction that is adjacently connected to a 2-port element. First, let us consider case (1). In BG language, any two adjacently connected junctions of the same type can be merged into one junction; therefore, it is assumed that any junction which is adjacently connected to a 0-controlled junction is a 1-junction. From the rules of causality, if the output variable of a 0-controlled junction is an input variable of a 1-port component, then the output variable of the 1-junction is not a variable of the bond that connects between the two junctions. Therefore, the output variable of the 1-junction is always well-defined, i.e., the 1-junction always has a valid causal form even when the 0-controlled junction is OFF. Now, consider case (2) where the controlled junction is adjacently connected to a 2-port component. The causality rules of 2-port component imply that when the 0-controlled junction is OFF, it imposes zero effort at the effort-input of the 2-port component (i.e., the input from the controlled-junction side) and the constitutive relation of the 2-ports component remains valid. This validity of the 2-port constitutive relations implies that the component has a valid causal form even when the controlled junction is OFF. Therefore, the OFF state of the controlled junction poses no invalid causal form to the HBG. The same argument applies to the case of 1-controlled junction.

Property 2.2 states a sufficient condition for a controlled-junctions causality that will not cause an invalid causal form to the HBG when the junction is OFF. One critical implication of this sufficient condition is that if the causalities of all controlled junctions are assigned in such a way that all their output variables are inputs of some 1-port components, then all active bonds maintain their causal forms, and the causality of the graph is valid at all operating modes.

One intuitive property can be useful in a special case is presented as follows.

Property 2.3 Consider two adjacently connected controlled junctions that have the same states at all operating modes. If one of the controlled junction's output variable is an input variable of the other controlled junction, then the OFF states of the controlled junction pose no invalid causal form to the HBG.

Property 2.3 is straightforward because the output variables of the two controlled-junctions are not input variable of any normal BG junction that is adjacently connected to them, and hence the OFF states of the controlled-junctions pose no invalid causal form to the HBG. Moreover, since the two controlled-junctions always have the same states, the OFF state of one implies the OFF state of the other; therefore, the two controlled-junctions do not pose any invalid causal form on themselves.

2.4.1.2 Causality Assignment for HBG

From the properties mentioned earlier, it is concluded that a controlled junction is said to be in preferred causality if the junctions output variable is an input variable of the following components: a 1-port component (R, C or I), a source element if the source is null when the junction is OFF, another controlled junction of different type which shares the identical state. For a given acausal HBG (HBG without causality assignment), it is desirable to assign a causality to the HBG such that all controlled junctions are in preferred causalities. In the case of the behavior model, it is desirable that all the storage components (C and I) of the HBG to be in integral causalities since integral causality is recommended for engineering simulation for ease of formulation and avoiding the numerical problems from differentiation. Here, a systematic procedure called the Sequential Causality Assignment Procedure for Hybrid Systems (SCAPH) is developed from an acausal HBG.

1. Choose any controlled junction that has no source component adjacently connected to it and assign the junction with its preferred causality.
2. Repeat step 1 until all the controlled junctions which have no source component connected to them have been assigned.
3. Choose any remaining controlled junction that has a source component connected to the junction. For the pair $\{S_e, 1_c\}$ or $\{S_f, 0_c\}$, the output variable of the controlled junction must not be an input variable of $\{0, 1\}$. For the pair $\{S_f, 1_c\}$ or $\{S_e, 0_c\}$, the output variable of the controlled junction must be the input variable of the connected source component.
4. Repeat step 3 until all the remaining controlled junctions have been assigned. After this step, all controlled junctions are treated as normal junctions $\{0, 1\}$ in the following procedures.
5. Choose any remaining source component and assign its causality. Immediately extend the causal implications through the HBG as far as possible using the constraint components $\{0, 1, GY, TF\}$.
6. Repeat step 5 until all sources have been assigned.
7. Choose any storage component and assign it with an integral causality. Immediately extend the causal implications through the HBG as far as possible using the constraint components $\{0, 1, GY, TF\}$.
8. Repeat step 7 until all storage components have been assigned with a causality.
9. Choose any unassigned R component and assign a causality to it. Immediately extend the causal implications through the HBG as far as possible using the constraint components $\{0, 1, GY, TF\}$.
10. Repeat step 9 until all remaining bonds have been assigned.

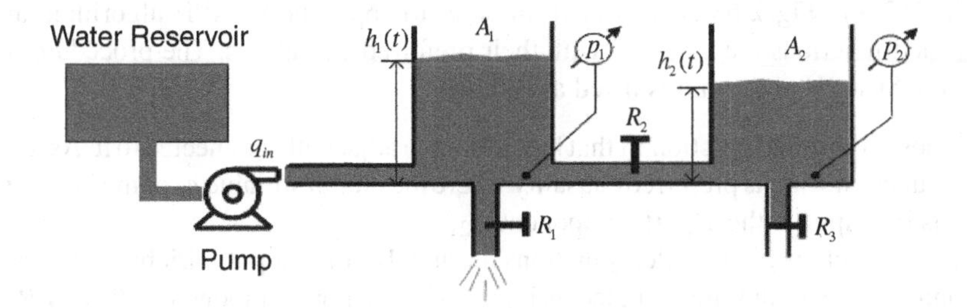

Fig. 2.62 A two-tank system

2.4.2 Illustrative Examples

A two-tank system is considered in this example (see Fig. 2.62). The two-tank system consists of two tanks, regulated centrifugal pump modeled as a source of flow q_{in} and three valves represented by R_1, R_2 and R_3. $A_1[m^2]$ and $A_2[m^2]$ are cross-section areas of the two tanks. The system is equipped with two pressure sensors ($p_1(t)$ and $p_2(t)$) to measure the pressure at the bottom of tank A_1 and tank A_2, respectively; this pressure is proportional to the liquid level, according to:

$$p_i(t) = \rho g h_i(t), \qquad i = 1, 2. \tag{2.28}$$

where ρ is liquid density, [kg/m^3]; g is the acceleration due to gravity, [m/s^2], $h_i(t)$ is liquid height in the tank, [m].

Each valve has two discrete states ON and OFF with negligible switching-time between open and closed states. The valves' dynamics is given by

$$f_j(t) = 0, \qquad j = 1, 2, 3 \qquad \text{when the valve is closed} \tag{2.29}$$

$$f_j(t) = \frac{\text{sign}(\Delta p(t))\sqrt{|\Delta p(t)|}}{R_j} = Cd_j \cdot \text{sign}(\Delta p(t))\sqrt{|\Delta p(t)|} \qquad \text{when the valve is open} \tag{2.30}$$

where f_j is the liquid-flow through the valve, [m^3/s]; $\Delta p(t)$ is the pressure difference across the valve, [Pa] and Cd_j is the coefficient of discharge, [$\sqrt{\text{kg} \cdot \text{m}}$].

The HBG of the two-tank plant is presented in Fig. 2.63. In this bond graph, the flow variable is the liquid volumetric flow (i.e., m^3/s), and the effort variable is pressure (i.e., [Pa]). The two tanks are modeled by the two storage components with coefficient $C_i = \frac{A_i}{g}$ for $i = 1, 2$. Each one of the three valves is modeled by a set of resistor with parameter $R_j = \frac{1}{Cd_j}$ for $j = 1, 2, 3$ and a controlled-junction.

The HBG in Fig. 2.63 is acausaled. In order to apply the SCAPH algorithm, all controlled-junctions are assigned with their preferred causalities. The procedure of applying SCAPH algorithm is listed as follows:

1. Choose controlled junction 1_3 that has no source adjacently connected to it. Assign the junction with its preferred causality where the output variable e_7 of the junction 1_3 is the input of the 1-port component R_2.
2. Similarly, choose controlled junctions 1_2 and 1_5, the output variables of these controlled junctions are assigned as inputs of the 1-port components (R_1 and R_3, respectively).
3. Since all controlled junctions have been assigned, all controlled junctions are treated as normal junctions $\{0, 1\}$ in the following procedures.
4. The source component Sf is assigned.
5. Since no other source, proceed to Step 6.
6. The two storage components C_1 and C_2 are assigned with preferred integral causality. Now, the causal implication using the components constraints to bonds 3, 6, 8 and 11 to complete the SCAPH algorithm.

The HBG is shown in Fig. 2.64. In Matlab® Simulink® environment, controlled junction is modeled with the aid of boolean variable $\alpha_j,\ j = 1, 2, 3$. Let us take

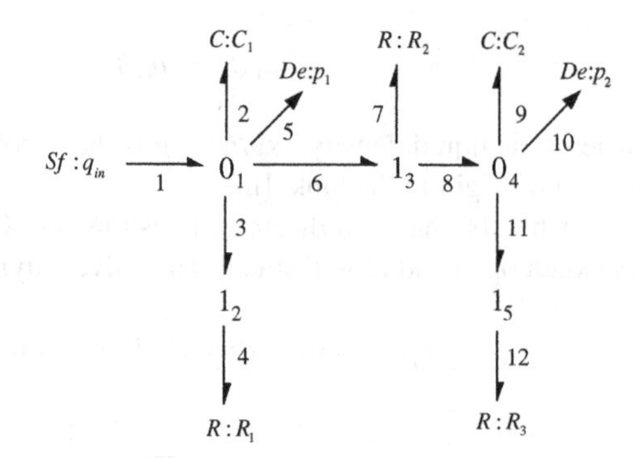

Fig. 2.63 The two-tank plant acausaled HBG

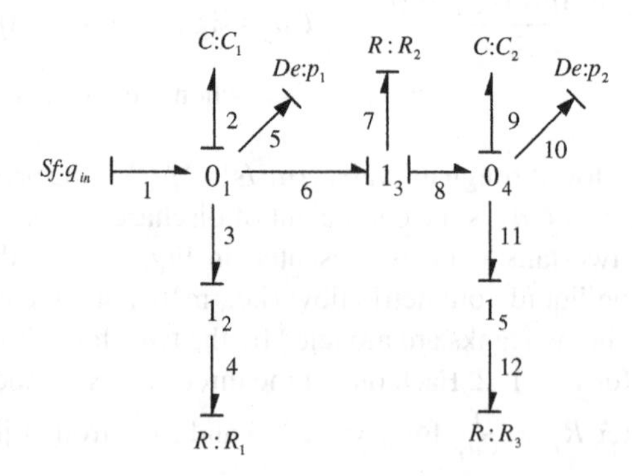

Fig. 2.64 The two-tank plant HBG

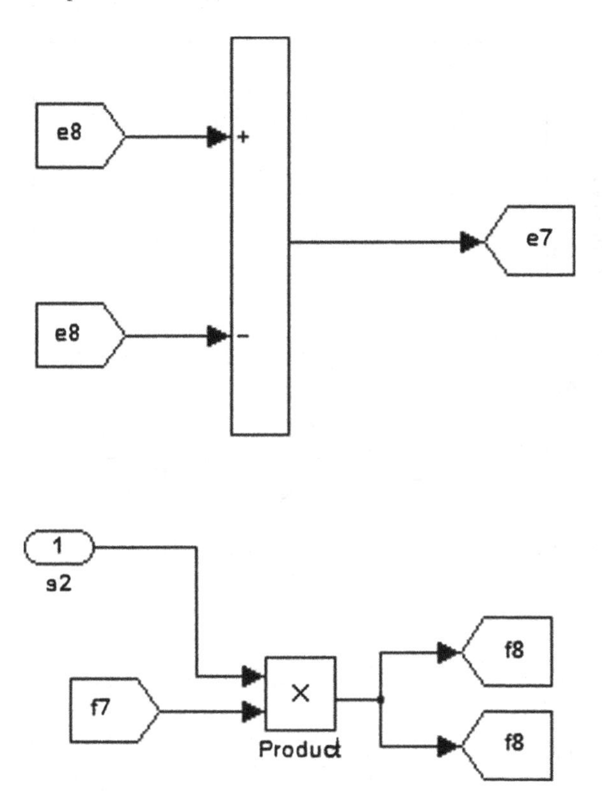

Fig. 2.65 Modeling of controlled junction 1_3 in Simulink®

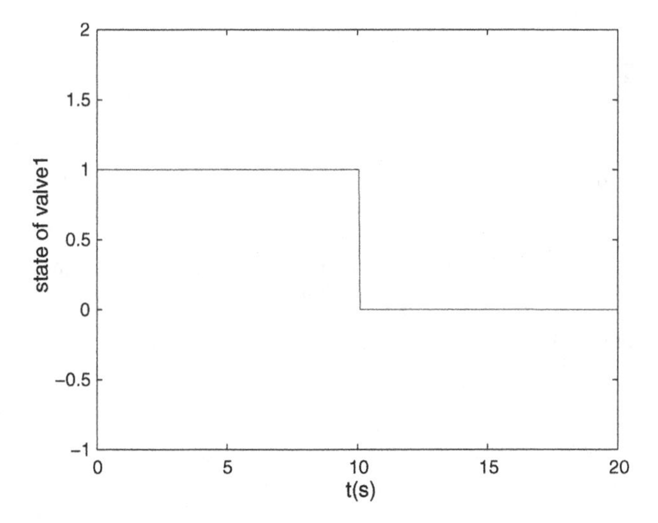

Fig. 2.66 State of valve1

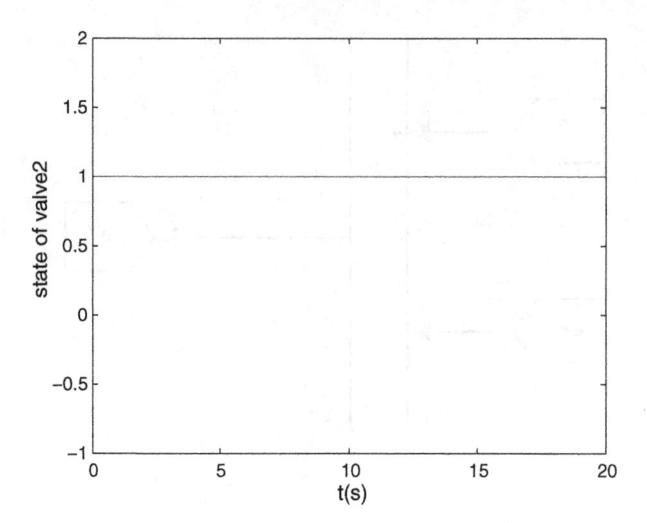

Fig. 2.67 State of valve2

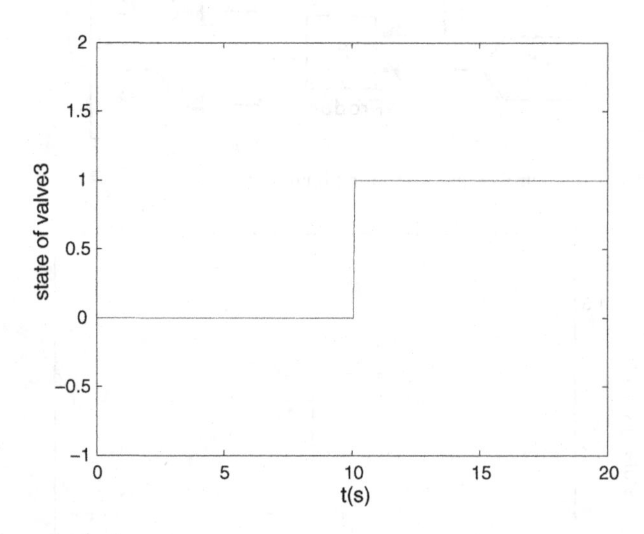

Fig. 2.68 State of valve3

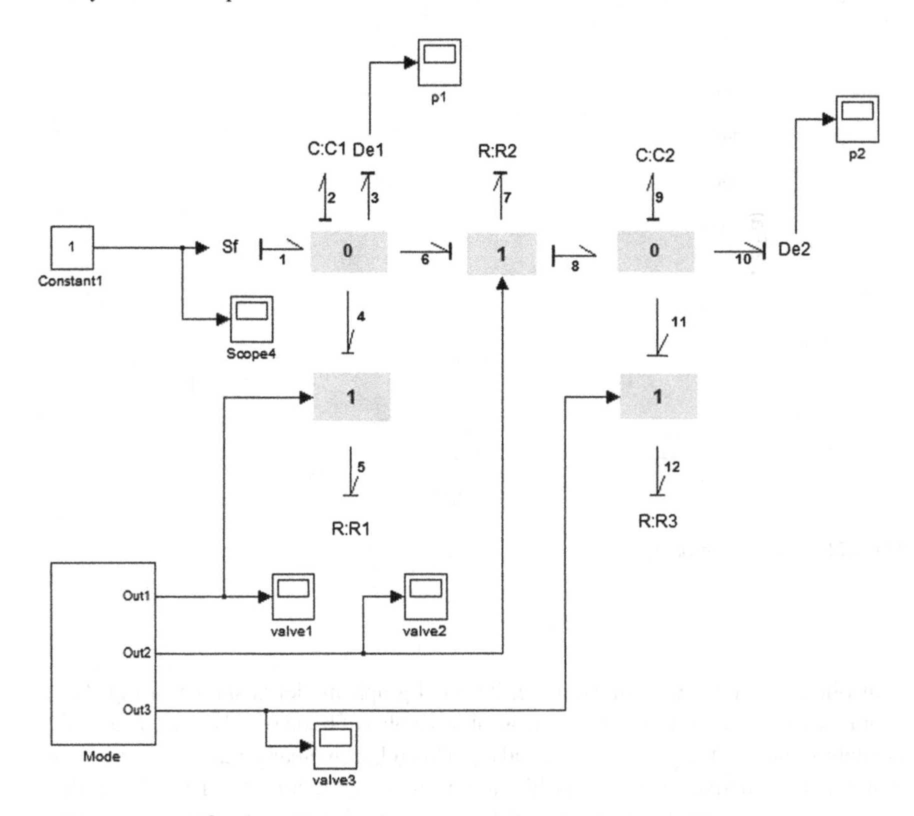

Fig. 2.69 Simulink® model of two tank system

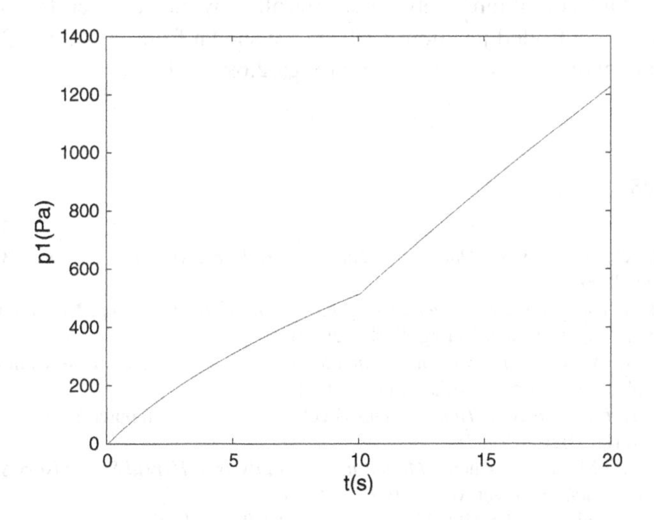

Fig. 2.70 Pressure sensor p_1

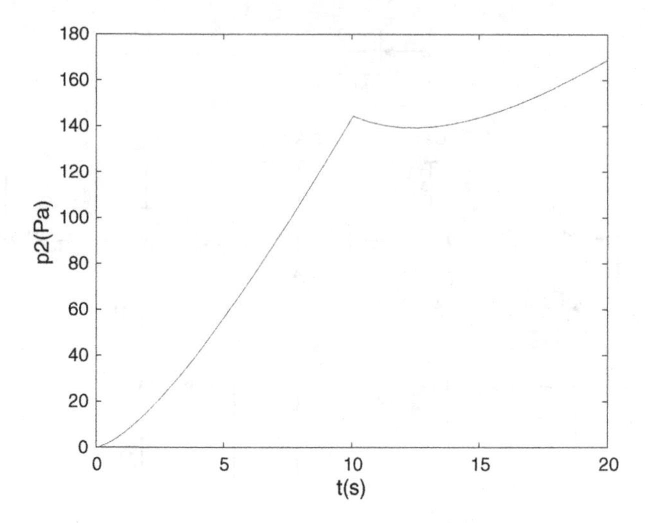

Fig. 2.71 Pressure sensor p_2

controlled junction 1_3 as an example, its bond graph model is shown in Fig. 2.65. From the figure, it is observed that the weak variables of bond 6 and 8 (i.e., f_6 and f_8) are determined by the product of f_7 and α_2. This relation means that if the controlled junction 1_3 is active, it functions like a common 1-junction and if the controlled junction 1_3 is deactivated all flow variables are forced to zero. Similarly, other two controlled junctions 1_2 and 1_5 can be modeled in the Simulink® environment. The states (ON or OFF) of all three valves are controlled by the designer. In simulation, the states of the controlled junctions (valves) are shown from Figs. 2.66–2.68. The whole BG model of the system is shown in Figs. 2.69–2.71.

References

1. A.S. Matveev, A.V. Savkin, *Qualitative Theory of Hybrid Dynamical Systems* (Birkhauser, Cambridge, 2000)
2. M. Rupak, T. Paulo, *Hybrid Systems: Computation and Control*. Lecture Notes in Computer Science, vol. 5469 (Springer-Verlag, Berlin, 2009)
3. B. Krogh, N. Lynch, *Hybrid Systems: Computation and Control*. Lecture Notes in Computer Science, vol. 1790 (Springer-Verlag, Berlin, 2000)
4. O. Maler, *Hybrid and Real-Time Systems*. Lecture Notes in Computer Science, vol. 1201 (Springer-Verlag, Berlin, 1997)
5. E. Villani, P.E. Miyagi, R. Valette, *Modelling and Analysis of Hybrid Supervisory Systems: A Petri Net Approach* (Springer-Verlag, Berlin, 2006)
6. J. Zander, I. Schieferdecker, P.J. Mosterman, *Model-Based Testing for Embedded Systems*. (CRC Press, Taylor & Francis, UK 2010)

7. P.J. Antsaklis, A. Nerode, IEEE Trans. Automat. Contr. (Spec. Issue Hybrid Control Syst.) **43** (1998)
8. A. Morse, C. Pantelides, S. Sastry, J. Schumacher, Automatica (Spec. Issue on Hybrid Systems). **35** (1999)
9. R. Evans, A.V. Savkin, Syst. Contr. Lett. (Spec. Issue on Hybrid Control Syst.) **38** (1999)
10. P.J. Antsaklis, Proc. IEEE. (Spec. Issue on Hybrid Syst. Theory App.) **88**(7) (2000)
11. F. Zhao, X. Koutsoukos, H. Haussecker, J. Reich, P. Cheung, Monitoring and fault diagnosis of hybrid systems. IEEE Trans. Syst. Man Cybern. B Cybern. **35**(6), 1225–1240 (2005)
12. K.M. Passino, U. Ozguner, Modeling and analysis of hybrid systems: examples, *in Proceedings of IEEE International Symposium on Intelligent Control*, (Arlington, 1991), pp. 251–256
13. T. Samad, G. Balas, *Software-Enabled Control: Information Technology for Dynamical Systems*. (Wiley-IEEE Press, New York, 2003)
14. C.A. Petri, W. Reisig, Petri net. Scholarpedia **3**(4), 6477 (2008)
15. P.J. Mosterman, G. Biswas, Behavior generation using model switching: a hybrid bond graph modeling technique. Trans. Soc. Simul. **27**(1), 177–182 (1995)
16. H.M. Paynter, Hydraulics by analog—an electronic model of a pumping plant. J. Boston Soc. Civil Eng. **46**(6), 197–219 (1959)
17. R.C. Rosenberg, D.C. Karnopp, A definition of the bond graph language. Trans. ASME J. Dyn. Syst. Measur. Control **94**(3), 179–182 (1972)
18. H.M. Paynter, An Epistemic Prehistory of Bond Graphs, ed. by P.C. Breedveld, G. Dauphin-Tanguy, in *Proceedings of the Bond Graphs for Engineers*, (Elsevier Science Publisher, Amsterdam, 1992), pp. 3–17
19. H.M. Paynter, *Analysis and Design of Engineering Systems* (MIT Press, Cambridge, 1961)
20. D.C. Karnopp, R.C. Rosenberg, *System Dynamics: A Unified Approach* (Wiley, New York, 1975)
21. D.C. Karnopp, D.L. Margolis, R.C. Rosenberg, *System Dynamics—Modeling and Simulation of Mechatronic Systems*, 3rd edn. (Wiley, New York, 2000)
22. D.C. Karnopp, D. Margolis, R.C. Rosenberg, *System Dynamics: Modeling and Simulation of Mechatronic Systems*, (Wiley, New York, 2006)
23. W. Borutzky, A. Orsoni, R. Zobel, eds. *in Proceedings of the 20th European Conference on Modelling and, Simulation* (2006)
24. J.U. Thoma, H.J. Halin (eds.), Special issue bond graphs for modeling and simulation. Simul. Prac. Theory **7**(5–6) (1999)
25. P.J. Gawthrop, S. Scavarda (eds.), Special issue on bond graphs. Proc. Inst. Mech. Eng. Part I: J. Syst. Control Eng. **216**(1) (2002)
26. I. Troch, W. Borutzky, P.J. Gawthrop (eds.), Special issue: bond graph modelling. Math. Comput. Model. Dyn. Syst. **2**(2–3) (2006)
27. W. Borutzky (ed.), Special issue: bond graph modelling. Simul. Model. Pract. Theory **17**(1) (2009)
28. P.J. Gawthrop, G.P. Bevan, Bond graph modeling. IEEE Control Syst. Mag. **27**(2), 24–45 (2007)
29. A.K. Samantaray, B. Ould Bouamama, *Model-based Process Supervision: A Bond Graph Approach* (Springer, London, 2008)
30. J.J. Granda, Computer Aided Modeling Program (CAMP), a Bond Graph Preprocessor for Computer-Aided Design and Simulation of Physical Systems Using Digital Simulation Languages, Thesis, University of California, Davis, 1982
31. Twentesim, *Users Manual of Twentesim (20sim)*. (Controlab Products Inc, Enschede, 1996)
32. A. Mukherjee, A.K. Samantaray, System modelling through bond graph objects on SYMNOLS 2000, in *Proceedings of the International Conference on Bond Graph Modeling and Simualtion (ICBGM'01)*. Simualtion Series, vol. 33 (2001), pp. 164–170
33. G.M. Asher, The robust modelling of variable topology circuits using bond graphs, in *Proceedings of International Conference on Bond Graph Modelling*, (1993), pp. 126–131
34. W. Borutzky, G. Dauphin-Tanguy, J.U. Thoma, Advances in bond graph modelling: theory, software and applications. Math. Comput. Simul. **39**(5–6), 465–475 (1995)

35. J.E. Stromberg, J. Top, U. Soderman, Variable causality in bond graph caused by discrete effects, in *Proceedings of International Conference on Bond Graph Modelling*, (1993), pp. 115–119
36. A.C. Umarikar, L. Umanand, Modelling of switched mode power converters using bond graph. IEEE. Proc. Electr. Power Appl. **152**(1), 51–60 (2005)
37. A.C. Umarikar, L. Umanand, Modelling of switching systems in bond graphs using the concept of switched power junctions. J. Franklin Inst. Eng. Appl. Math. **342**(2), 131–147 (2005)
38. P.J. Mosterman, G. Biswas, Behavior generation using model switching: A hybrid bond graph modeling technique. Trans. Soc. Comput. Simul. **27**(1), 177–182 (1995)
39. P.J. Mosterman, G. Biswas, Behaviour generation using model switching-a hybrid bond graph modelling technique, in *Proceedings of the International Conference on Bond Graph Modelling*, (1995), pp. 177–182
40. P.J. Mosterman, G. Biswas, A theory of discontinuities in physical systems model. J. Franklin Inst. **335**(3), 401–439 (1998)
41. I. Roychoudhury, M. Daigle, G. Biswas, X. Koutsoukos, P.J. Mosterman, A method for efficient simulation of hybrid bond graph, in *Proceeding of International Conference Bond Graph Modeling and Simulation*, vol. 335 (2007), pp. 177–184
42. I. Roychoudhury, M. Daigle, G. Biswas, X. Koutsoukos, Efficient simulation of hybrid systems: an application to electrical power distribution systems, in *Proceeding of the 22nd European Conference Modeling, Simulation* (2008), pp. 471–477

Chapter 3
Quantitative Hybrid Bond Graph-Based Fault Detection and Isolation

3.1 Introduction

A Fault Detection and Isolation (FDI) system detects a fault or failure so that safety can be ensured and replacement of essential components can be performed timely and effectively to ensure smooth operations. In general, a model-based FDI algorithm evaluates a set of equations named Analytical Redundancy Relations (ARRs). ARRs represent the discrepancy between information obtained from the actual system and that generated by a model of the system. In short, an ARR is a constraint relation that contains known terms of the system. These known terms usually include inputs, model parameters and measurements. This constraint relation is derived by eliminating its unknown variables based on other analytic equations so that some analytical redundancies are embedded in the relation. As a result, this relation is able to detect a change when a fault has occurred. The numerical evaluation of ARRs produces residuals. FDI is performed by evaluating residuals in a real time manner. Besides fault detection, the residuals generated by these ARRs can also be used to isolate the causes of faults.

The objective of this chapter is to introduce ARR based FDI approach which lays the essential foundation for the study on model-based FDI for hybird systems. The tasks of designing and constructing model-based FDI for hybird systems will be formulated and presented in this chapter.

3.2 Bond Graph-Based Fault Diagnosis

The performance of a model-based FDI method depends on the quality of the system's model. Modelling of a physical system is a demanding step where the difficulty increases with the complexity of the system. Fortunately, BG modelling provides an approach to deal with a complex system which possesses large number of subsystems and components [1]. Additionally, BG model provides a systematic and convenient

D. Wang et al., *Model-based Health Monitoring of Hybrid Systems*,
DOI: 10.1007/978-1-4614-7369-5_3, © Springer Science+Business Media New York 2013

tool to design FDI algorithms for complex systems. From the FDI point of view, the causal properties of the bond graph model are used to determine the origin of the faults. In the BG diagnosis, it is important to have a BG model that includes all the parameters that are related with the faults under analysis.

Recently, BG-based FDI algorithms have been proposed [2–5] with applications [6–8]. These BG-based methods can be classified as *quantitative* [3–8] and *qualitative* methods [2, 9]. Quantitative BG-based approaches utilize BG model to derive quantitative ARRs for FDI [3–8]. These quantitative ARRs are evaluated to generate quantitative residuals to assess system fault status in both transient and steady-state operations. Additionally, quantitative approaches allow both incipient and abrupt parametric faults to be detected. On the other hand, qualitative approach utilizes qualitative descriptions of the system's BG for FDI [2, 9, 10].

Quantitative BG-based methods can be further divided into *symbolic* and *numerical* methods. For symbolic methods [3, 4], symbolic ARRs are derived from BG to generate residuals for FDI. In contrast, numerical BG-based method does not require symbolic ARRs to generate residuals for diagnosis [5]. The numerical BG-based method formulates each quantitative ARR implicitly based on simpler constraint equations that allow residuals generation in graphical-programming environments, e.g., MATLAB® SIMULINK®. In the absence of nonlinear algebraic causal loop, both approaches would lead to the same number of residuals. However, if there exists a nonlinear algebraic causal loop in the BG, then the numerical method may offer a solution to generate the maximum number of residuals for the system. Although the numerical method offers possibility to generate maximum number of residuals in the presence of a nonlinear algebraic causal loop, the approach has two limitations. First, implementation of the numerical method is restricted to graphical-based programming systems. Second, the numerical method does not exploit symbolic ARR equations for FDI designs. Symbolic ARRs provide useful insights on residual behaviors which can be exploited for better FDI designs. For example, analytical bounds of residuals with respect to a parametric modelling error, may be deduced analytically with the help of the symbolic ARRs. In short, symbolic and numerical BG-based FDI methods have their advantages and limitations, and the choice between the two approaches depend on the characteristics of the monitored system, and the available computational resources. In this chapter, symbolic ARRs are adopted for FDI design.

In general, quantitative model based fault diagnosis method consists of two main stages: residual generation and residual evaluation. The residual generation in brief is a procedure for constructing ARRs using system model. In the residual evaluation, the trend of the generated residuals are evaluated to determine whether a fault has occurred. A well-designed residual simplifies the residual evaluation process.

3.2.1 Analytical Redundancy Relationships

ARRs are model constraints, expressed by known variables such as inputs, sensor measurements and physical parameters. In symbolic ARRs derived from BG model, an ARR has the general form of

$$F_l(\theta, De, Df, u) = 0 \qquad \text{for } l = 1, \ldots, m. \tag{3.1}$$

where m denotes the number of ARRs derived from the BG model, $\theta = [\theta_1, \ldots, \theta_p]^T$ represents the parameters of the components, where p represents the number of parameters used in BG to describe the system. u denotes the system's inputs, De and Df denote the effort and flow sensors of the BG.

Usually, ARRs are derived from physical laws, such as Newton law and Kirchoff law. However, an ARR itself may not always have a physical meaning; nevertheless an ARR still represents constraint which needs to be satisfied during system normal operation.

3.2.2 Residual Evaluation and Fault Signature Matrix

Numerical evalution of an ARR is a residual. Since in most cases the formulation of ARRs require the use of derivatives or high order derivatives of sensor measurements, usually a pre-filter is required to filter out the noise influence on the residuals. In general, residual r_l, defined by an ARR, is said to be sensitive to a fault in a component if the ARR depends on the component's parameters. When the system is fault free, every computed residual will be consistent with the system's behavior. This means that the absolute value of every residual r_l falls below a small threshold value ε_l. To apply the set of residuals for FDI, a binary coherence vector $C = [c_1 \ \ldots \ c_m]$ is defined. Each component c_l of C is obtained by the following simple decision rule:

$$c_l = \begin{cases} 1 & \text{if } |r_l| > \varepsilon_l; \\ 0 & \text{otherwise} \end{cases} \qquad \text{for } l = 1, \ldots, m \tag{3.2}$$

If the system is fault-free, then the binary coherence vector C will be zero. On the other hand, if the system is faulty, then the coherence vector will be a nonzero vector.

To study the fault detectability and fault isolability, a Fault Signature Matrix (FSM) is generated from the m ARRs of the system [6]. A typical FSM is shown in Table 3.1. In Table 3.1, the column headers represent the residuals $r_1, \ldots, r_m$, the

Table 3.1 Fault signature matrix (FSM)

	r_1	...	r_m	D_b	I_b
θ_1	1 or 0				
.					
.					
.					
θ_p					

fault detectability (D_b), and the fault isolability (I_b). Each entry of the table holds a boolean value. For each row, the boolean entries under the columns $r_1, \ldots, r_m$ form the fault signature of the parameter θ_i that corresponds to a fault in θ_i. The fault is represented by changing the value of θ_i from a nominal to faulty value, either abruptly or incipiently. Under a residual column, a 1 in an entry indicates that the residual is sensitive to the fault of the corresponding parameter that lies in the matching row. On the other hand, a 0 in the entry represents that the residual is insensitive to the fault of the parameter. If at least a 1 appears in the fault signature of the parameter θ_i, then the parameter is said to be fault detectable. This ability is represented by $D_b = 1$ in the matrix. When the fault signature of the parameter is unique, then the parameter is said to be fault isolable. This is denoted by $I_b = 1$. It is beneficial to note that fault detectability is a necessary but not sufficient condition for fault isolability.

3.2.3 Generation of ARRs

There are two existing BG-based methods to generate symbolic ARRs [3, 4]. The ARR generation method presented in [3] is called the *Covering Path method*. Roughly speaking, the method exploits the causality information of BG to derive symbolic ARRs by scrutinizing the constitutive relations of all BG junctions and the causal paths associated with the junctions. This procedure provides an intuitive and direct way to derive ARRs from BG; however, it will be shown later that this approach alone may not generate the optimum number of symbolic ARRs. Additionally, this approach requires all storage elements of the graph to be assigned in *derivative causality*. The storage elements are preferred to be in *derivative causality* to avoid unknown initial states in the ARR equations. The second symbolic BG-based ARR generation method, which is called *Causality Inversion* method, generates symbolic ARR equations from a BG junction that is equipped with a sensor. This method requires a causality inversion on a sensor attached to the junction to generate an ARR. If the sensor's causality is irreversible, then the ARR generation algorithm is not able to derive an ARR from the junction's constitutive relation.

An ARR is generated by eliminating unknown variables of a system's constraint equation. However, the elimination of these unknown variables is not necessarily a trivial task, especially for complex systems. For linear systems, the *parity space technique* developed in [11] can be viewed as elimination of the unknown variables using projection method [12]. Similarly, this concept has been extended to nonlinear systems for generation of ARRs [13]. In these methods, the generation of ARRs are not systematic. For large complex systems, the elimination of these unknown variables becomes a daunting task.

The parity space method is a well-known ARR based FDI technique, but the way to derive ARR is different with the one used in BG based FDI. The parity space is well suited for sensors and actuators FDI. While the identified model is given under state equation form, the generated ARRs parameters do not have a physical perception and cannot be associated with physical component fault. In other words, by comparing a BG based FDI with a parity space FDI method using state-space model, BG provides component-level information that allows the FDI algorithm to isolate a fault at component level easily.

3.2.3.1 Covering Path Method

Covering path method is a systematic approach that generates symbolic ARRs from a BG [3]. This approach utilizes the causal paths of BG to eliminate all unknown variables to generate ARRs. The method is presented algorithmically in [6] which can be stated as follows.

1. Choose a BG junction.
2. Find the corresponding ARR by writing the constitutive (structural) relation (or the sum relation) of the considered junction. The goal is to reduce the total unknown variables of the relation so that the ARR can be obtained. The unknown variables are solved from the BG model using the causal paths that connect to the variables.
3. Consider the next junction.
4. If the second ARR is independent of the other derived ARRs, then keep it, else consider another junction.
5. Repeat step 2 until all the junctions are considered, and the independent signatures are obtained.

The above procedure provides an orderly way to derive ARRs by eliminating unknown variables of constitutive relation of every junction using causal paths of the graph.

A system under monitoring, *S*, may be described by a set of constraints, *F*, a set of variables, *Z* and a set of parameters θ [14]. The set of constraints *F* basically represents the algebraic constrains and constitutive (structural) relations of elements in the system model. The system can be represented as

$$S = S(F, Z, \theta) \tag{3.3}$$

(a) Constraints

The constraints *F* describe the relations which link the system variables and parameters. In general, in BG model, the constraints *F* include the structure constraints (F_S), the behavior constraints (F_B), the measurement constraints (F_M), and the control constraints (F_C):

$$F = F(F_S, F_B, F_M, F_C) \tag{3.4}$$

In BG model, the structure constraints (F_S) represent a set of conservation laws deduced from the junctions (0-junction and 1-junction) and 2-port elements (*TF* and *GY*) with $F_S \in R^{n_s}$, and n_s is equal to the sum of the number of power bonds connected to each junction (0-junction or 1-junction) and twice the number of 2-port elements. The behavior constraints (F_B) describe the physical (constitutive) relations in lumped-parameter BG elements (I, C and R) with $F_B \in R^{n_B}$, and n_B is equal to the total number of power bonds connected to the lumped-parameter elements (I, C and R). The measurement constraints (F_M) represent the measurements from sensors (De and Df) with $F_M \in R^{n_M}$, and n_M is equal to the total number of sensors in BG model. The control constraints (F_C) describe the controllers, the modulated sources (MSe and MSf) and simple sources (Se and Sf) with $F_C \in R^{n_C}$, and n_C is equal to the sum of the number of controllers and the number of the sources. As a result, $F \in R^{n_s+n_B+n_M+n_C}$.

(b) Variables

The set of variables Z include known (K) and unknown (U):

$$Z = K \cup U \tag{3.5}$$

The known variables includes input variables and measured variables with $K \in R^{n_M+n_C}$. The unknown variables contain power variables (effort and flow) associated with the power bonds. Thus, $U \in R^{2n_D}$, and n_D is equal to the total number of power bonds in BG model.

(c) Parameters

In BG model, parameters refer to the physical quantities of lumped-parameter 1-port elements (I, C and R). Define the vector of parameters as $\theta \in R^{n_p}$, and n_p is equal to the total number of lumped-parameter 1-port elements. It is obvious that $n_p = n_B$.

Consider a BG model with one junction having n bonds connected to it. Each bond is assumed to be connected to one element. Thus, the total number of power variables is $2n$. The junction has n-1 number of equality constraints (i.e., efforts are same for 0-junction and flows are same for 1-junction) and one sum equation [14]. In addition, the n number of bonds connected to n number of external elements give n number of constitutive constraints. As a result, there are $2n$ number of constraints for $2n$ number of variables. If the sensor measurements are known which will bring additional known variables, then the system becomes over-constrained. Therefore, ARRs can be derived from such system.

In order to illustrate the procedure of covering path method, an example of circuit shown in Fig. 3.1 is considered. It comprises of two effort sensors $\{De_1, De_2\}$, two capacitors C_1 and C_2, four resistors $\{R_1, R_2, R_3, R_4\}$, and one voltage source V_1. The BG model of the circuit with preferred derivative causality for storage elements (two capacitors C_1 and C_2) is shown in Fig. 3.2.

The set of unknown variables is:

$$U = [f_1, e_1, f_2, e_2, f_3, e_3, f_4, e_4, f_5, e_5, f_6, e_6, f_7, e_7, f_8, e_8, f_9, e_9, f_{10}, e_{10}] \tag{3.6}$$

The structure constraints (F_S) include constitutive laws of four junctions 1_1, 0_2, 1_3 and 0_4 as follows:

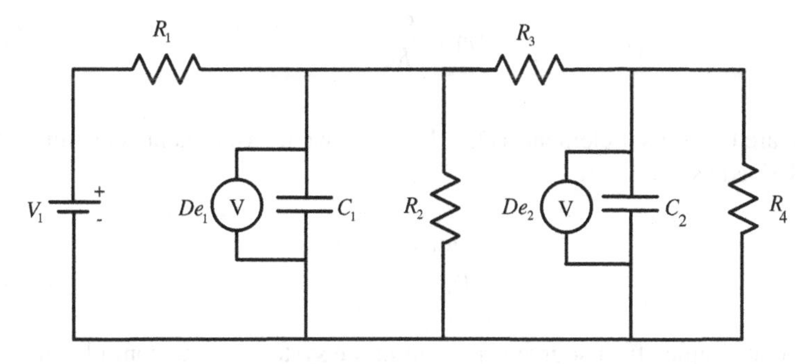

Fig. 3.1 An electrical circuit

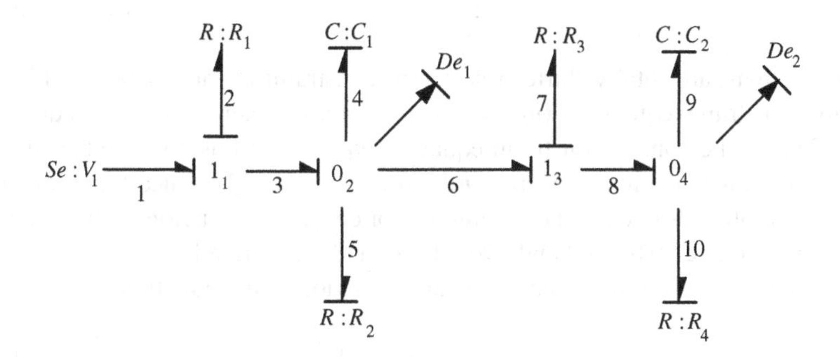

Fig. 3.2 Bond graph model of the circuit

$$f_1 = f_3, f_2 = f_3, e_1 - e_2 - e_3 = 0 \tag{3.7}$$

$$e_4 = e_3, e_5 = e_3, e_6 = e_3, f_3 - f_4 - f_5 - f_6 = 0 \tag{3.8}$$

$$f_6 = f_8, f_7 = f_8, e_6 - e_7 - e_8 = 0 \tag{3.9}$$

$$e_9 = e_8, e_{10} = e_8, f_8 - f_9 - f_{10} = 0 \tag{3.10}$$

As for behavior constraints (F_B), there are six power bonds connected to the lumped-parameter elements (bond 2, bond 4, bond 5, bond 7, bond 9, bond 10), thus six constitutive equations are generated

$$e_2 = R_1 f_2 \tag{3.11}$$

$$f_4 = C_1 \frac{de_4}{dt} \tag{3.12}$$

$$f_5 = \frac{e_5}{R_2} \tag{3.13}$$

$$e_7 = R_3 f_7 \tag{3.14}$$

$$f_9 = C_2 \frac{de_9}{dt} \tag{3.15}$$

$$f_{10} = \frac{e_{10}}{R_4} \tag{3.16}$$

There are two sensor elements (De_1, De_2), so two measurement constraints (F_M) can be derived as

$$De_1 = e_3 \tag{3.17}$$

$$De_2 = e_8 \tag{3.18}$$

Only one simple effort source is present in the system, so the control constraints (F_C) is represented as

$$e_1 = V_1 \tag{3.19}$$

Finally, there are totally thirteen structure constraint equations ($n_s = 13$), six behavior constraint equations ($n_B = 6$), two measurement constraint equations ($n_M = 2$) and one control constraint equation ($n_C = 1$). Thus, the number of constraint equations is twenty-two ($n_s + n_B + n_M + n_C = 22$). Since the number of unknown variables is twenty and the number of constraint equations is twenty-two, the system is over-constrained and two ARRs can be generated.

First, consider junction 1_1, the constitutive relation of this junction is

$$e_1 - e_2 - e_3 = 0 \tag{3.20}$$

The variable e_3 is determined from the measurement Eq. (3.17), and e_1 is known according to (3.19). The variable e_2 is given by the constitutive relation (3.11). Covering the causal path 2-3 leads to

$$f_2 = f_3 \tag{3.21}$$

The variable f_3 is calculated by covering the causal paths 3-4, 3-5 and 3-6

$$f_3 = f_4 + f_5 + f_6 \tag{3.22}$$

The variable f_4 is derived from the constitutive relation (3.12) with $e_4 = De_1$. The variable f_5 is calculated from (3.13) with $e_5 = De_1$.

$$f_4 = C_1 \frac{De_1}{dt} \tag{3.23}$$

$$f_5 = \frac{De_1}{R_2} \tag{3.24}$$

As for f_6, covering the causal paths 6-8 gives $f_6 = f_8$. The variable f_8 is calculated by covering the causal paths 8-9 and 8-10 with (3.15) and (3.16)

$$f_6 = f_8 = f_9 + f_{10} = C_2 \frac{De_2}{dt} + \frac{De_2}{R_4} \tag{3.25}$$

Finally, an ARR can be generated

$$\begin{aligned} ARR_1 &= V_1 - R_1 \left(C_1 \frac{De_1}{dt} + \frac{De_1}{R_2} + C_2 \frac{De_2}{dt} + \frac{De_2}{R_4} \right) - De_1 \\ &= ARR_1(V_1, De_1, De_2, R_1, R_2, R_4, C_1, C_2) \end{aligned} \tag{3.26}$$

For junction 0_2, the constitutive relation can be written as

$$f_3 - f_4 - f_5 - f_6 = 0 \tag{3.27}$$

However, an ARR cannot be generated from this junction since the causal paths of the unknown f_3 (3-4, 3-5 and 3-6) indicates $f_3 = f_4 + f_5 + f_6$, which is Eq. (3.27) itself. Therefore, the resultant residual is always zero regardless of faulty or fault-free condition.

Next, the constitutive relation of junction 1_3 leads to the following equation

$$e_6 - e_7 - e_8 = 0 \tag{3.28}$$

Since $e_6 = De_1$ and $e_8 = De_2$, Eq. (3.28) becomes

$$De_1 - e_7 - De_2 = 0 \tag{3.29}$$

Using (3.14) the unknown variable e_7 is solved $e_7 = R_3 f_7$. Covering causal paths 7-8-9 and 7-8-10, one obtains

$$e_7 = R_3 f_7 = R_3 f_8 = R_3(f_9 + f_{10}) \tag{3.30}$$

According to (3.15) and (3.16), Eq. (3.30) becomes

$$e_7 = R_3 f_7 = R_3 f_8 = R_3(f_9 + f_{10}) = R_3 \left(C_2 \frac{De_2}{dt} + \frac{De_2}{R_4} \right) \tag{3.31}$$

The second ARR can be obtained

$$ARR_2 = De_1 - R_3 \left(C_2 \frac{De_2}{dt} + \frac{De_2}{R_4} \right) - De_2 = ARR_2(De_1, De_2, R_3, R_4, C_2) \tag{3.32}$$

Finally, the constitutive relation of junction 0_4 is considered

$$f_8 - f_9 - f_{10} = 0 \tag{3.33}$$

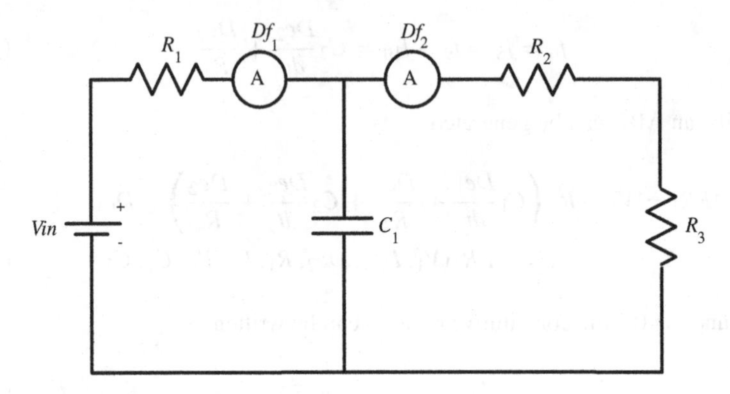

Fig. 3.3 A simple circuit

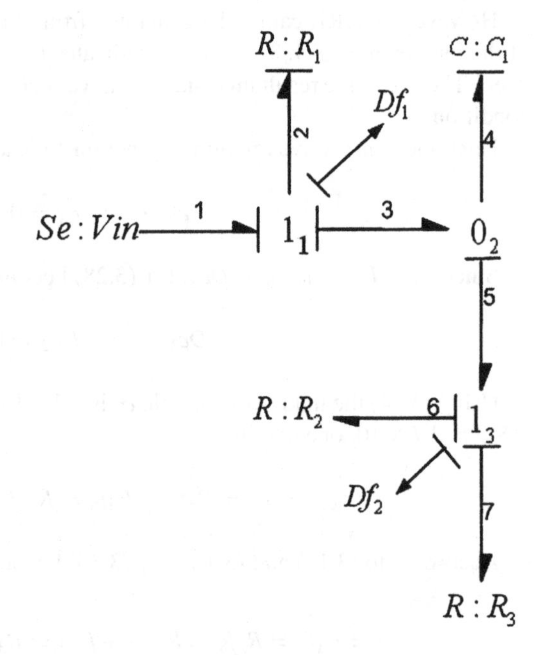

Fig. 3.4 A BG with causality assigned

Similarly, Eq. (3.33) cannot generate an ARR because the unknown variable f_8 requires Eq. (3.33) itself to solve. Thus, totally two independent ARRs are obtained and each of them is sensitive to different set of faults. This result also matches the previous analysis.

In order to explore several properties of covering path method, another example is considered as shown in Fig. 3.3 and its BG model is shown in Fig. 3.4. The system has two flow sensors $\{Df_1, Df_2\}$. Df_1 is connected to junction 1_1 and Df_2 is connected to junction 1_3. Flow sensor Df_1 measures the flow variables $\{f_1, f_2, f_3\}$; hence,

$f_1 = f_2 = f_3 = Df_1$. Similarly, Df_2 measures the flow variables $\{f_5, f_6, f_7\}$, and $f_5 = f_6 = f_7 = Df_2$.

First, the constitutive relation (3.34) of junction 1_1 is considered for ARR.

$$e_1 - e_2 - e_3 = 0 \tag{3.34}$$

The unknown variables that are associated with this relation are $\{e_1, e_2, e_3\}$. In this case, an ARR cannot be generated from this relation since the causal path of the unknown e_2 indicates $e_2 = e_1 - e_3$, which is Eq. (3.34) itself; hence, the resultant residual is always zero regardless of faulty or fault-free condition.

Next, the constitutive relation (3.35) of junction 0_2 is considered for ARR.

$$f_3 - f_4 - f_5 = 0. \tag{3.35}$$

This relation may lead to an ARR since there is no unknown variable that requires Eq. (3.35) itself to solve. Since $f_3 = Df_1$ and $f_5 = Df_2$, Eq. (3.35) becomes

$$Df_1 - f_4 - Df_2 = 0. \tag{3.36}$$

To derive an ARR from Eq. (3.3), the unknown variable f_4 is required to be solved using the causal paths of the graph.

Figure 3.5 depicts the causal paths of the unknown variable f_4. The causal paths indicate that $f_4 = C_1 \frac{d(e_4)}{dt}$, $e_4 = e_5$, $e_5 = e_6 + e_7$, $e_6 = R_2 Df_2$, and $e_7 = R_3 Df_2$. These relations lead unknown variable f_4 to

Fig. 3.5 Solving f_4 using Causal Paths

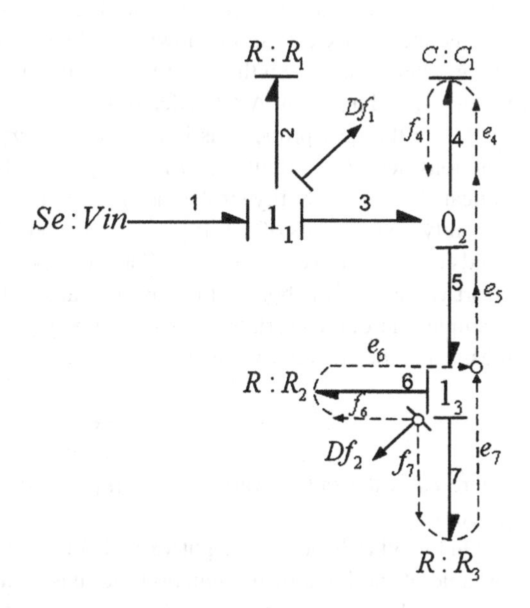

$$f_4 = C_1 \frac{d}{dt}((R_2 + R_3)Df_2). \tag{3.37}$$

By substituting (3.37) into Eq. (3.36),

$$ARR_1 = Df_1 - C_1 \frac{d}{dt}((R_2 + R_3)Df_2) - Df_2. \tag{3.38}$$

Finally, the constitutive relation (3.39) of junction 1_3 is considered.

$$e_5 - e_6 - e_7 = 0 \tag{3.39}$$

Similarly, Eq. (3.39) cannot generate an ARR using the covering path method because unknown variable e_5 requires Eq. (3.39) itself to solve.

This case study shows that although covering path method provides a direct and systematic approach to generate symbolic ARR by scanning all BG junctions, the total number of ARRs derived by this method may not be optimum. From FDI perspective, it is desirable to attain as many independent ARRs as possible for a FDI design. Hence, it is desirable to analyze the properties of the covering path method.

One direct but important property observed from the covering path algorithm can be stated as follows.

Property 3.1 Every ARR that is derived by the covering path method is based on BG junction's constitutive relation.

Property 3.1 can be observed from the covering path procedure. The procedure attempts to eliminate all unknown variables of every junction's constitutive relation to generate an ARR. If all unknown variables of the constitutive relation are solved by the causal paths, then the constitutive relation can be expressed in term of known variables, and hence an ARR is derived.

Before the next property is introduced, the *input* and *output* notions of the BG junctions need to be defined. In BG language, the output variable of a junction is indicated by the causality strokes assigned to the junction. Figures 3.6a, b depict the causality assignments of a 1-Type junction and a 0-Type junction with n number of bonds connected to each junction. For a 1-Type junction, its output is defined as the effort variable of the bond which has no causal stroke assigned to the junction. For example, the output variable of the 1-Type junction shown in Fig. 3.6a is e_n. This means that the constitutive relation of the junction is expressed as

$$e_n = e_1 - e_2 - \cdots e_{n-1}, \tag{3.40}$$

where e_n is the output variable and $\{e_1, \ldots, e_{n-1}\}$ are the input variables of the junction.

On the other hand, the output variable of a 0-Type junction is defined as the flow variable of the bond of the junction that has a stroke assigned to the junction. For instance, the output variable of the 0-Type junction shown in Fig. 3.6b is f_n. This

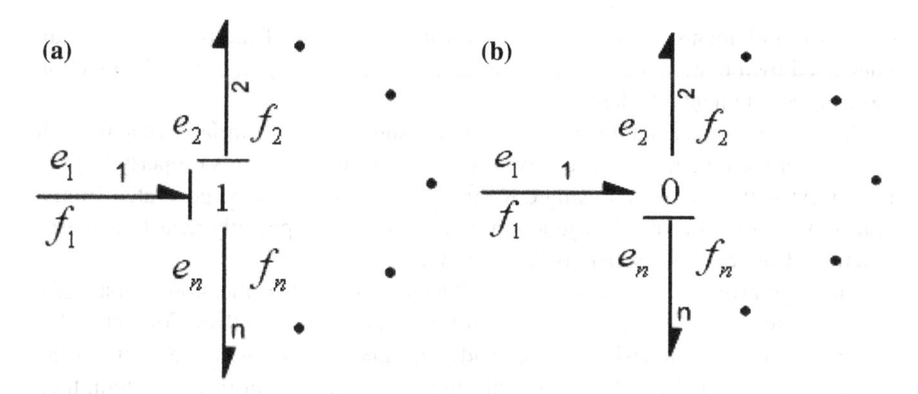

Fig. 3.6 Causalities of 1-Type and 0-Type junctions

implies that the junction constitutive relation is expressed as

$$f_n = f_1 - f_2 - \cdots f_{n-1}, \tag{3.41}$$

where f_n is the output variable and $\{f_1, \ldots, f_{n-1}\}$ are the input variables of the 0-Type junction.

With this input-output notion, a necessary condition for a junction's constitutive relation to be an ARR candidate can be stated as follows.

Property 3.2 A junction's constitutive relation is an ARR candidate only if the output variable of the junction is measurable by a sensor.

Proof Without loss of generality, let us consider the constitutive relation of a 1-Type junction with n bonds, which is $e_1 + e_2 + \cdots + e_n = 0$. The covering path method attempts to solve all the unknown variables of the equation to derive an ARR using the causal paths connected to the variables. If the output variable of the relation is not measurable (known), then it is necessary to apply the causal paths to solve the variable. Since the solution of the output variable is obtained from the causal paths which correspond to the constitutive equation of the 1-Type junction, the constitutive relation of the junction cannot be an ARR since it is always zero. The same arguments also apply to the 0-Type junction; and this completes the proof of Property 3.2.

Based on Property 3.2, the eligibility for a junction's constitutive relation to be an ARR can be evaluated without solving all the unknown variables of the relation. With help of Property 3.2, the ARR generation efforts can be drastically reduced by just considering BG junctions whose output variables are measurable. One consequence of Property 3.2 can be stated as follows.

Property 3.3 A junction's constitutive relation generates an ARR only if the junction is adjacently connected to a sensor junction.

Two BG elements are said to be *adjacently connected* if the two elements are connected by a bond. And a junction is said to be a *sensor junction* if the junction has a sensor element attached to it.

Property 3.2 states that it is necessary to measure the output variable of a junction in order for the junction's constitutive relation to be an ARR. This property implies that an ARR that is derived using covering path method must be generated from a constitutive equation of a BG junction that is located graphically next to a sensor junction. This feature is stated as Property 3.3.

Although Property 3.3 looks intuitive, it leads to several implications. First, there is no need to consider all junctions' constitutive relations for ARRs. For a complex system, it would be undesirable to consider all junctions for ARRs. Second, if the FDI designer has the freedom to modify the sensor placement of the system, then Property 3.3 provides a guideline on how a sensor should be located on the BG to ensure that some critical elements are fault detectable. Third, the property suggests the possible junctions that may generate ARRs can be identified by inspecting the junctions that are located near the sensor junctions. Even though the covering path method is a systematic algorithm, the correlation between the ARRs and the sensors is not well-understood. It is clear that an increase in the number of sensors increases the maximum number of independent ARRs; therefore, it would be beneficial if the correlation between the ARRs and the sensors of BG can be established. The following property reveals this association.

Property 3.4 Every ARR that is derived by the covering path method is associated with a unique sensor of the BG.

Proof Figure 3.7a, b depict the general BG topology where a junction's output variable is measurable by a sensor connected to its adjacent junction. Without loss of generality, let us consider the D_f case shown in Fig. 3.7a. The figure shows that the output variable (f_1) of the 0 junction is measurable by the D_f sensor. More importantly, the causalities of the graph indicate that there is only one output variable of BG junction that is measurable by the sensor Df. Hence, the sensor which associates with the ARR is unique. Same argument applies to the De case which is shown in Fig. 3.7b and this completes the proof.

3.2.3.2 Causality Inversion Method

This section describes and analyzes the second BG-based ARR generation method, which is called causality inversion method [4]. The causality inversion method is an ARR generation method that generates ARRs from constitutive relations of some sensor junctions. The core of this method is to modify the causality strokes of the junction such that an ARR can be generated from the junction's constitutive equations without violating any causality rule of the BG.

To describe the method, the following 0-Type and 1-Type sensor junctions are considered. Figure 3.8a depicts a 0-Type junction with a BG element X and an effort

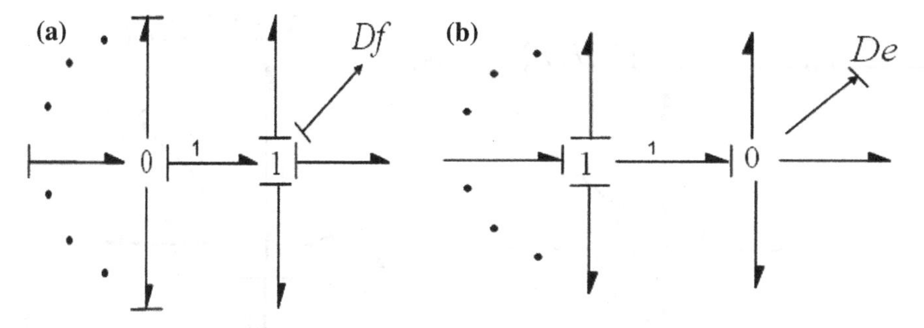

Fig. 3.7 **a** 0-Type junction case: output variable measured by *Df* sensor. **b** 1-Type junction case: output variable measured by *De* sensor

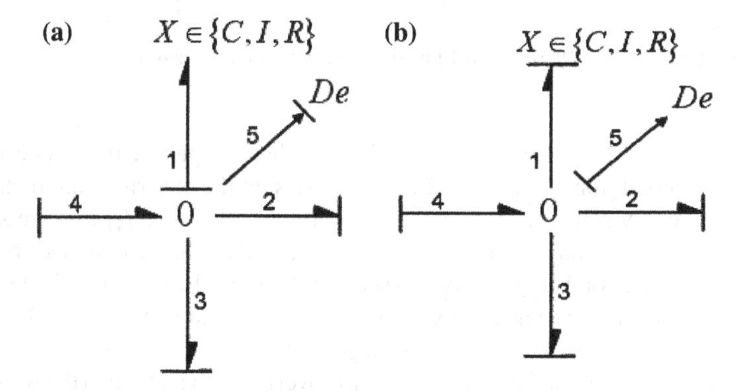

Fig. 3.8 **a** 0 junction with *De* sensor. **b** 0 junction with inverted causality

sensor attached to it. The bonds of the junction are enumerated for reference purposes. As shown in the graph, the causality strokes assigned to the junction indicate that the junction's output variable (f_1) is the input of the element X. This causality inversion method shifts the output variable of the junction to the input of the effort sensor (*De*) connected to the junction by modifying the causal strokes of the junction. This modification is achieved by reversing the causal strokes of the BG element X and the sensor as shown in Fig. 3.8b. With this inversion, the constitutive relation of the junction can be written as

$$f_4 - f_1 - f_2 - f_3 - f_5 = 0. \tag{3.42}$$

In the same spirit of the covering path method, all unknown variables in Eq. (3.42) are then solved using the causal paths of the graph. In this case, the output variable f_5 needs not to be measurable by a sensor since the flow is zero, i.e., $f_5 = 0$. This feature leads Eq. (3.42) to

$$f_4 - f_1 - f_2 - f_3 = 0, \tag{3.43}$$

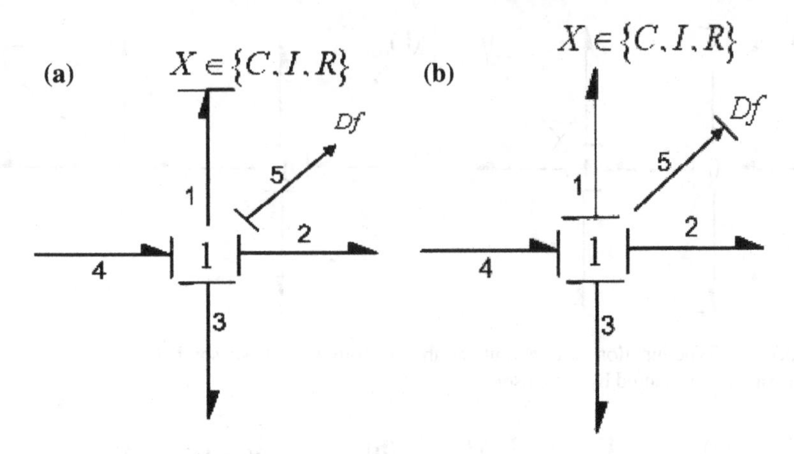

Fig. 3.9 **a** 1 junction with *Df* sensor. **b** 1 junction case with inverted causality

where f_i for $i = 1, \ldots, 4$ are input variables of the 1-Type junction. As a result, the necessary condition (Property 3.2) that imposes on the covering path method is avoided, i.e., every solution of the unknown variables $\{f_1, f_2, f_3, f_4\}$ that are solved by the causal paths is not Eq. (3.43) itself; hence, an ARR can be obtained from the equation. Likewise for 1-Type junction (see Fig. 3.9a, b), the output variable of the causality inverted 1-Type junction is zero, i.e., $e_5 = 0$, and hence, an ARR can be derived from the constitutive relation of the junction.

The causality inversion method is similar to covering path method in the sense that it also utilizes junction's constitutive relation and causal paths to generate an ARR equation. The main difference between the two methods is the necessary condition for a junction's constitutive relation to be an ARR. For the covering path method, the output variable of the constitutive relation must be measurable by a sensor in order to derive an ARR from the relation. This necessary condition is relaxed by performing a causality inversion to the element which the considered junction is outputting to.

The causality inversion method implicitly expresses the junction output variable in term of measured sensor value via the constitute relation of the causality-inverted element. That explains why the analytical redundancy relation can still be derived despite the fact that the junction output variable is not measurable by a sensor. These insights are summarized as properties in the remainder of this section.

For illustration purpose, still consider the example in Fig. 3.1 of BG with inverted sensor as shown in Fig. 3.10.

For junction 0_2, the constitutive relation is

$$f_3 - f_4 - f_5 - f_6 = 0, \tag{3.44}$$

The causal path 3-2 indicates that the unknown variable $f_3 = f_2$. Since $f_2 = \frac{e_2}{R_1}$ and $e_2 = e_1 - e_3 = V_1 - De_1$, so $f_3 = \frac{V_1 - De_1}{R_1}$. As for f_4 and f_5, the causal paths

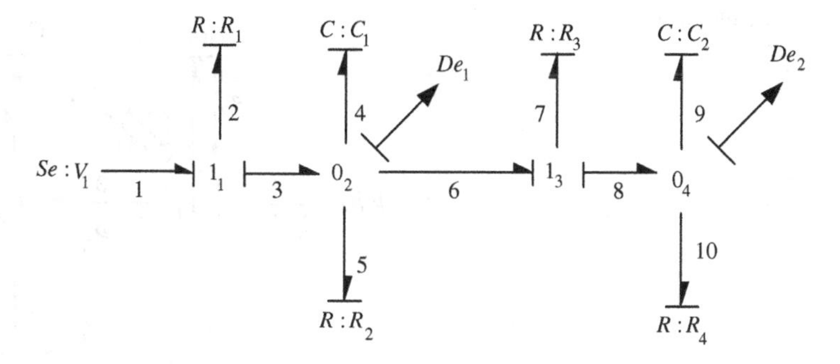

Fig. 3.10 Bond graph model of the circuit with inverted sensor

indicate $f_4 = C_1 \frac{De_1}{dt}$ and $f_5 = \frac{De_1}{R_2}$. According to causal path 6-7, $f_6 = f_7 = \frac{e_7}{R_3} = \frac{e_6 - e_8}{R_3} = \frac{De_1 - De_2}{R_3}$. By substituting all the solved variables into Eq. (3.44)

$$ARR_1 = \frac{V_1 - De_1}{R_1} - C_1 \frac{De_1}{dt} - \frac{De_1}{R_2} - \frac{De_1 - De_2}{R_3} \\ = ARR_1(V_1, De_1, De_2, R_1, R_2, R_3, C_1) \tag{3.45}$$

Next, consider junction 0_4 with the following constitutive law

$$f_8 - f_9 - f_{10} = 0, \tag{3.46}$$

From causal path 8-7, $f_8 = f_7 = \frac{e_7}{R_3} = \frac{e_6 - e_8}{R_3} = \frac{De_1 - De_2}{R_3}$. As for f_9 and f_{10}, the causal paths indicate $f_9 = C_2 \frac{De_2}{dt}$ and $f_{10} = \frac{De_2}{R_4}$. Finally, the second ARR can be obtained by substituting all the solved variables into Eq. (3.46)

$$ARR_2 = \frac{De_1 - De_2}{R_3} - C_2 \frac{De_2}{dt} - \frac{De_2}{R_4} = ARR_2(De_1, De_2, R_3, R_4, C_2) \tag{3.47}$$

It is not hard to find that these two ARRs are structurally independent because they are sensitive to different sets of faults.

In order to gain more insights on causality inversion method and to reveal the properties, consider the example of Fig. 3.11. First, the causality strokes of the flow sensors of BG junctions 1_1 and 1_3 are inverted. Unfortunately, the causality strokes of the graph suggest that only the causalities of the flow sensor Df_1 and R_1 can be inverted without violating the causality rules. The inverted causal strokes are represented as the dash lines on bond 2 and bond $s1$ of Fig. 3.11.

To generate an ARR from junction 1_1, the unknown variables of the junction's constitutive relation (3.48) need to be solved.

$$e_1 - e_2 - e_3 = 0 \tag{3.48}$$

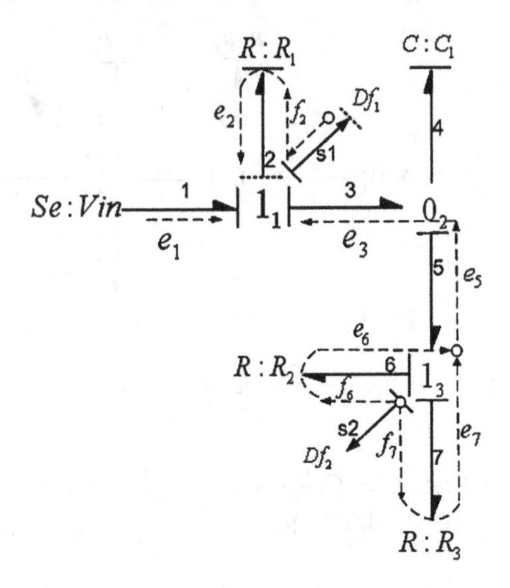

Fig. 3.11 BG with inverted sensor

The causal paths indicate that the unknown variables $\{e_1, e_2\}$ are $e_1 = V_{in}$, $e_2 = R_1Df_1$. As for e_3, the causal paths indicate $e_3 = e_5$ and $e_5 = e_6 + e_7$. By covering the paths of bonds 6 and 7, $e_6 = R_2Df_2$ and $e_7 = R_3Df_2$. Hence, $e_3 = R_2Df_2 + R_3Df_2$. By substituting all the solved variables into Eq. (3.48)

$$ARR = V_{in} - R_1Df_1 - R_2Df_2 - R_3Df_2 \tag{3.49}$$

is established.

From this case study, it can be observed that the causality inversion method avoids the necessary condition that is imposing on the covering path method. However, this case study also highlights that the total number of ARRs derived by the causality inversion method may also be less than the total number of sensors. In this example, only one ARR is derived by this method although there are two sensors. The limitations that are associated with this method remains unclear. Some properties of the causality inversion method are presented as follows.

Property 3.5 Every ARR equation that is derived by the causality inversion method is based on a BG junction's constitutive relation.

Although the causal inversion is not presented in an algorithmic manner as the covering path method, this approach also utilizes the constitutive relation of a junction and causal paths to generate an ARR. Property 3.5 points out that this approach ignores constitutive relations of other BG elements as ARR candidates.

Property 3.6 Every ARR that is derived by the causality inversion method is uniquely associated to a sensor junction.

Proof Property 3.6 is shown by the following argument. Property 3.5 suggests that the causality inversion method derives an ARR from a sensor junction's constitutive relation; hence, the derived ARR can be associated with the sensor junction. It is clear that in BG langauge, there is only one sensor attached to a junction and this completes the proof of Property 3.6.

Property 3.6 associates an ARR to a sensor junction, and hence to the junction's sensor. Although Property 3.6 is intuitive, it suggests that the sensors which have not been well-exploited by the method can be identified. Another property can be stated as follows.

Property 3.7 A junction's constitutive relation is an ARR candidate only if the junction's output variable is an input of an causality invertible BG element.

Proof This property can be revealed from Figs. 3.8a, 3.9b for 0-Type and 1-Type junctions. In order to generate an ARR from the junction's constitutive relation, the junction's causalities must be assigned in such a way that the junction's output is a input to a causality invertible element; hence, Property 3.7 is established.

Property 3.7 states that not every sensor junction's constitutive relation can become an ARR based on the causality inversion method. In some systems, the causality of an element cannot be inverted if the element is already assigned with a preferred causality. In FDI applications, storage elements are preferred to be in derivative causalities to avoid unknown initial states in the ARRs. If the causality assignment of a storage element is already in derivative causality, then it would be undesirable to apply a causality inversion to the storage element.

3.2.3.3 Comparison Between the ARR Generation Methods

This section highlights some key similarities and differences between the two symbolic ARR generation methods.

First, Property 3.1 and Property 3.5 state that every ARR can only be generated based on constitutive relation of a junction. This similarity suggests that these methods are rather restrictive, since they only consider junctions' constitutive relations as *ARR candidates*. An ARR candidate is a constitutive relation that becomes an ARR when all the unknown variables of the equation are solved. There are also other constitutive relations that may generate an ARR, e.g., the constitutive relation of $\{I, R, C, TF, GY\}$, which are not considered as ARR candidates in these methods. Second, Property 3.4 and Property 3.6 suggest that any sensor that has been explicitly exploited in generating an ARR can be uniquely identified and associated with the ARR. The third similarity is that the performance of the two methods is sensitive to the causalities assigned to the BG. This observation means that any of these methods may not be able to maximize the sensors for the ARRs generation for a given causality assignment.

Besides the similarities, the two methods also possess some different and complementary properties. First, the covering path method considers all BG junctions in the

ARR generation process. In contrast, the causality inversion method derives ARRs only from sensor junctions. This observation suggests that the exhaustive approach of ARRs generation can be avoided if the BG junctions that are eligible to generate ARRs can be identified. By selecting suitable BG junctions for ARRs generation, the FDI design process can become more efficient when handling a complex system. Second, from the properties of the two methods, it can be observed that covering path method may be applied to derive an ARR in cases where the causality inversion method fails. Likewise, causality inversion method can also help to generate an ARR when covering path method fails to generate an ARR associated to a sensor. These complementing properties are attractive and will be exploited to develop an integrated ARR generation strategy that combines the benefits of the two existing methods in an efficient and effective manner.

3.2.3.4 An Integrated Strategy for Bond Graph-Based Symbolic ARR Generation

The insights gained from the preceding section suggest that the covering path method can be extended so that the ARR derivation can be less sensitive for a given causality assignment. In this section, *Extended Covering Path* is proposed to enhance the applicability of the BG-based ARR generation.

(a) Extended covering path method

In general, covering path method relies only on junctions' constitutive relations for ARR candidates. In this part, other potential ARR candidates are considered in the ARR generation process. Additionally, an improved ARR generation method is developed to avoid the exhaustive search inherent in the covering path method.

An ARR is basically the difference between a measured and an analytically computed terms where the difference is theoretically null when the system is fault-free. A careful examination of the BG structure suggests that besides the BG junctions, constitutive relations of $\{TF, GY, I, C, R\}$ elements can also be considered as ARR candidates. Here, these relations are considered as the ARR candidates.

First, let us consider the constitutive relations of *TF* element as an ARR candidate. Figure 3.12a, b show the possible causality strokes assigned to the *TF* element where the constitutive relations are as follows

$$e_2 = ne_1, \quad f_1 = \frac{1}{n}f_2 \tag{3.50}$$

$$f_2 = \frac{1}{n}f_1, \quad e_1 = ne_2. \tag{3.51}$$

where n is the parameter of the *TF* element, f_2 and e_2 are the respective output variables.

Figures 3.13a, b show the BG where Eqs. (3.50) and (3.51) can be considered as ARR candidates since the output variables (f_2, e_2) are measurable. Figure 3.13a depicts that the output variable of *TF* (f_2) is measurable by sensor Df; hence equation

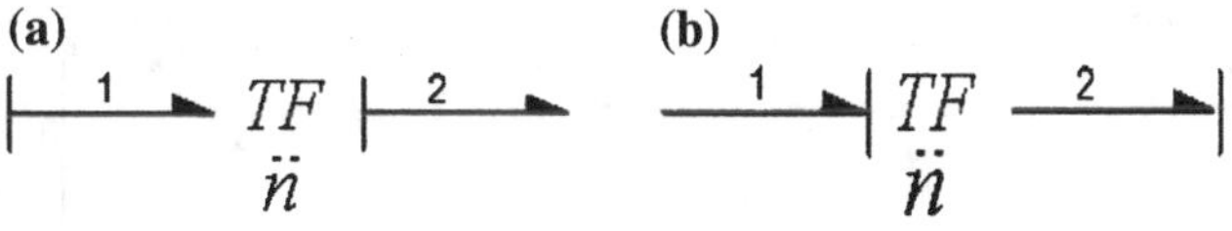

(a) **(b)**

Fig. 3.12 Possible causalities of *TF* element

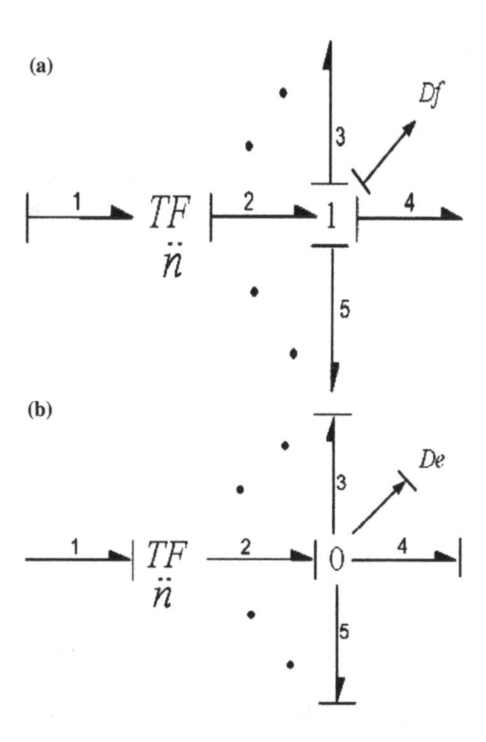

Fig. 3.13 *TF* constraint as ARR candidate

($f_1 = \frac{1}{n} f_2$) can be re-arranged into

$$Df - \frac{1}{n} f_1 = 0 \tag{3.52}$$

where the left hand side of the equation is an ARR once the unknown variable f_1 is solved by the causal paths of the graph. Similarly, Fig. 3.13b shows that the measurable output variable e_2 leads equation ($e_1 = ne_2$) to

$$De - ne_1 = 0 \tag{3.53}$$

where the left hand side of the equation is an ARR once e_1 is solved by causal paths.

The same principle applies to the *GY* element as shown in Fig. 3.14a, b.

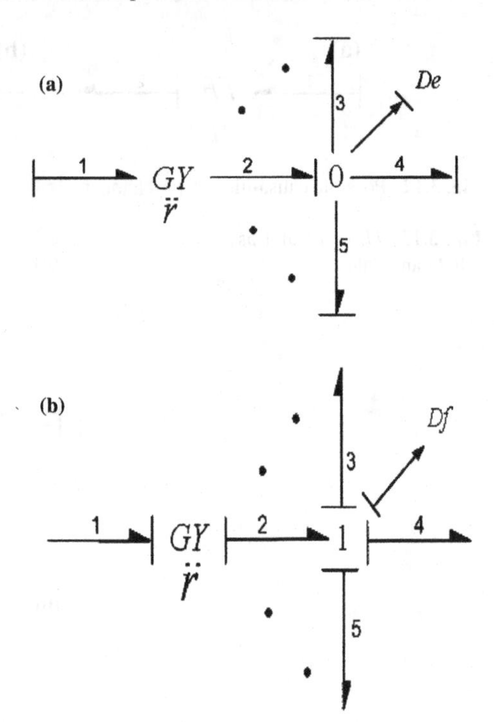

Fig. 3.14 *GY* constraint as ARR candidate

$$e_2 = rf_1 \tag{3.54}$$

$$\Rightarrow De - rf_1 = 0 \tag{3.55}$$

Equation (3.54) is the constitutive relation of the *GY* element with the causalities shown in Fig. 3.14a. Its output variable e_2 is measurable by *De*, hence, Eq. (3.54) can be expressed into an ARR candidate (3.55). Likewise for the causality assignments shown in Fig. 3.14b, Eq. (3.56) can be presented into ARR form as

$$f_2 = \frac{1}{r}e_1 \Rightarrow Df - \frac{1}{r}e_1 = 0. \tag{3.56}$$

The above discussion demonstrates that the constitutive relations of the BG element $\{TF, GY\}$ can also be ARR candidates.

BG elements $\{0, 1, TF, GY\}$ are connective elements that provide the basic BG topology of a physical system. Besides these connective elements, the constitutive relations of other BG elements also play important roles in BG modelling. Here, it is shown that the constitutive relation of a 1-port BG element $X \in \{R, I, C\}$ can also be an ARR candidate.

Without loss of generality, a BG element $X \in \{R, I, C\}$ that is connected to a sensor junction is considered (see Fig. 3.15a, b).

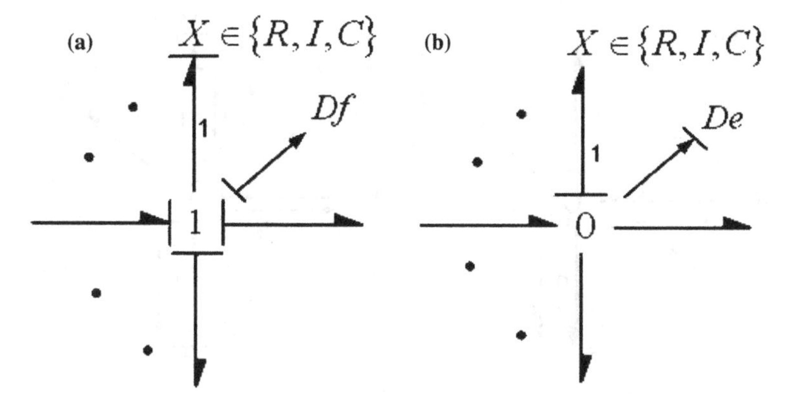

Fig. 3.15 $\{R, I, C\}$ constitutive relation as ARR candidate

Let Φ_X denotes the constitutive relation of the BG element X shown in Fig. 3.15a. The causal stroke of bond 1 implies that the input and output variables of the element X is $\{e_1, f_1\}$. Therefore, the constitutive relation is written as

$$f_1 = \Phi_X(e_1). \tag{3.57}$$

The BG indicates that the output variable of Eq. (3.57) (f_1) is measurable by the *Df* sensor, hence the equation can be written as

$$Df - \Phi_X(e_1) = 0 \tag{3.58}$$

where the left hand side of Eq. (3.58) becomes an ARR equation once the unknown variable e_1 is solved using the causal paths.

Likewise for 0-Type junction (see Fig. 3.15b), the constitutive relation can be written as

$$e_1 = \Phi_X^{-1}(f_1)De - \Phi_X^{-1}(f_1) = 0 \tag{3.59}$$

where the left hand side of Eq. (3.59) becomes an ARR once f_1 is solved by the causal paths.

From this investigation, it is clear that besides BG junctions, the constitutive relations of the BG elements $\{TF, GY, R, I, C\}$ can be utilized as ARR candidates to derive ARRs. Before these new candidates can be incorporated for ARR generation, a property that is useful in formalizing an ARR generation procedure is stated as follows.

Property 3.8 Consider a BG with a valid causality assignment. For any sensor element, there exists one and only one output variable that is measurable by the sensor.

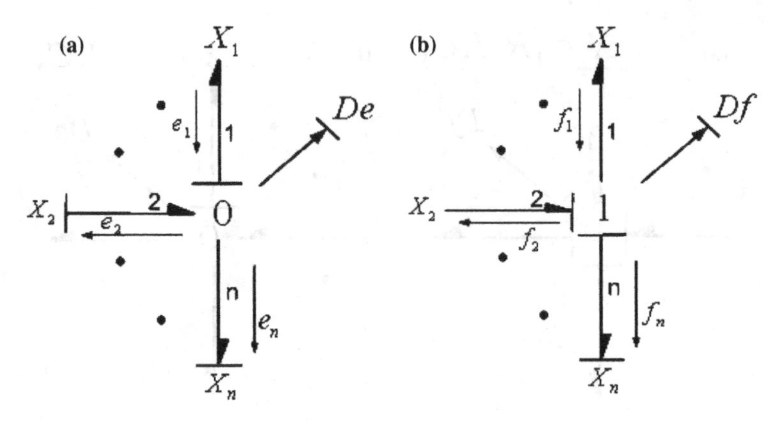

Fig. 3.16 $\{R, I, C\}$ constitutive relation as ARR candidate

Proof The proof of the property can be deduced directly from the BG topology shown in Fig. 3.16a. The figures depict a n-bond 0- and 1-Type sensor junctions. For the 0-Type junction, X_i for $i = 1, \ldots, n$ denote some BG elements connected to the junction. Figure 3.16a shows the effort variables of the n bonds. The orientation of the causal paths that are defined by the causalities of the 0-Type junction indicates that there exists one and only one output variable (e_1) that is measurable by the sensor *De*. Likewise for 1-Type junction (see Fig. 3.16b), there exists one and only one output variable (f_1) measurable by sensor *Df*.

One consequence of Property 3.8 is that there exists an *one to one* mapping between an ARR candidate and a sensor connected to the BG. This property means that instead of considering all 0-Type and 1-Type junctions for ARR candidates, the N number of ARR candidates for the N ARR equations can be accurately identified. N denotes the total number of sensors. In this manner, the ARR derivation process can be more efficient, and also able to consider more potential candidates for ARRs. With these points in mind, the *Extended Covering Path algorithm* is proposed and described as follows.

Consider a given BG that has N number of sensors and a valid causality assignment,

1. Choose a sensor.
2. Find the corresponding ARR candidate of the given sensor. This candidate is the constitutive relation whose output variable is measurable by the sensor.
3. Eliminate all unknown variables of the ARR candidate using causal paths of the BG.
3. Consider the next sensor.
4. Repeat step 2 until all the N sensors of the BG are considered.

The extended covering path method directly seeks an ARR candidate from the BG based on the causalities of any sensor junction. This concept of ARR derivation is more efficient compared with the classical covering path method where all BG

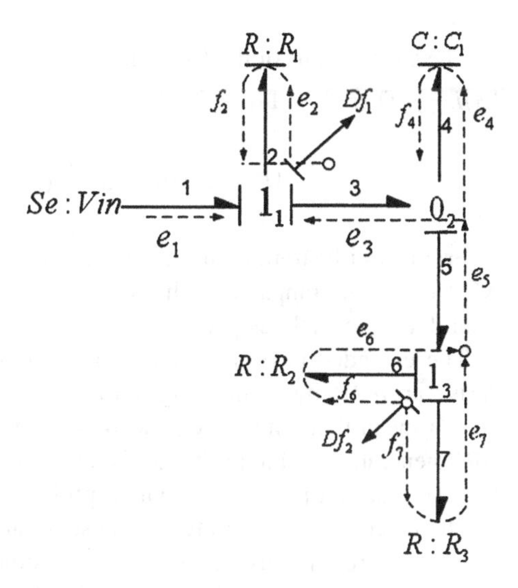

Fig. 3.17 BG with causal paths

junctions are considered. This strategy allows the extended covering path method to maximize the given number of sensors for symbolic ARR generation. The extended covering path method is illustrated on the BG that is shown in Fig. 3.4.

Figure 3.17 depicts the BG that is considered in this study. The graph has two sensors $\{Df_1, Df_2\}$. First, let us consider sensor Df_1. By inspecting the causality of junction 1_1, it can be observed that the sensor is measuring the output variable (f_2) of the constitutive relation $f_2 = \frac{1}{R_2}e_2$. Hence, the relation is chosen as the ARR candidate of sensor Df_1 which can be expressed as

$$Df_1 - \frac{1}{R_2}e_2 = 0, \tag{3.60}$$

where e_2 is the unknown variable to be solved. The causal paths of the graph indicate that $e_2 = e_1 - e_3$ where $e_1 = V_{in}$. By covering causal paths 3-5-6-7, $e_3 = e_5 = e_6 + e_7$. Since $e_6 = R_2Df_2$, and $e_7 = R_3Df_2$, the unknown variable e_2 is

$$e_2 = V_{in} - R_2Df_1 - R_3Df_2. \tag{3.61}$$

By substituting Eq. (3.61) into Eq. (3.60), the ARR candidate becomes

$$ARR_1 = Df_1 - \frac{1}{R_2}(V_{in} - R_2Df_1 - R_3Df_2). \tag{3.62}$$

Next, sensor Df_2 is considered. By examining the causalities of the sensor junction 1_3, the output variable of 0_2 (f_5) is measurable by sensor Df_2. The variable f_5 is the output variable of junction 0_2's constitutive relation

$$f_3 - f_4 - f_5 = 0. \tag{3.63}$$

The causal paths of the BG imply that $f_3 = Df_1, f_5 = Df_2$, and $f_4 = C_1 \frac{d}{dt}((R_2 + R_3)Df_2)$. Therefore, Eq. (3.22) becomes

$$ARR_2 = Df_1 - C_1 \frac{d}{dt}((R_2 + R_3)Df_2) - Df_2. \tag{3.64}$$

From this illustration, the extended covering path method obtains one additional ARR equation compared with the classical covering path method. This benefit is desirable from FDI perspectives.

The extended covering method associates an ARR candidate to a sensor and derive an ARR by eliminating the unknown variables of the ARR candidate. So does it mean that with the extended covering path method, the causality inversion ARR generation technique becomes redundant? A study reveals that if a BG element $\{I, C\}$ of a sensor junction is not in its preferred causality, the causality of the element can be inverted into its preferred causality and derive an ARR from the junction's constitutive relation using causality inversion method. This positive feature of the method can be integrated seamlessly with the extended covering path method to achieve a unified symbolic BG-based ARR generation method.

(b) An integrated strategy: causality inversion aided extended covering path method

This section presents the integrated ARR generation methods. For a given BG that has N number of sensors and a valid causality assignment,

1. Consider a sensor and its corresponding sensor junction.
2. Find the ARR candidate of the sensor. This ARR candidate is simply the constitutive of a BG element's relation whose output variable is measurable by the sensor. If the ARR candidate is the constitutive relation of a BG element (I, R, C) which is not in its preferred causality, then inverts the causality of the BG element and choose the junction's constitutive relation as the ARR candidate of the sensor.
3. Eliminate all unknown variables of the ARR candidate using causal paths of the BG.
4. Consider the next sensor.
5. Repeat step 2 until all the sensors of the BG are considered.

The main advantage of combining causality inversion method with extended covering path method is that if the ARR candidate is a constitutive relation of an element which is not in its preferred causality, e.g., integral causality of a storage element, the combined strategy allows user to invert the element's causality into its preferred causality and selects the junction's constitutive relation as the ARR candidate. In this case, the causality inversion aids the extended covering path method to generate an ARR equation.

Here, the integrated ARR generation scheme is applied to an electro-mechanical system (see Fig. 3.18). Figure 3.18 depicts the schematic of a DC motor whose input

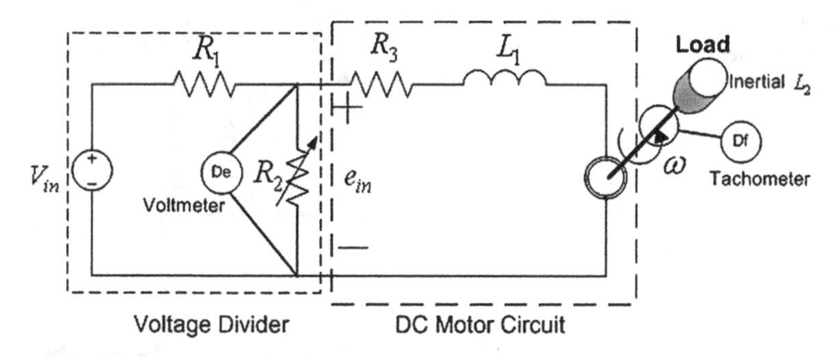

Fig. 3.18 Electro-mechanical system a DC-motor interfaced with voltage divider

is interfaced with a voltage divider circuit. This circuit is an electro-mechanical system whose objective is to rotate a mechanical load that is attached to the motor shaft. The angular velocity of the motor shaft is adjusted by the mechanism of the voltage divider.

The voltage divider consists of components $\{V_{in}, R_1, R_2\}$. V_{in} represents a constant DC voltage source with V_{in} volt. R_1 is a constant resistor with a resistance value R_1 ohm, where R_2 is a variable resistor with a resistance value R_2 ohm. R_1 and R_2 form the voltage divider circuit that changes the input voltage of the DC motor that is denoted by e_{in}. This change of input voltage is achieved by varying the resistance value of R_2. D_e denotes a voltmeter that measures e_{in}.

The DC motor consists of components $\{R_3, L_1\}$, and a mechanical load attached to the motor's shaft. Similarly, R_3 is a resistor with a resistance value R_3 ohm, L_1 represents an inductor with an inductance value of L_1 Henries, ω denotes the angular velocity of the motor shaft, and L_2 represents the inertia of the motor shaft with the load attached. In this system, there is a tachometer attached to the motor's shaft, measuring the angular velocity of the shaft.

The electro-mechanical system can be modeled by the BG shown in Fig. 3.19. In the BG, the electrical and mechanical components of the system are modeled by the generalized BG components $(I, L, R, GY, Se, 0, 1, De, Df)$. BG component $I : L_1$ represents the inductor L_1, $I : L_2$ represents the inertial value of the motor shaft, $R : R_1$, $R : R_2$, and $R : R_3$ denote the resistors of the system, $R : R_4$ represents the rotation friction across the motor shaft, $GY : r$ models the conversion between electrical and mechanical energy in the system, $Se : V_{in}$ denotes the constant DC source, $\{0, 1\}$ represents the connections of the components, De denotes the measured value of voltmeter, and Df represents the measured angular rotation of the motor's shaft.

Based on the properties of the Covering Path and Causality Inversion, it is clear that either method will not lead to the optimum number of ARRs, which is 2 in this case, since there are two sensor components. On the other hand, the integrated strategy allows us to derive this optimum number of ARRs from the BG. This is illustrated as follows.

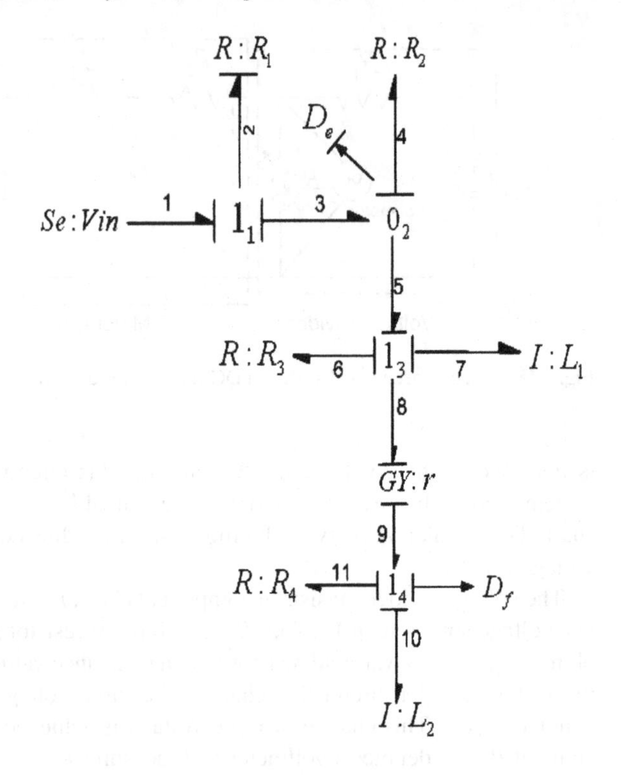

Fig. 3.19 BG of the electro-mechanical system

Figure 3.20 depicts the BG with the causal paths. The selected ARR candidates of the sensors $\{De, Df\}$ are constitutive relations of BG components $\{R : R_2, GY : r\}$. Note that the components of the ARR candidates are highlighted by dotted circles, whereas the causal paths are represented by the dotted lines.

First, the ARR candidate of sensor De is considered for ARR derivation.

The ARR candidate is written as

$$e_4 - R_2 f_4 = 0, \tag{3.65}$$

where f_4 is an unknown variable. The causal paths of the BG indicate that $f_4 = f_3 - f_5$, where $f_3 = f_2 = e_2/R_1$. Since $e_2 = V_{in} - e_3$ and $e_3 = De$, $f_3 = (V_{in} - De)/R_1$. As for f_5, the causal paths show that $f_5 = f_8 = e_9/r$ and $e_9 = e_{10} + e_{11}$. Because $e_{10} = L_2 \frac{d(Df)}{dt}$, and $e_{11} = R_4 f_{11}$ with $f_{11} = Df$, $f_5 = (L_2 \frac{d(Df)}{dt} + R_4 Df)/r$. These solutions lead Eq. (3.65) to ARR equation

$$ARR_1 = De - R_2 \left\{ \frac{V_{in} - De}{R_1} - \frac{L_2 \frac{d(Df)}{dt} + R_4 Df}{r} \right\} \tag{3.66}$$

Next, the ARR candidate of sensor Df is considered. The equation is

Fig. 3.20 BG with causal paths

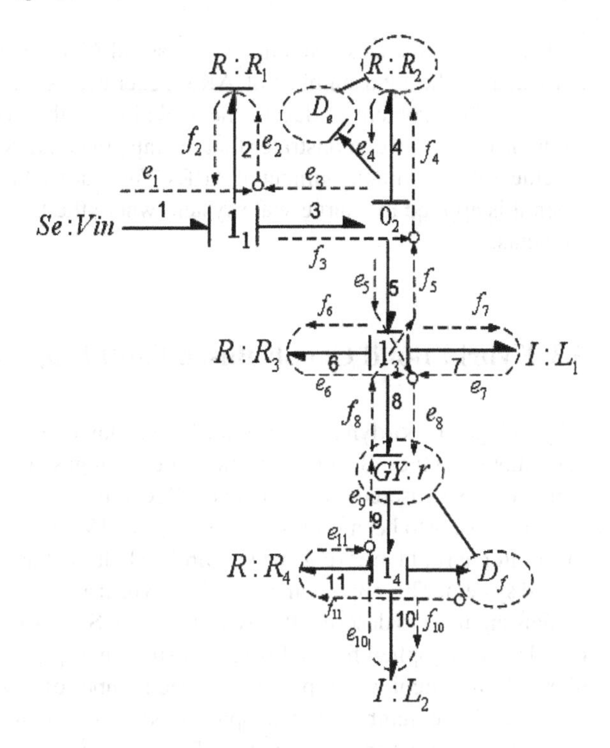

$$f_9 - \frac{e_8}{r} = 0, \tag{3.67}$$

where variable f_9 is measurable by Df; hence, it left unknown e_8 to be solved. The causal paths show that $e_8 = e_5 - e_7 - e_6$, where e_5 is measurable by De, $e_7 = L_1 \frac{d(f_7)}{dt}$, and $e_6 = R_3 f_6$. Since $f_5 = f_6 = f_7 = f_8$,

$$e_8 = De - \frac{L_1}{r}\left\{L_2\frac{d^2(Df)}{dt^2} + R_4\frac{d(Df)}{dt}\right\} - \frac{R_3}{r}\left\{L_2\frac{d(Df)}{dt} + R_4 Df\right\}. \tag{3.68}$$

By substituting (3.68) into ARR candidate (3.67),

$$ARR_2 = Df - \frac{1}{r}\left\{De - \frac{L_1}{r}\left\{L_2\frac{d^2(Df)}{dt^2} + R_4\frac{d(Df)}{dt}\right\} - \frac{R_3}{r}\left\{L_2\frac{d(Df)}{dt} + R_4 Df\right\}\right\} \tag{3.69}$$

This study illustrates that if the two classical ARR generation methods are applied individually, the total number of ARR generated is less than the total number of sensors. This problem is due to the violation of the necessary condition of each method. This study demonstrates that the integrated ARR generation strategy is able to achieve the maximum number of ARRs. These advantages will be more perceptible when it is applied to a large scale system where the BG involves a large number of elements.

3.3 Hybrid Bond Graph-Based Fault Diagnosis

A hybrid system is a dynamic system whose behavior evolution combines discrete and continuous changes. Besides the continuous changes of system's states, a hybrid system undergoes discrete mode changes. Recently, fault diagnosis systems for hybrid systems have also been proposed [15–19]. In [15, 16], FDI algorithms are designed to monitor abrupt faults (e.g., printer broken belt) and incipient parametric faults of a hybrid system. This approach is based on hybrid automaton model where state-space models are used to describe the hybrid system. Similarly, Hybrid Automaton model has also been applied for FDI based on structured parity residuals [19]. For a complex hybrid system which possesses large number of modes, this hybrid automaton requires a large number of state-space models; consequently, it imposes difficulties on the FDI designs and real-time implementation. By comparing a BG with a state-space model, BG provides component-level information that allows a FDI algorithm to isolate a fault at component level easily.

Fortunately, an extended BG-based modeling approach called Hybrid Bond Graph (HBG) is proposed to extend the benefits of BG to hybrid systems [20], and new method has been proposed to utilize HBG for efficient simulation [21, 22]. Recently, HBG has been used to develop a qualitative and quantitative diagnosis framework for hybrid systems with abrupt parametric faults [17, 18]. The fault detection is based on a hybrid observer, which integrates a Kalman filter (for continuous tracking) and a mode-change detector. The discrete events are either known controlled mode changes or autonomous mode changes (triggered by continuous state variables). Once the mode-change conditions are detected, a Finite State Machine is utilized to determine the new mode. The state-space model of the new mode is dynamically computed (from the HBG) and applied to the Kalman filter. If a fault is detected, the hybrid observer is unable to track autonomous mode changes. Consequently, fault-hypotheses are generated in all previous modes up to the level of system diagnosability. A qualitative approach based on Temporal Causal Graph (TCG) is utilized to narrow the set of fault hypotheses, and only those hypotheses which are consistent with the systems observations are considered for fault parameter identification. In this work, when data is collected for fault parameter identification, the mode is assumed to be persistent. In [23], a parameterized causality concept is proposed to address the causality change issue faced in [17, 18]. This work improves the efficiency of the FDI design framework.

3.3.1 Causality Assignment from FDI Perspective

To exploit the rich information of a HBG for FDI, it is desirable to have a consistent causality description for the hybrid system at all modes. This consistent causality description facilitates FDI designs by eliminating the need for causality reassignment which leads to derivation of unified constraint relations. In this section, the causality properties of the HBG is studied to gain insights on this issue. In this work, the focus is on FDI applications; hence, it is desirable that all the storage components (C, I) of the HBG to be in derivative causalities. This is because ARR equations that are derived based on derivative causality are free from unknown initial state.

3.3.1.1 SCAPHD with Allowing Inversion of Sensor Causalities

In Chap. 2, a consistent causality method (i.e., SCAPH) is developed to hybrid system for behavior model. In this part, this method is modified from FDI perspective to Diagnostic Hybrid Bond Graph (DHBG) that possesses the potential for efficient and effective FDI design. A HBG with proper causality assignments is named as Diagnostic Hybrid Bond Graph. In short, a DHBG is a HBG with a causality assignment that all its controlled junctions and storage components are assigned with preferred causalities. From the FDI perspectives, it is beneficial to design FDI algorithms based on a DHBG since the consistent causality of the graph provides a convenient way to describe the behavior of the hybrid system at all modes. Here, a systematic procedure called Sequential Causality Assignment Procedure for Hybrid System Diagnosis (SCAPHD) is developed to achieve a DHBG from an acausal HBG. The SCAPHD extends the classical SCAP by introducing the concept of preferred causality of controlled junctions. The procedure is listed as follows.

1. Choose any controlled junction that has no source component adjacently connected to it and assign the junction with its preferred causality.
2. Repeat step 1 until all the controlled junctions which have no source component connected to them have been assigned.
3. Choose any remaining controlled junction that has a source component connected to the junction. For the pair $\{S_e, 1_c\}$ or $\{S_f, 0_c\}$, the output variable of the controlled junction must not be an input variable of $\{0, 1\}$. For the pair $\{S_f, 1_c\}$ or $\{S_e, 0_c\}$, the output variable of the controlled junction must be the input variable of the connected source component.
4. Repeat step 3 until all the remaining controlled junctions have been assigned. After this step, all controlled junctions are treated as normal junctions $\{0, 1\}$ in the following procedures.
5. Choose any remaining source component and assign its causality. Immediately extend the causal implications through the HBG as far as possible using the constraint components $\{0, 1, GY, TF\}$.
6. Repeat step 5 until all sources have been assigned.

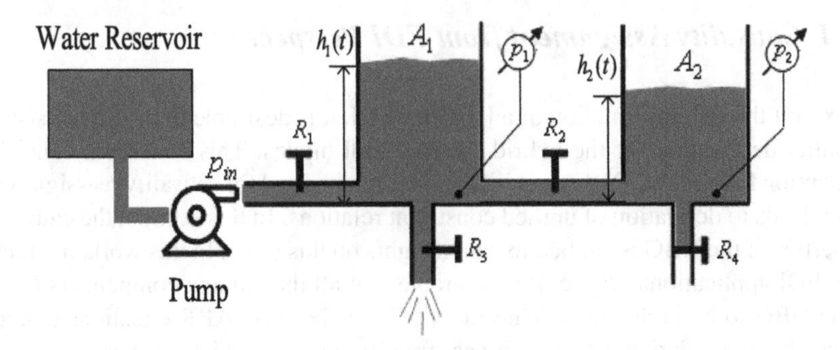

Fig. 3.21 A two-tank system

7. Choose any storage component and assign it with a derivative causality. Immediately extend the causal implications through the HBG as far as possible using the constraint components $\{0, 1, GY, TF\}$.
8. Repeat step 7 until all storage components have been assigned with a causality.
9. Choose any unassigned R component and assign a causality to it. Immediately extend the causal implications through the HBG as far as possible using the constraint components $\{0, 1, GY, TF\}$.
10. Repeat step 9 until all remaining bonds have been assigned.

It worth to note that the preferred causalities of the controlled junctions impose additional restrictions to the causality assignment of the HBG; as a result, it is possible that some controlled-junctions or some storage elements may not be assigned with their preferred causalities due to the restrictions. Fortunately, the sensor causality inversion will facilitate the causality assignment for a given HBG to achieve a DHBG.

A two-tank system is considered, as given in Fig. 3.21. The two-tank system consists of two tanks, regulated centrifugal pump modeled as a source of pressure $P_{in}(t)$[Pa] and four valves represented by R_1, R_2, R_3 and R_4. The cross-section areas of the two tanks are $A_1[m^2]$ and $A_2[m^2]$. The system is equipped with two pressure sensors (i.e., $p_1(t)$ and $p_2(t)$) to measure the pressure at the bottom of tank A_1 and tank A_2, respectively. The pressure is proportional to the liquid level, according to:

$$p_i(t) = \rho g h_i(t), \qquad i = 1, 2. \tag{3.70}$$

where ρ is liquid density, [kg/m^3]; g is the acceleration of gravity, [m/s^2], $h_i(t)$ is liquid height in the tank, [m].

Each valve has two discrete states ON and OFF. A negligible switching-time between open and closed states is assumed. The valves' dynamics is given by

$$f_j(t) = 0, \quad j = 1, 2, 3, 4 \quad \text{when the valve is closed} \tag{3.71}$$

$$f_j(t) = \frac{\text{sign}(\Delta p(t))\sqrt{|\Delta p(t)|}}{R_j} = Cd_j \cdot \text{sign}(\Delta p(t))\sqrt{|\Delta p(t)|}$$
$$\text{when the valve is open} \quad (3.72)$$

where f_j is the liquid-flow through the valve, [m^3/s]; $\Delta p(t)$ is the pressure difference across the valve, [Pa] and Cd_j is the coefficient of discharge, [$\sqrt{\text{kg} \cdot \text{m}}$].

The HBG of the two-tank plant is presented in Fig. 3.22. In this bond graph, the flow variable is the liquid volumetric flow (i.e., m^3/s), and the effort variable is pressure (i.e., [Pa]). The two tanks are modeled by the two storage components with coefficient $C_i = \frac{A_i}{g}$ for $i = 1, 2$. Each one of the four valves is modeled by a set of resistor with parameter $R_j = \frac{1}{Cd_j}$ for $j = 1, 2, 3, 4$ and controlled-junction with boolean variable α_j, $j = 1, 2, 3, 4$.

The HBG in Fig. 3.22 is acausaled. In order to apply the SCAPHD algorithm, all controlled-junctions are assigned with their preferred causalities. The procedure of applying SCAPHD algorithm with allowing inversion of sensor causalities is listed as follows:

1. Choose controlled junction 1_3 that has no source adjacently connected to it. Assign the junction with its preferred causality where the output variable e_6 of the junction 1_3 is the input of the 1-port component R_3.
2. Similarly, choose controlled junctions 1_4 and 1_6, the output variables of these controlled junctions are assigned as inputs of the 1-port components (R_2 and R_4, respectively).
3. Choose controlled junction 1_1 that has a source Se connected to it. This pair indicates that the output of the controlled junction must not be an input variable of normal junction 0_2. Hence the output of the controlled junction 1_1 is assigned as the input variable of 1-port component R_1.
4. Since all controlled junctions have been assigned, all controlled junctions are treated as normal junctions {0, 1} in the following procedures.

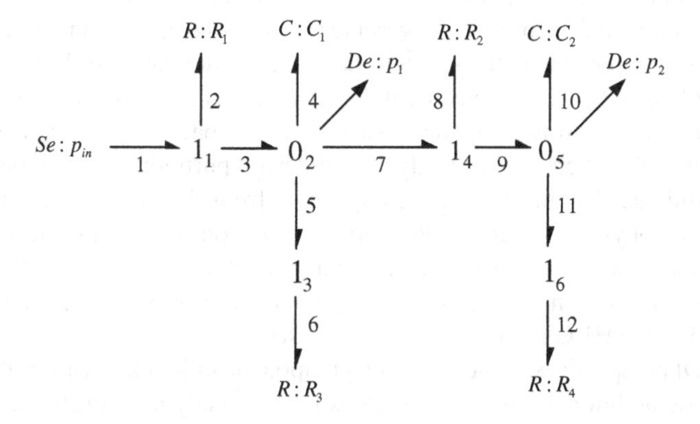

Fig. 3.22 The two-tank plant acausaled HBG

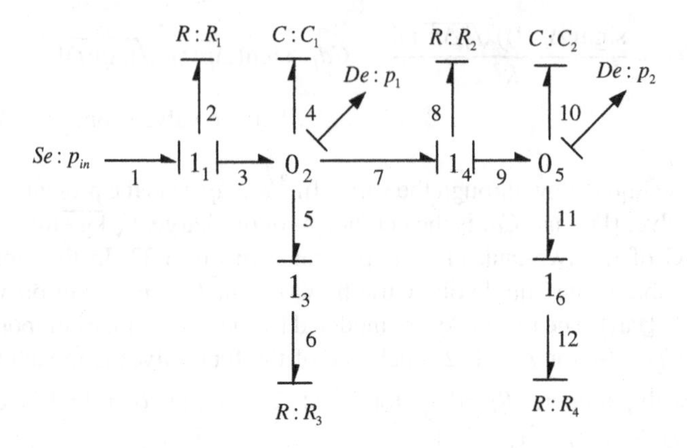

Fig. 3.23 The two-tank plant HBG

5. The source component *Se* is assigned and the causal implication is extended to bond 3.
6. Since no other source, proceed to Step 7.
7. The two storage components C_1 and C_2 are assigned with preferred derivative causality and the sensor causality is inverted. Now, the causal implication using the components constraints to bonds 7, 5, 9 and 11 to complete the SCAPHD algorithm with allowing inversion of sensor causalities when necessary.

The complete DHBG is shown in Fig. 3.23.

3.3.1.2 SCAPHD with Model Approximation

In BG modeling, a *R* component has no preferred causality. This property suggests that a *R* component could facilitate the causality assignment for a given HBG. However, the number of *R* components is fixed in a physical system. In this section, a model approximation technique is presented that allows user to increase the total number of *R* components in the graph to facilitate the derivation of DHBG [24].

Without loss of generality, electrical domain is considered to gain some insights. In electrical systems, stray components such as stray capacitors are usually neglected from the circuit's schematic for analysis and design purposes. This model approximation eliminates the insignificant components from the model and captures the essence of the physical system's behaviors. Here in contrast, the physical system is approximated by adding stray resistors (*R* components) to the model without losing the essence of the system's behavior. This approximation allows more *R* components to be added to the HBG model of a physical system.

From FDI perspectives, an additional *R* component adds unknown variables to the model. These additional unknown variables require analytical constraints to solve, and hence they reduce the analytical redundancy of the model which undermines

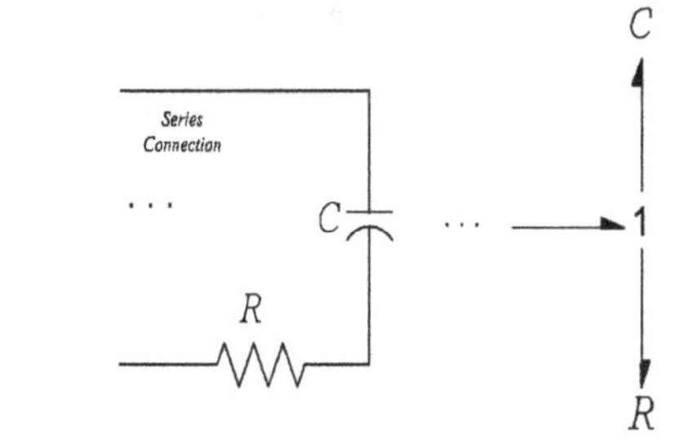

Fig. 3.24 Series connection

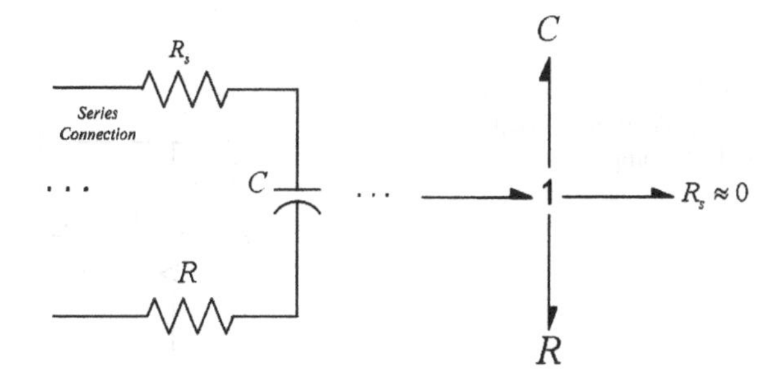

Fig. 3.25 Approximated model: Series connection with R_s component

the monitoring ability of the system. Moreover, how this concept can be applied effectively and systematically to the HBG remains unclear. In this section, a model approximation method is proposed to add *R* components to the graph without degrading the monitoring ability of the system. Additionally, a procedure is developed to combine the model approximation method with the SCAPHD algorithm to yield a DHBG systematically.

A series connection (1-junction) of an electric circuit and its corresponding HBG are shown in Fig. 3.24. For this part of electric circuit, it can be approximated by assuming a low-resistive stray resistor $R_s \approx 0$ resides in the series circuit. In this manner, the physical system is approximated by the original system with an additional R_s component in the series connection. The corresponding approximated circuit and the HBG are shown in Fig. 3.25. The figure illustrates how an additional R_s component is generated in a series circuit by approximating the physical system with a stray resistor. The same approximation technique is also applicable to parallel connections of a circuit. For a parallel connection (see Fig. 3.26), we approximate the original physical system by adding a high-resistive *R* component $R_s \approx \infty$ to the parallel connection. The approximated model is shown in Fig. 3.27.

From HBG perspective, each additional *R* component adds a pair of variables $\{e, f\}$ to the HBG. If these variables are unknown, then the approximation degrades the system's monitoring ability. Fortunately, this problem can be avoided if the approximation is carried out at suitable parts of the system. To gain better insights on how this problem can be avoided, the case of series connection (see Fig. 3.25) is considered. Let $\{e_s, f_s\}$ be the effort and flow variables associated to the $R_s \approx 0$. Since R_s is assumed to be small, $e_s = 0$. Moreover, if f_s is measurable by a flow sensor D_f

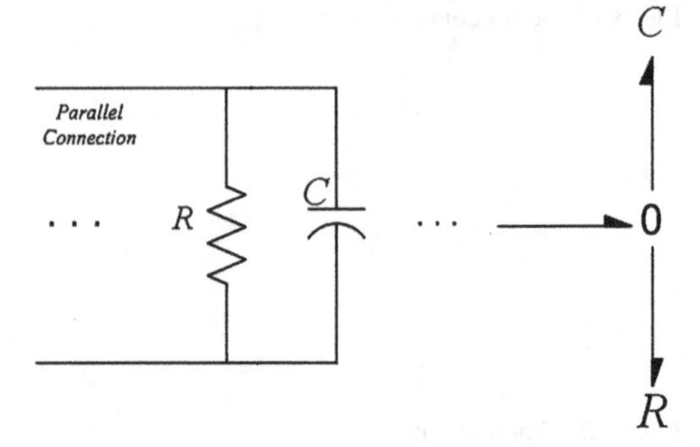

Fig. 3.26 Parallel connection

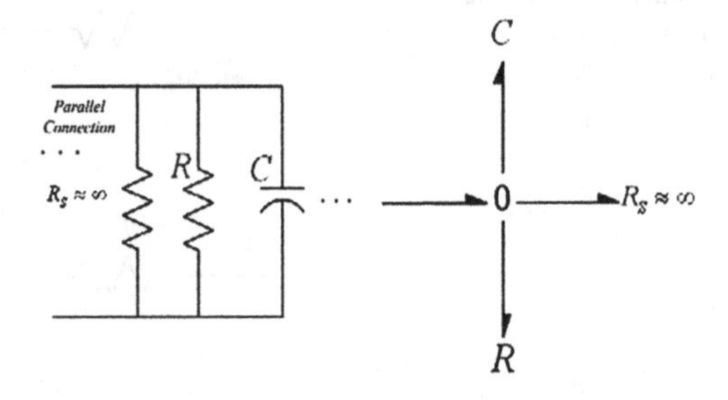

Fig. 3.27 Approximated model: Parallel connection with R_s component,

that is attached to the series connection, then $f_s = D_f$. This observation suggests that if $R_s \approx 0$ is only added at a series connection that has a flow sensor, then unknown variables due to this R_s component will not be added to the model. Likewise for parallel connection. If the $R_s \approx \infty$ is added to a parallel junction that has an effort sensor D_e, then $f_s = 0$ since the R_s is highly resistive, and $e_s = D_e$. Note that the locations where this model approximation is applicable can be easily identified, since the locations of the physical sensors are explicitly shown in a HBG. These junctions are called *sensor junctions*. This explicit description of HBG allows the idea of model approximation to be implemented easily.

In the preceding discussion, it has been shown that the unknown variables $\{e_s, f_s\}$ due to an additional R_s can be overcome by adding it at a sensor junction. Till now, the causality implications of adding a R_s to the graph has not been considered. Here, the causality implications of the R_s components is studied. Additionally, a procedure is developed to combine the model approximation method with the SCAPHD algorithm for systematic derivation of a DHBG from an acausal HBG.

Figure 3.28a, b depict a 1-junction with a R_s component added to it. These figures show two possible causal forms of the $R_s \approx 0$ component. The bonds of the original graph are denoted by the solid lines in the figures. From Fig. 3.28a, it can be observed that the causal form of the R_s has no contribution to the freedom of causality assignment since the elimination of the R_s maintains valid causalities of the original bonds $\{1, 2, 3\}$. In other word, the R_s is redundant. The causal form of the R_s shown in Fig. 3.28a is said to be *indifferent*. In contrast, Fig. 3.28b shows the same graph with another causality assignment. In this case, the component R_s cannot be eliminated from the graph in order to maintain the valid causalities of the original bonds. The

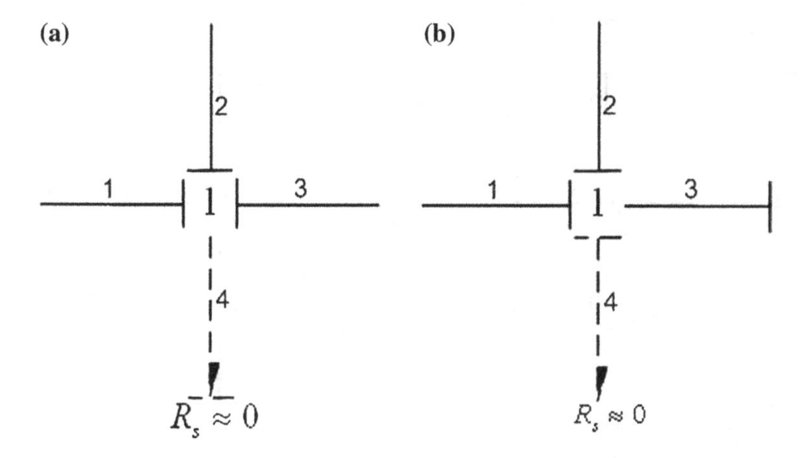

Fig. 3.28 **a** Causal form 1: Indifferent form. **b** Causal form 2: functional form

causal form of the R_s in Fig. 3.28b is said to be *functional*. One implication is that the causal strokes of the original bonds $\{1, 2, 3\}$ cannot be assigned in the way as shown in the Fig. 3.28b if the R_s is not added and assigned with the functional causal form. This means that a R_s that has a functional causal form allows a causality assignment to the original bonds, which is initially impossible without the R_s component. This illustration shows a R_s component that has a functional causality contributes to the freedom of causality assignment.

This representation of causal relationships has a strong similarity with other constraint imposition formalisms available in the literature, namely inversion of sensor causality [25], use of modulated virtual sources and sensors [5], and use of bicausalled source sensors [14]. Despite of the strong similarity, this representation leads to new insights in exploiting HBG for FDI of hybrid systems.

Let us consider the case of a 0-junction. Figure 3.29a, b show the possible causalities of the graph. Similarly, the causal form of R_s shown in Fig. 3.29a is indifferent, and Fig. 3.29b depicts the R_s with a functional causal form. Figure 3.30a, b summarize the functional and indifferent causal forms of the R_s components for the 1- and 0-junctions.

This study suggests that every R_s component that has indifferent causal form can be eliminated from the approximated HBG since it does not contribute any freedom of causality assignment to the graph. In contrast, the R_s components are equipped with functional causalities should remain in the graph.

Note that a functionally causalled R_s component is the preferred location to break the causal path and derive the ARRs through its constitutive relation. At a 0-junction, the expression for the flow variable in the functionally causalled R_s element is an ARR. Similarly, at a 1-junction, the expression for the effort variable in the functionally causalled R_s element is an ARR.

Finally, Fig. 3.31a, b show the causal paths of the additional variables introduced by the R_s components. The functional causal forms of the R_s components determine

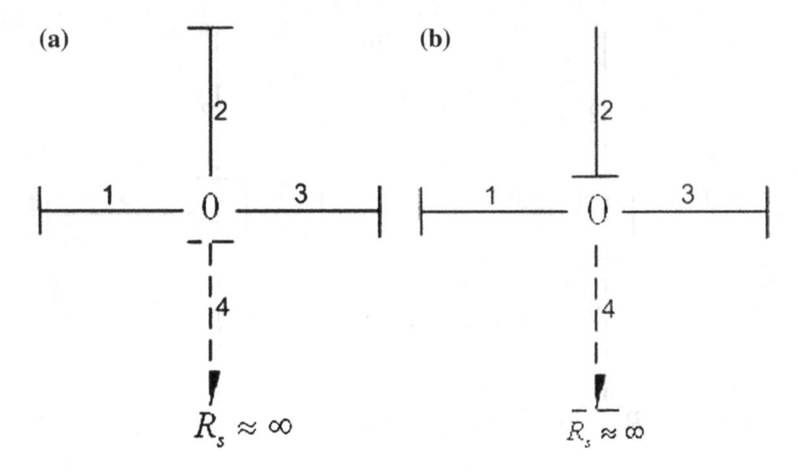

Fig. 3.29 **a** Causal form 1: Indifferent form. **b** Causal form 2: functional form

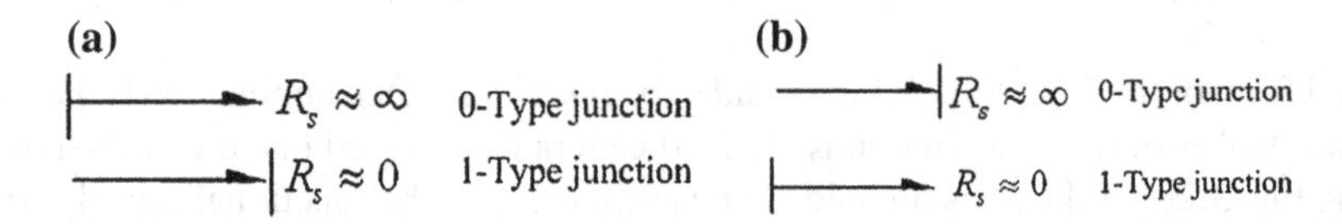

Fig. 3.30 **a** Functional causal forms. **b** Indifferent causal forms

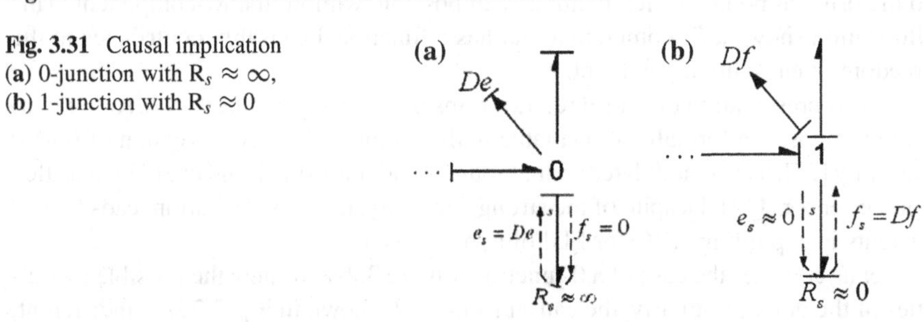

Fig. 3.31 Causal implication (**a**) 0-junction with $R_s \approx \infty$, (**b**) 1-junction with $R_s \approx 0$

the causal paths of the graph as shown in the figures. From the figures, it can be observed that the introduction of R_S components pose no problem in solving any unknown variable using causal paths since the sub-paths that lead to the R_S components are terminated by the sensors of the junctions. This causal property shows that the additional variables $\{e_S, f_S\}$ introduced by a R_S component at a sensor junction does not pose any problem when solving an unknown variable.

The insights gained from this study suggests that we can combine the SCAPHD algorithm with the model approximation method via the following steps. For a given acausal HBG,

1. Add a R_S component to every sensor junction of the HBG.
2. Apply the SCAPHD algorithm to assign the causality of the approximated HBG.
3. Eliminate every R_S that is in indifferent causality.

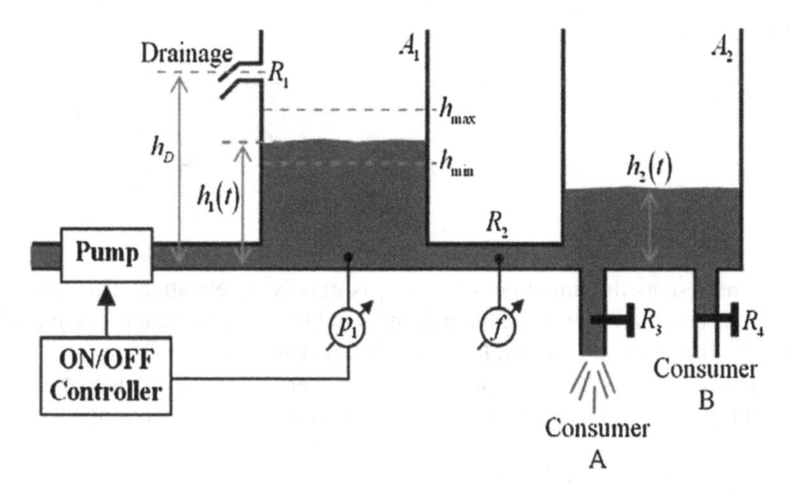

Fig. 3.32 The two-tank plant

This procedure integrates the model approximation technique with the SCAPHD algorithm effectively to derive a DHBG from a given acausal HBG. The procedure exploits the properties gained in this study to facilitate the application of SCAPHD algorithm. With the help of the model approximation method, a DHBG can be easily obtained from the given acausal HBG.

Note that additional R_s components can be obtained by having more sensors in the system. This implies that by adding more sensors to the system, user may be able to derive a DHBG when the procedure fails to derive it from the initial HBG.

The SCAPHD and the Model Approximation (MA) method are demonstrated in the following example. The aim of the example is to give an illustration on how the developed results can be used to derive unified ARRs for FDI application. It will be demonstrated that the stray resistors introduced by the MA method do not pose any numerical problem in this FDI application. Let us consider a two-tank plant as given in Fig. 3.32:

The plant is composed of two tanks (A_1, A_2), liquid-level controller, and two valves (R_3, R_4). The goal is to provide a continuous liquid supply to two customers (A and B). A_1 and A_2 represent the two tanks' cross section area. The liquid level at the tank A_1 is controlled by an ON/OFF controller to maintain the liquid level between two limits (h_{min}, h_{max}). The controller turns ON or OFF a pump. When the pump is ON, the flow through the pump is q_{in}, when the pump is OFF, the flow is zero. The liquid-level at the tank A_1 is measured by the pressure sensor p_1, which is installed at the bottom of the tank. The relation between liquid-level and liquid-pressure is given as follows:

$$p(t) = \rho g h(t), \tag{3.73}$$

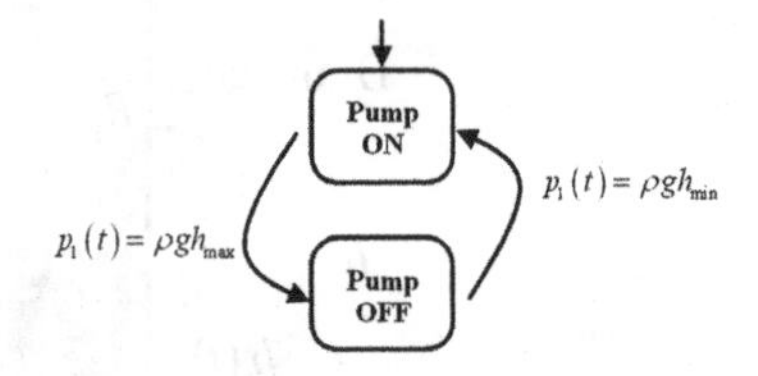

Fig. 3.33 The ON/OFF controller

where ρ represents the liquid density and g is gravity acceleration. The controller dynamics is based on the automaton presented in Fig. 3.33 (where it is assumed that the two tanks are initially empty, hence the initial state is ON).

The two valves (R_3, R_4) are operated by two users. Each valve has two discrete states OPEN and CLOSED, and the valves' hybrid dynamics is given by:

$$\begin{aligned} &\text{The Valve is CLOSED} \Rightarrow f_{Ri}(t) = 0 \\ &\text{The Valve is OPEN} \Rightarrow f_{Ri}(t) = \frac{1}{R_i}\sqrt{p_2(t)} \quad \text{for } i = 3, 4 \end{aligned} \tag{3.74}$$

where $f_{Ri}(t)$ is the volumetric flow of the liquid through the valves, $p_2(t)$ is the pressure at the bottom of the tank A_2, and R_i is the valve discharge coefficient.

In this system, the plant is equipped with a flow sensor which measures the liquid flow between the two tanks. This flow is proportional to square root of the pressure difference $\Delta p = p_1 - p_2$ which is given as follows:

$$f(t) = \frac{1}{R_2}\text{sign}(\Delta p(t))\sqrt{|\Delta p(t)|}. \tag{3.75}$$

If the liquid-level controller fails, a liquid overflow would occur. To prevent an overflow, tank A_1 is equipped with an additional outlet to drain the liquid when $h_1(t)$ is higher than h_D. The liquid-flow which travels through the drainage is described as follows:

$$\begin{aligned} &h_1(t) \le h_D \Rightarrow f_D(t) = 0 \\ &h_1(t) > h_D \Rightarrow f_D(t) = \frac{1}{R_1}\sqrt{\Delta p_D(t)} \end{aligned} \tag{3.76}$$

where $\Delta p_D = p_1 - p_D = \rho g(h_1(t) - h_D)$.

The HBG of the two-tank plant is presented in Fig. 3.34. In this graph, the flow variable is the liquid volumetric flow (i.e., [m^3/s]), and the effort variable is the liquid pressure (i.e., [pa]). The hybrid nature of the pump and the ON/OFF controller is modeled by the flow-source q_{in} and the controlled-junction 1_{c1}. The storage components C_1 and C_2 represent the two tanks, where the coefficients are given by $C_i = A_i/g$ for $i = 1, 2$. The equation of the drainage is given in (3.75), which is represented in the bond graph by the controlled-junction 1_{c2} and the components p_D and R_1. The resistor R_2 and the junction 1_1 represent the center tube that connects the two

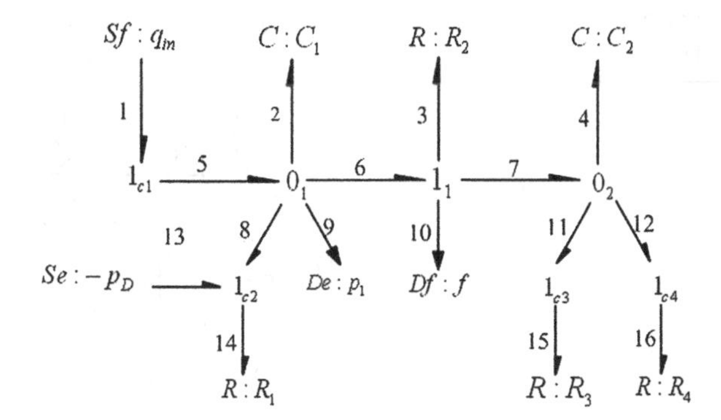

Fig. 3.34 The two-tank plant acausaled HBG

tanks, and resistors R_3, R_4 together with the two controlled-junctions 1_{c3}, 1_{c4} model the two customers' valves. The two detectors p_1 and f are the two sensors modeled in the hybrid bond graph model. Note that the ON/OFF controller is not presented in the bond graph model as its presence is irrelevant for the causality assignment.

The HBG in Fig. 3.34 is acausaled, and the three steps procedure of SCAPHD with MA is applied to form a DHBG. First step, a R_s component is added to every sensor junction; in this example a very-high resistive component is added to the junction 0_1 and a very-low resistive component is added to the junction 1_1 (with their bond numbers 17 and 18). The second step is to apply the SCAPHD algorithm. In this step, all controlled-junctions are assigned with their preferred causality. The output variables of the controlled-junctions 1_{c3} and 1_{c4} are assigned as inputs of the 1-port components (R_3 and R_4 respectively). The model includes two controlled-junctions that have a source connected to the junction. For the pair $\{p_D, 1_{c2}\}$ the output variable of the controlled junction must not be an input variable of a standard junction, hence the output variable of 1_{c2} is assigned as the input variable of R_1 (which is a 1-port component). For the pair $\{q_{in}, 1_{c1}\}$, the output variable of the controlled junction must be the input variable of the connected source component (such configuration is valid only if the source is inactive when the junction is OFF). The source component p_D is assigned according to step 5 and the causal implication is extended to bond 8. Since the DHBG is required for the plant health diagnosis (e.g. ARR derivation), the two storage components C_1 and C_2 are assigned with their preferred derivative causality. Now, we extend the causal implication using the components constraints to bonds 5, 11, 12, 7, 3, 10, 18, 9, 17 to complete the SCAPHD algorithm for the approximated HBG. The final step is to eliminate every R_s that is in indifferent causality.

From Fig. 3.28a, b, it can be observed that R_{s2} is redundant and therefore is removed from the DHBG with its bond 18 (this is marked by the dotted x in Fig. 3.35). On the other hand R_{s1} is functional, hence it is remained in the graph. The DHBG with the two storage components in derivative causality would not be possible if the HBG is not equipped with the additional R_{s1} via the modeling approximation. This benefit is presented in Fig. 3.35. Additionally, we demonstrated that the concept of stray resistors is not unique to electrical systems and can be implemented in other domains. In this hydraulic domain, R_{s1} represents a small hole at the bottom of

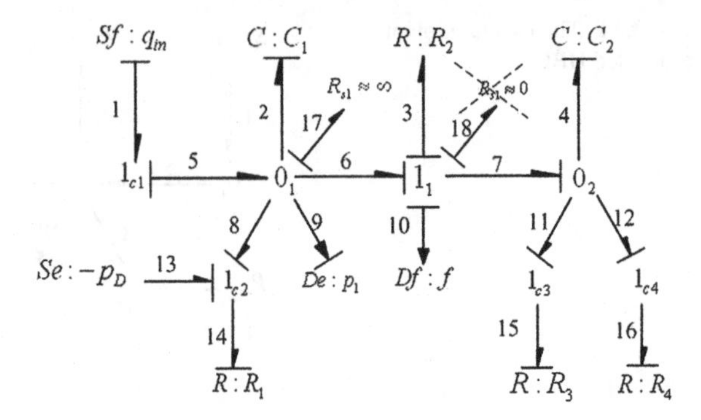

Fig. 3.35 The two-tank plant DHBG

the tank A_1, where the liquid-pressure across the hole is p_1. Note that R_{s1} has high resistance, therefore, the liquid flows through the hole is negligible.

As pointed out previously, the MA method has a strong similarity with other constraint imposition formalisms, e.g., the sensor causality inversion [25]. A single sensor junction attached with an R_s element is equivalent to the same sensor junction without R_s, if the sensor causality is inverted; however the classical application of sensor inversion method, encounters difficulties if it is applied to hybrid systems. These difficulties can be avoided using the proposed SCAPHD+MA method. This point is illustrated by this example.

Consider the acausal HBG of the two-tank plant given in Fig. 3.34. The classical causality inversion method developed in [4] is formulated for continuous systems (or to individual modes of a hybrid system [25]). It is well known that the constitutive relations of sensor junctions with inverted sensor causality lead to ARRs. To apply this method directly to this hybrid system without the aids of the proposed SCAPHD+MA approach, we first assign the two sensors of the two-tank plant with their inverted causality. Since the system is hybrid and a DHBG is required, it is desirable to assign all controlled junctions with their preferred causality. Here, it is desirable to assign the two storage components with their preferred causal description for FDI (i.e. derivative causality). However, one of the two storage elements is required to be in integral causality description as depicted in Fig. 3.36 by the dashed circle. This leads to an additional unknown variable which deteriorates analytical redundancy.

The reason for the above-mentioned problem is that in some modes (e.g. when all controlled junction are OFF), the flow sensor is redundant and its causality is not invertible. While in other modes (e.g. when all controlled junction are ON), the flow sensor is not redundant and can be inverted. On the other hand, for a DHBG, these sensors' causalities are required to be consistent at all modes. The SCAPHD+MA algorithm eliminates these difficulties by letting the causality of the R_s elements to be the last step of the assignment procedure. This advantage motivates us to apply SCAPHD+MA algorithm for hybrid bond graph instead of using the classical causality inversion.

As a matter of fact, the SCAPHD algorithm can be implemented with the aid of sensor causality inversion to achieve a DHBG. In this case, sensors causality

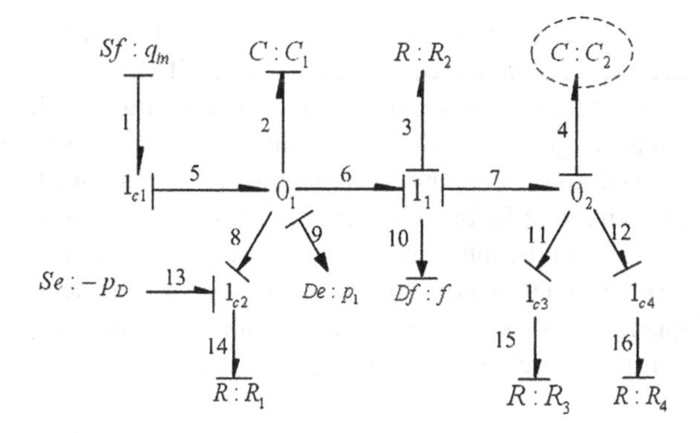

Fig. 3.36 Incorrect implementation of sensor causality inversion

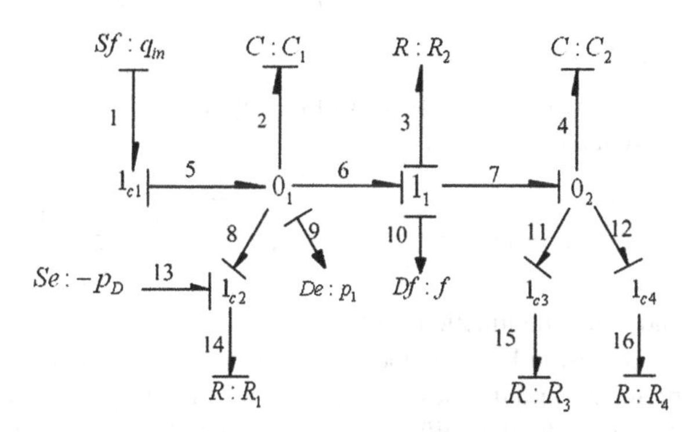

Fig. 3.37 Correct implementation of sensor causality inversion

assignment has the same priority of the R_s elements of the original SCAPHD+MA algorithm (and junctions with inverted sensor are analogous to junctions with functional R_s). The result is presented in Fig. 3.37.

Despite the similarities between the MA method and the causality inversion method, the MA method is derived from model approximation viewpoint. On the other hand, the classical causality inversion is proposed from the inverse dynamics perspective to derive ARR equation for continuous systems. These different viewpoints lead to different FDI design frameworks to different system domains.

The FDI applications of DHBG can be described using the same example. Note that the two-tank plant is a hybrid system where its discrete dynamics is represented by 16 different modes (the mode is determined by the controlled-junction's state). Each mode is characterized by a continuous dynamics and different modes correspond to different continuous models. The HBG allows us to represent all these modes under a single modeling framework. In the DHBG, every active component has a valid and consistent causality at all operating modes. In other words, it is not required to reassign a different causality for different operating modes. One important implication is that the causal paths of the graph remain the same structure, except that some of the sub-paths are eliminated due to the OFF states of the controlled-junctions. This feature of DHBG allows derivation of unified equations that govern the hybrid system behavior at all modes. The following illustrates the concept by deriving one unified equation from the DHBG.

ARR derivation for FDI is based on constitutive relations of bond graph components and junctions. Some elements of the graph are preferred, because the ARRs derived from their constitutive relations can be shown to be structurally independent [5]. An example for such preferred element is the functionally causalled R_s component. The constitutive relation of the R_s component connected to the junction 0_1 is $f_{17} = 0$. This relation is an ARR, if the unknown variable f_{17} can be expressed by known quantities (such as inputs, known parameters and measurements). The junction 0_1 constitutive relation leads to

$$f_{17} = f_5 - f_2 - f_8 - f_6. \tag{3.77}$$

Based on the DHBG (Fig. 3.35), the unknown flow f_8 can be solved in the following way using the causal paths of the graph.

When the junction is ON,

$$f_8 = f_{14} = \frac{\sqrt{e_{14}}}{R_1} = \frac{\sqrt{e_8 - e_{13}}}{R_1} = \frac{\sqrt{e_9 - e_{13}}}{R_1} = \frac{\sqrt{p_1 - p_D}}{R_1}, \tag{3.78}$$

and when the junction is OFF, $f_8 = 0$.

To design FDI methods for hybrid systems, a unified description of variables is preferred, and it is desirable to obtain a unified representation of ARRs in all modes. For example, the unified description of f_8 at all modes can be defined as

$$f_8 = a_2 \frac{\sqrt{p_1 - p_D}}{R_1}, \tag{3.79}$$

where $a_2 \in \{0, 1\}$ is a discrete variable representing the state of controlled junction 1_{c2} (i.e., $a_2 = 1$, when the junction is ON, and $a_2 = 0$ when the junction is OFF). Any causal path containing the variable f_8 is cut when the junction 1_{c2} is OFF (and the variable $f_8 = 0$); however all remaining terms of (3.78) remain unchanged. This suggests that a unified ARR is achievable by substituting (3.79) into (3.78). Using the same approach, the unknown f_2, f_5, and f_6 are solved, and the unified ARR is (3.79)

$$a_1 q_{in} - C_1 \dot{p}_1 - a_2 \frac{\sqrt{p_1 - p_D}}{R_1} - f_6 = 0. \tag{3.80}$$

Note that the unified ARR (3.80) describes the system at different modes by using different corresponding values of a_1 and a_2. This representation implies that this equation can be easily implemented to monitor a hybrid system at different modes. This unified ARR for FDI requires knowledge of the system operating mode and autonomous mode changes are detected by our mode tracker [26].

This example demonstrates the concept of utilizing a unified description to describe a hybrid system at different modes. Other detailed useful applications of DHBG to FDI are presented in [27].

3.3.2 Global Analytical Redundancy Relationships

From the previous section, it is observed that it is possible to utilize a unified description to describe a hybrid system at different modes. In HBG modeling language, an OFF controlled junction enforces the respective power variables of its adjacently connected bonds to zero. This property suggests that the covering path method can be extended based on the DHBG via the following procedures.

1. Define a boolean variable a_i that represents the state of every controlled junction $i = \{1, \ldots, q\}$, and let N be the total number of sensors $\{De, Df\}$ modeled in the HBG.
2. Let n_i be the total number of bonds of the controlled junction i. For each controlled junction i, there are n_i bonds adjacently connected to it. Every flow variable f_i^j (for $j = 1, \ldots, n_i$) of the n_i bonds is substituted with $a_i.f_i^j$ if the considered junction is 1-Type. Similarly, every effort variable e_i^j (for $j = 1, \ldots, n_i$) of the n_i bonds is substituted with $a_i.e_i^j$ if the considered junction is 0-Type.
3. Repeat step 2 until all the q controlled junctions are considered.
4. Choose a BG junction. The junction can be a standard BG junction or controlled-junction.
5. Based on the constitutive relations of the considered BG junction, all unknown variables of these relations are to be eliminated using the causal paths of the DHBG. If all unknown variables of any of these relations are solved, i.e., expressed only in terms of known variables (input, sensors variables), then the GARR is the relation with its solved unknown variables. Repeat solving until all the considered constitutive relations have been attempted.
6. Choose another BG junction.
7. Repeat step 5 until all the junctions of the DHBG have been considered, or until the number of generated GARRs equals N.

The main idea of introducing the boolean variables is to model the eliminations of the flow or effort variables due to the OFF states of the controlled junctions. Since a DHBG has a uniform causal structure, boolean variables can be used to model the flow and effort variables of the system efficiently based on the DHBG.

The rationale that the GARR generation procedure stops when N GARRs are derived is from the fact that the maximum number of ARRs that can be generated for a single mode is equal to the total number of sensors, modeled in the bond graph [5, 25]. This property leads to possible early termination of the GARR generation procedure without considering all the components of the DHBG.

With this unified representation, the unified constraint equations (GARRs) of the hybrid system can be derived from the procedure.

In general, a GARR has the form of

$$F_l(\theta, a, De, Df, u) = 0 \qquad \text{for } l = 1, \ldots, m. \tag{3.81}$$

where m denotes the number of GARRs derived from the DHBG, $\theta = [\theta_1, \ldots, \theta_p]^T$ represents the parameters of the HBG components, where p represents the number of HBG parameters used to describe the hybrid system. $a = [a_1, \ldots, a_n]^T$ denotes the operating mode of the hybrid system. u denotes the system's inputs, De and Df denote the effort and flow sensors of the graph.

Equation (3.81) has the same form as the ARR of continuous systems except that Eq. (3.81) has included the mode information α of the hybrid system. This similarity suggests that FDI algorithms can be developed based on the GARRs. In the remainder of this chapter, these GARRs are exploited for fault detectability analysis and fault diagnosis design. Notice that if a GARR is derived based on the constitutive relation of a controlled junction or from the relations of its adjacently connected BG components, then the GARR would be deactivated if the controlled junction's state is OFF. Therefore, the GARR would not be able to provide meaningful information about the behavior of the hybrid system at the operating modes where the junction is OFF. In this case, the total number of GARRs would not be constant at all modes. Nevertheless, this type of GARR is still able to provide valuable information at the operating modes where the junction is active.

3.3.3 Fault Detectability and Isolability Analysis

This chapter considers parametric faults, where their occurrence are modelled as a change in some parameters of the faulty components. For continuous systems, the ARRs have been applied to study fault detectability and fault isolability [12, 13]. This study is structural in the sense that the values of the ARRs' arguments are not required to deduce the system's fault detectability and fault isolability. In this section, the GARRs are exploited to study the fault detectability and fault isolability [28].

Unlike continuous systems, hybrid systems are multiple modes in nature. This feature suggests that the system's fault detectability and fault isolability are required to be evaluated at different operating modes for effective FDI analysis and designs. Fortunately, the unified characteristic of the GARRs provide a convenient way to generate the Fault Signature Matrix (FSM) of each operating mode. Before illustrating on how FSM can be generated from the GARRs, some notions that are useful to describe the different degrees of fault detectability and fault isolability are presented.

Definition 3.1 A parameter θ_i is said to be *all-mode detectable* if the parameter is detectable at all operating modes.

This definition refers to a parameter which is detectable at all modes. Assuming a hybrid system contains some critical components where their health status are important, then it is desirable that these components' parameters are all-mode fault detectable. Definition 3.1 defines a strong fault diagnosis property. Some weaker notions can be stated as follows.

Definition 3.2 A parameter θ_i is said to be *weakly detectable* if there exists a mode such that the parameter is detectable.

Definition 3.3 A parameter θ_i is said to be *all-mode non-detectable* if the parameter is undetectable at all operating modes.

Definition 3.2 refers to parameters that are fault detectable at some modes but not all modes. On the other hand, Definition 3.3 refers to parameters that are non-detectable for all modes. In the same manner, definitions for fault isolability are defined as follows.

Definition 3.4 A parameter θ_i is said to be *all-mode isolable* if the parameter is isolable at all operating modes.

Definition 3.5 A parameter θ_i is said to be *weakly isolable* if there exists a mode such that the parameter is isolable.

Definition 3.6 A parameter θ_i is said to be *all-mode non-isolable* if the parameter is non-isolable at all operating modes.

With these definitions, the fault detectability and fault isolability of each parameter can be characterized. Here, the GARRs and the structure of the HBG are exploited to deduce the fault detectability and fault isolability of each parameter.

Property 3.9 If a component is adjacently connected to a controlled junction, then all the component's parameters are not all-mode detectable and not all-mode isolable.

Proof proof is straightforward. Since the component is adjacently connected to a controlled junction, then one of its two power variables is enforced to be zero, in modes, where the junction is deactivated. As there is no energy flow in the component, there is no way to observe the behavior of the component based on sensors. Hence, the component's parameters are also not all-mode detectable. Since fault detectability is a necessary condition for fault isolability, the component's parameters are not all-mode isolable. Property 3.9 provides a necessary condition for a component's parameters to be all-mode detectable and isolable but not weakly detectable. Based on the characteristic of the controlled junction, another property can be deduced from Property 3.6.

Property 3.10 If a component is adjacently connected to a controlled junction, then the component is non-detectable at operating modes where the controlled junction is OFF.

Proof The proof follows the same argument used in the proof of Property 3.9.

Table 3.2 Fault signature matrix (FSM) at mode a

Mode a	r_1	...	r_m	D_b	I_b
θ_1	1 or 0				
.					
.					
.					
θ_p					

Property 3.10 provides a sufficient condition to determine non-detectable parameters using the HBG without inspecting the GARRs. Property 3.9 implies that $D_b = 0$ for those parameters of the components that are adjacently connected to the controlled junctions at the respective operating modes. To conclude other fault detectability and fault isolability based on Definitions 3.1–3.6, GARRs are required to provide the information. In HBG modeling language, each operating mode of a hybrid system is represented by the corresponding states of all controlled junctions of the system's HBG. The fault detectability and fault isolability of the hybrid systems are evaluated at all modes by inspecting the GARRs. To derive the FSM of mode a, the parameters θ of the m GARRs are inspected at that mode.

In the same spirit of the FSM generation for continuous systems, the FSM for hybrid systems is derived based on the GARRs. Consider the case where m GARRs are governing the behavior of a hybrid system. The FSM of mode a can be formulated as follows (see Table 3.2).

Based on the D_b and I_b of Table 3.2, the detectability and isolability of the parameters can be determined for each mode. If a component is weakly detectable or isolable at all modes, then the component is said to be all-mode detectable or all-mode isolable. It is worth to note that the *all-mode detectability* of a parameter can be directly deduced by inspecting F_l for $l = 1, \ldots, m$. A sufficient condition for this test is as follows.

Property 3.11 A parameter θ_i is said to be all-mode detectable if for any mode, there exists a $l \in \{1, \ldots, m\}$ such that F_l depends on θ_i.

Proof If for any mode, there exists a $l \in \{1, \ldots, m\}$ such that F_l is dependent on θ_i, then a change of the parameter is always detectable by some residuals. In other words, the D_b of the parameter is always 1 for every FSM for any mode.

Note that the GARRs that are generated from constitutive relations of controlled junctions can be ignored at the corresponding modes where the junctions are inactive.

3.3.4 Case Study

For illustration, a 2-mode hybrid system is utilized (see Fig. 3.38). The system has a switch (Sw), 7 components, and 3 sensors. The input is represented by the source

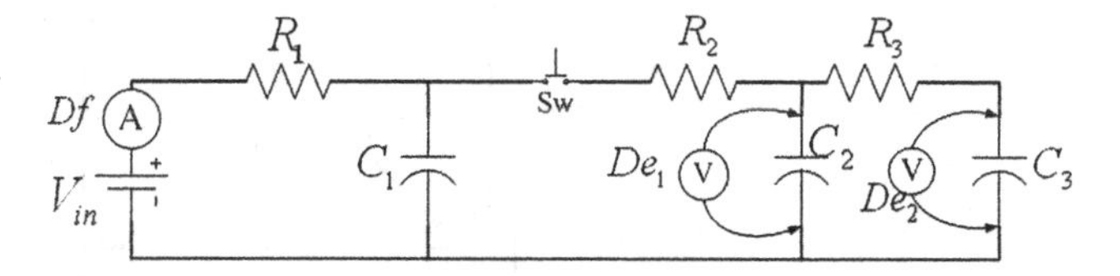

Fig. 3.38 An electrical hybrid system

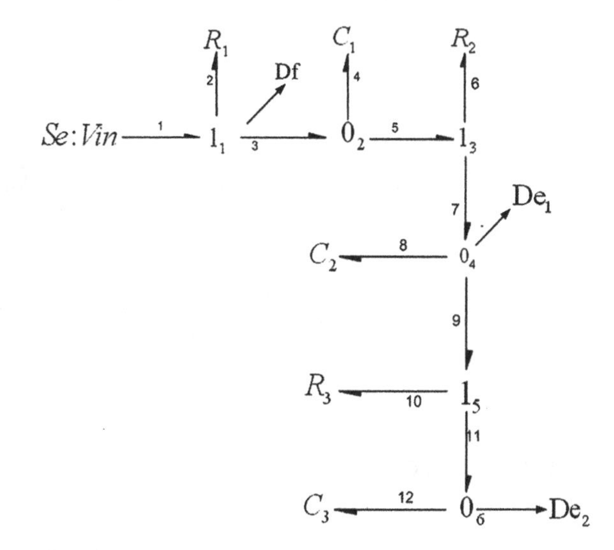

Fig. 3.39 The system's acausal HBG

element *Se*. D_f denotes the current (flow) sensor, $\{De_1, De_2\}$ represent the voltage (effort) sensors that measure the respective voltages of the circuit shown in the figure. The acausal HBG of the hybrid system is shown in Fig. 3.39. Controlled junction 1_3 models the switching behavior of the physical system (Sw) in HBG language. For this system, $a \in \{0, 1\}$ denotes the operating mode of the hybrid system. a is the boolean variable that represents the state of the controlled junction 1_3.

Before the GARRs are derived, the 3-step integrated procedure presented in the previous section is applied to obtain a DHBG of the system's acausal HBG. First, the sensor junctions are identified, and the stray R elements are added to the HBG. In this case, $\{R_{s1}, R_{s2}, R_{s3}\}$ are added to sensor junctions $\{1_1, 0_4, 0_6\}$. Next, the SCAPHD algorithm is used to assign the causality of the approximated HBG. The derived DHBG of the hybrid system is shown in Fig. 3.40. Notice that the storage component (C) and the controlled junction (1_3) are assigned with their preferred causalities. This assignment wouldn't be feasible if the HBG is not equipped with the additional R_{s2} via the modeling approximation. And since the added stray resistors $\{R_{s1}, R_{s3}\}$ are assigned with indifferent causal form, they are redundant and are eliminated from the approximated HBG. This elimination is depicted in Fig. 3.40.

With the DHBG, the GARR generation procedure can be applied to derive the GARRs. Based on the boolean variable a, the flow variables of bond $\{5, 6, 7\}$ are

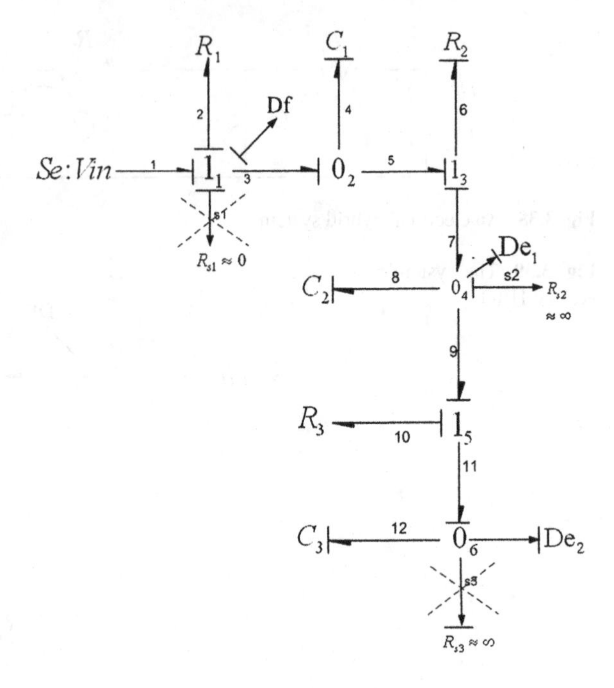

Fig. 3.40 The system's DHBG

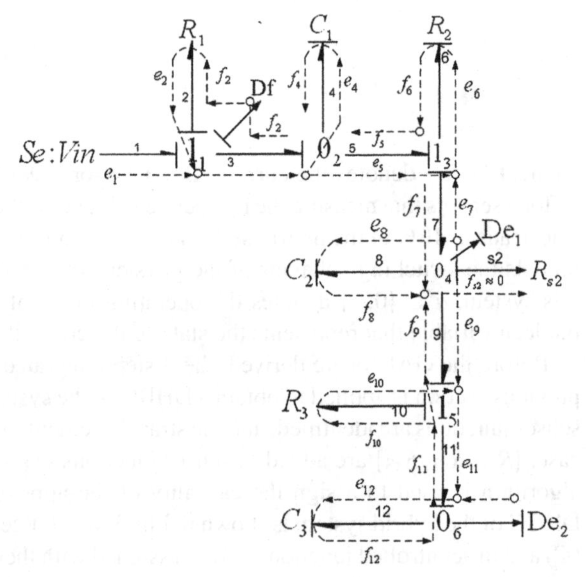

Fig. 3.41 The DHBG with causal paths

replaced with $\{af_5, af_6, af_7\}$. With the help of the causal paths (see Fig. 3.41), the generation procedure is applied to derive the GARRs. The following GARRs are derived from the constitutive relations of junctions $\{0_2, 0_4, 1_5\}$.

For junction 0_2, its constitutive relation is

$$f_3 - f_4 - af_5 = 0, \tag{3.82}$$

where f_4 and f_5 are unknown variables. By covering casual paths 4-3 and 5-6, $f_4 = C_1 \frac{d}{dt} e_3$, and $f_5 = f_6 = \frac{1}{R_2} e_6$, and hence Eq. (3.82) becomes

$$f_3 - C_1 \frac{de_3}{dt} - \frac{e_6}{R_2} a = 0. \tag{3.83}$$

where f_3 is measurable by sensor D_f, hence $f_3 = D_f$. The unknown variable e_3 is given by the 1_1 junction's constitutive relation,

$$e_3 = e_1 - e_2 = V_{in} - R_2 D_f. \tag{3.84}$$

Now it left unknown variable e_6 to be solved. Based on the constitutive relation of junction 1_3, $e_6 = e_5 - e_7$. Since $e_7 = De_1$ and $e_5 = e_3$, e_6 can be written as

$$e_6 = (V_{in} - R_2 f_2) - De_1. \tag{3.85}$$

With Eqs. (3.84) and (3.85), the constitutive relation (3.83) becomes

$$GARR_1 = D_f - C_1 \frac{d(V_{in} - R_1 D_f)}{dt} - \frac{1}{R_2}\{V_{in} - R_1 D_f - De_1\}a. \tag{3.86}$$

Next, the constitutive relation of junction 0_4 is considered. Since $f_{s2} \approx 0$, the equation can be written as

$$af_7 - f_8 - f_9 = 0. \tag{3.87}$$

In this case, $\{f_7, f_8, f_9\}$ are unknown variables. For f_7, the path 7-6 is covered to obtain $f_7 = f_6 = e_6/R_2$. This leads f_7 to

$$f_7 = \frac{1}{R_2}(V_{in} - R_1 D_f - De_1). \tag{3.88}$$

By constitutive relation of C_1, $f_8 = C_2 d(De_1)/dt$. As for unknown f_9, the path 9-11-12 is covered to yield $f_9 = f_{11} = f_{12} = C_3 d(De_2)/dt$. These relations lead Eq. (3.87) to

$$GARR_2 = \frac{1}{R_2}(V_{in} - R_1 D_f - De_1)a - C_2 \frac{d(De_1)}{dt} - C_3 \frac{d(De_2)}{dt}. \tag{3.89}$$

The third GARR is derived from the constitutive relation of junction 1_5

$$e_9 - e_{10} - e_{11} = 0. \tag{3.90}$$

Table 3.3 Fault signature matrix (FSM) at mode $a = 1$

$x = 1$	r_1	r_2	r_3	D_b	I_b
R_1	1	1	0	1	0
C_1	1	0	0	1	1
R_2	1	1	0	1	0
C_2	0	1	0	1	1
R_3	0	0	1	1	1
C_3	0	1	1	1	1
V_{in}	1	1	1	1	1

Variables e_{11} and e_9 are measurable by sensors De_2 and De_1; hence $e_{11} = De_2$ and $e_9 = De_1$. It left unknown variable e_{10} to be solved. By covering path 10-11-12,

$$e_{10} = R_3 f_{10} = R_3 f_{12}. \tag{3.91}$$

Based on the constitutive relation of element C_3, $f_{12} = C_3 d(De_2)/dt$. Hence Eq. (3.90) becomes

$$GARR_3 = De_1 - R_3 C_3 \frac{d(De_2)}{dt} - De_2. \tag{3.92}$$

For this hybrid system, three GARRs are obtained. To deduce the fault detectability and fault isolability from the GARRs, the FSMs of the two modes are derived. $\{r_1, r_2, r_3\}$ represent the residuals that are evaluated from $\{GARR_1, GARR_2, GARR_3\}$.

Table 3.3 shows the FSM of the hybrid system at mode $a = 1$ and Table 3.4 depicts the FSM of the system at mode $a = 0$. The fault detectability and fault isolability of each parameter is gained from the $\{D_b, I_b\}$ values of the two FSMs. This information can be summarized in Table 3.5.

Table 3.5 shows that all parameters except R_2 are all-mode detectable. R_2 is not all-mode detectable but is weakly detectable at mode $a = 1$. This result shows the consistency of Property 3.8 with respect to the controlled junction 1_3. Components $\{C_2, R_3, C_3\}$ are all-mode detectable and isolable, whereas V_{in} is all-mode detectable and weakly isolable at mode $a = 1$.

Table 3.4 Fault signature matrix (FSM) at mode $a = 0$

$x = 0$	r_1	r_2	r_3	D_b	I_b
R_1	1	0	0	1	0
C_1	1	0	0	1	0
R_2	0	0	0	0	0
C_2	0	1	0	1	1
R_3	0	0	1	1	1
C_3	0	1	1	1	1
V_{in}	1	0	0	1	0

Table 3.5 Fault detectability and fault isolability of components

θ	Detectability	Isolability
R_1	All-mode	Nil
C_1	All-mode	Mode $a = 1$
R_2	Mode $a = 1$	Nil
C_2	All-mode	All-mode
R_3	All-mode	All-mode
C_3	All-mode	All-mode
V_{in}	All-mode	Mode $a = 1$

This illustration shows that the fault detectability and fault isolability of each component can be well-presented in a compact fashion based on the notions defined in this section. Additionally, the fault detectability and fault isolability of each component can be systematically deduced based on the GARRs. With this information, a FDI designer would have a preliminary measure on the performance of the monitoring system.

Since the GARRs govern the hybrid system's behavior efficiently at all-mode, the equations can also be exploited to develop a comprehensive real-time FDI monitoring system for hybrid systems. This motivation leads us to propose a Quantitative Hybrid Bond Graph-based (QHBG) FDI system for a high-performance real-time FDI system.

Figure 3.42 depicts the architecture of the QHBG diagnosis framework. The framework utilizes the unified information of the HBG to monitor a hybrid system efficiently and effectively. This framework consists of a *GARR alarm generator*, a *fault detection module*, a *fault isolation module*, a *fault estimator*, and a *mode tracer*. These modules are based on the HBG of the hybrid system. The mode tracer constantly determines the instantaneous operating mode using sensors and input information obtained from the hybrid system. This instantaneous mode information allows users to evaluate the GARRs for residuals effectively at all operating modes. With these residuals, the QHBG framework is able to detect, isolation, and estimate the fault occurred in the system. The fault detection module decides whether or not a fault has occurred when any of the residual signals is non-zero. For those fault isolable parameters, the GARRs generate a set of unique residuals that allows the faulty parameter to be identified. For those non-isolable but detectable parameters, the module will select a set of potential fault candidates. Finally, the parameter estimation module estimates the selected fault parameters to access the health status of the system. This information can also be used to refine the non-isolable parameter. In this chapter, the fault detection and isolation modules are presented.

The application of the quantitative residuals for fault diagnosis is also applicable for hybrid systems. To apply these residuals, a binary coherence vector $C = [c_1, \ldots, c_m]$ is defined. Each component c_l of C is a boolean variable whose value is obtained by the following simple decision rule:

$$c_l = \begin{cases} 1 & \text{if } |r_l| > \varepsilon_l^a; \\ 0 & \text{otherwise} \end{cases} \quad \text{for } l = 1, \ldots, m \tag{3.93}$$

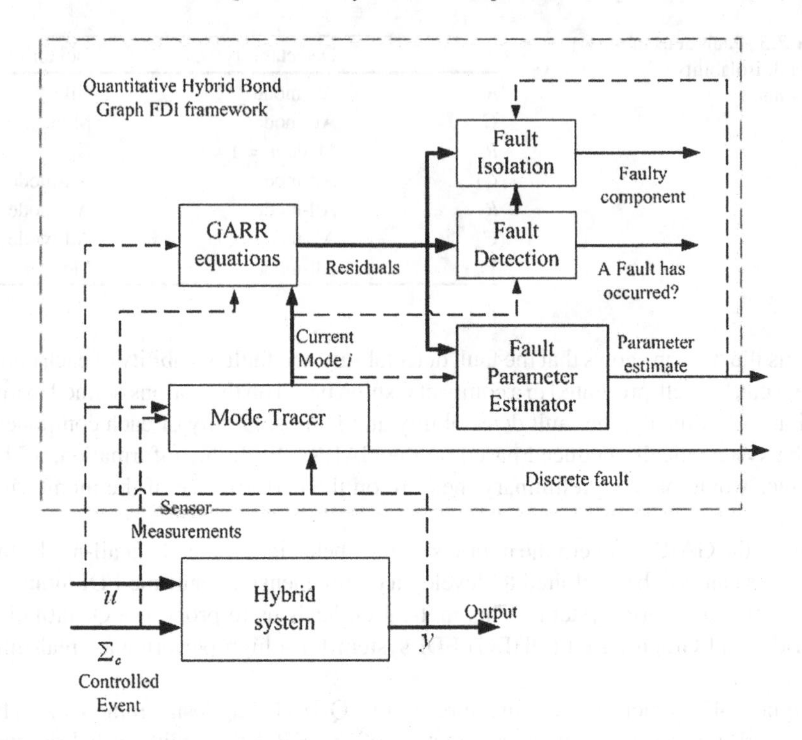

Fig. 3.42 Architecture of QHBG FDI system

where a denotes the instantaneous operating mode.

If the hybrid system is fault-free, then the binary coherence vector C will be a zero vector. On the other hand, if the system is faulty, then the coherence vector will be a nonzero vector.

Suppose a fault occurs at time k. To isolate the faulty parameter, the binary coherence vector C is matched with the fault signature of the FSM of the operating mode at time k. Under the assumption that the system can have only one fault, the fault can be detected and uniquely isolated based on its fault detectability and fault isolability.

For hybrid systems, the degree of discrepancy (nonzero value) for each residual due to modeling error may vary from mode to mode. Therefore, the thresholds may need to be defined carefully for each mode to avoid conservative thresholds. Alternatively, more sophisticated decision making procedures that have been utilized in [29, 30] may be used here for better decision performance.

(a) Simulation Study

The electrical circuit considered in Fig. 3.38 is used for simulations to evaluate the QHBG FDI framework proposed in this chapter. In this simulation, the QHBG is implemented to detect and isolate faults simulated in the hybrid system. For illustration, the fault-free components' parameters θ are chosen as $\{R_1 = 100\,\text{k}\Omega,\ C_1 = 10\,\mu\text{F},\ R_2 = 100\,\text{k}\Omega,\ C_2 = 110\,\mu\text{F}, R_3 = 100\,\text{k}\Omega,\ C_3, = 10\,\mu\text{F}, V_{in} = 7\ \text{v}\}$. In

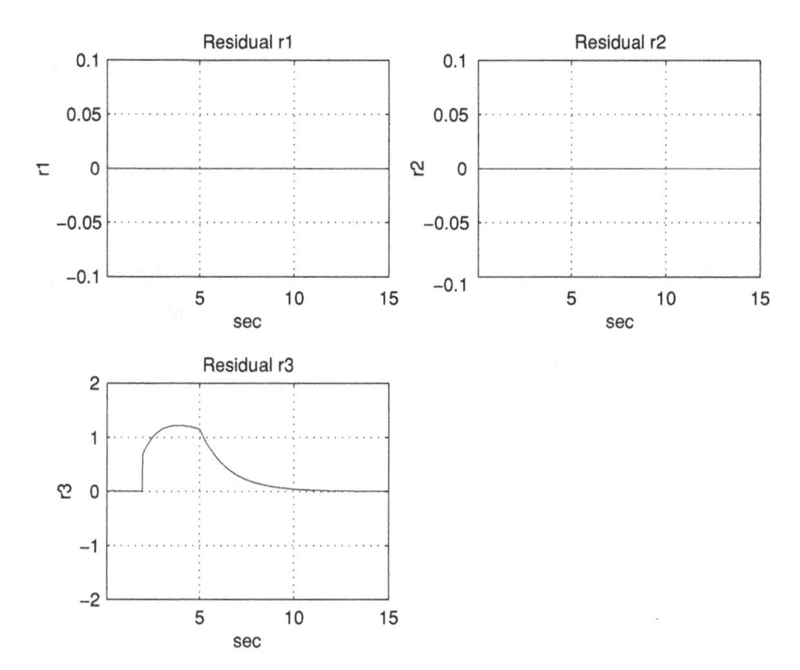

Fig. 3.43 Residuals in run 1

this case study, the switching signal of the hybrid system is known; therefore, the mode tracer module of the framework is replaced by one to one mapping between the state of the switch to the operating states of the hybrid system $a \in \{0, 1\}$. For simplicity, the thresholds ε_l^a for $l = 1, \ldots, 3$ are fixed at 0.02. A fault is simulated in R_3 component where its resistive value changes abruptly from 100 kΩ to 200 kΩ in two runs. In both runs, the hybrid system has an initial mode $a = 1$ (Sw ON) and switches to mode $a = 0$ (Sw OFF) at $t = 5$ s. In the first run, the fault is simulated at $t = 2$ s, and in the second run, the same fault was simulated at $t = 6$ s. The residuals of the two runs are presented in Figs. 3.43 and 3.44.

From the FSMs of the system, the component R_3 is an all-mode isolable component where its fault signature is $[r_1\ r_2\ r_3] = [0\ 0\ 1]$ for all modes. That implies that a deviation from the zero level is expected in residual r_3. Figure 3.43 depicts the simulated responses of the residuals in run 1. From the figure, residual r_3 is deviated from zero at $t = 2$ s. This is due to the R_3 fault simulated at $t = 2$ s. The coherence binary vector at the time instant is $C = [0\ 0\ 1]$ which indicates that R_3 is the faulty parameter.

Figure 3.44 depicts the residuals of run 2 where the fault is simulated at $t = 7$ s when the system is at mode $a = 0$. Likewise in run 1, a deviation in residual r_3 is expected upon the R_3 fault occurrence. The simulation results shown in Fig. 3.44 confirms this behavior. One observation to note is that the magnitude of the deviation in r_3 due to the fault is significantly different at the two modes. This difference

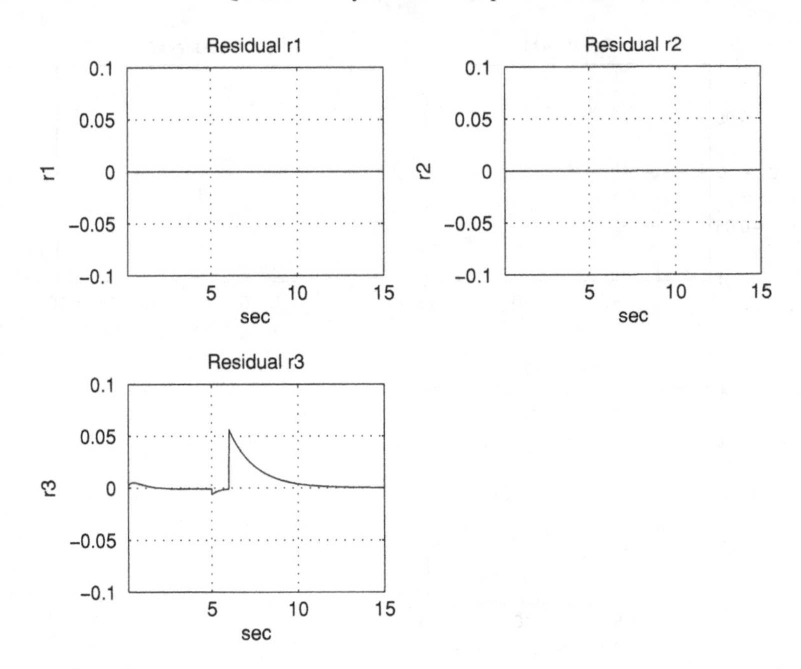

Fig. 3.44 Residuals in run 2

suggests that the threshold ε_3^0 can be chosen smaller than its counterpart ε_3^1 to have a less conservative and more reliable coherence binary vector C for fault detection.

From this simulation, an all-mode fault detectable and isolable component can be monitored at all operating modes based on the GARRs. This result is consistent with the fault detectability and fault isolability shown in Table 3.4, and more importantly, it demonstrates the efficiency of the QHBG framework. One physical interpretation of the all-mode detectable ability is that its corresponding component possesses energy interaction at all times. This interpretation matches with our intuition since there is no way the behavior of the component can be monitored based on its constitutive equation if the component has no energy. Energy within a component is essential for the model-based fault diagnosis.

(b) Experiment Study

In this experiment, the key concept of utilizing GARRs equations for FDI of hybrid systems is validated.

Figure 3.45 depicts the schematic of an electrical hybrid system. This system comprises of two main loops, one flow sensor Df, two efforts sensors $\{De_1, De_2\}$, a capacitor C_1, an input DC voltage V_{in}, three resistors $\{R_1, R_2, R_3\}$, and two autonomous switches {Sw1, Sw2}. In this circuit, two autonomous switches {Sw1, Sw2} control the voltage of the capacitor V_{c1}, such that the voltage stays within a predefined range $0 < b_1 \leq V_{c1} \leq b_2$. The control logics control the states of the two switches.

The basic operations of the system can be described as follows. First, the initial states of the switches are Sw1 ON and Sw2 OFF. When the DC input voltage V_{in}

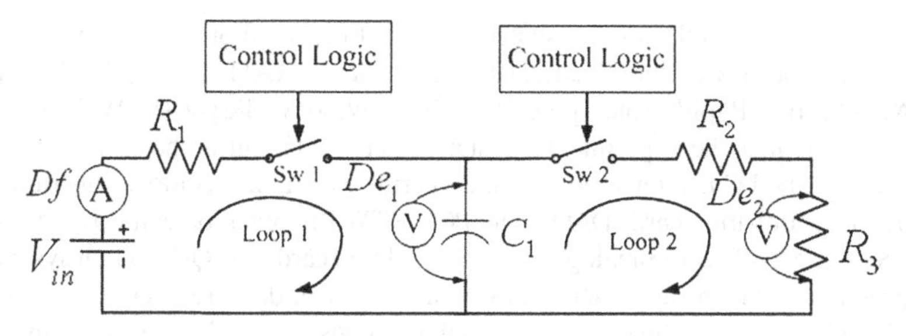

Fig. 3.45 Circuit of a hybrid system

Fig. 3.46 A Real-time Model-based FDI experimental system

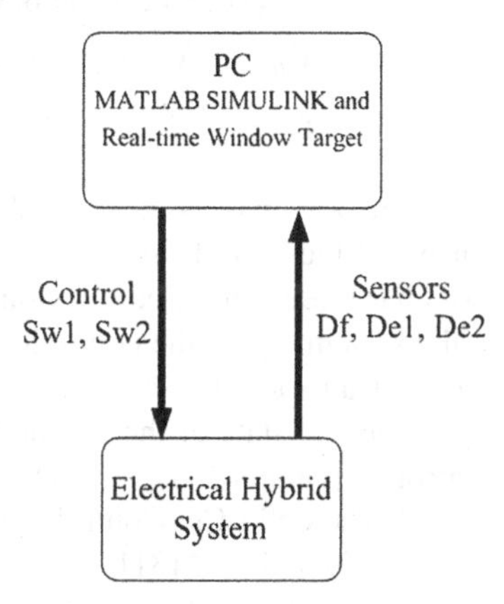

Fig. 3.47 Layout of the experimental system

is applied to the circuit, $V_{c1} \to V_{in}$ with a time constant $\tau_1 = R_1C_1$. When $V_{c1} > b_2$, Sw1 is OFF, and Sw2 is ON. At this operating mode, Loop 1 is disabled and the charges stored in the capacitor discharges onto resistive loads R_2 and R_3 via Loop 2. This discharge results $V_{c1} \to 0$ in the 1st-order manner with a time constant $\tau_2 = C_1\{R_2 + R_3\}$. As $V_{c1} < b_1$, Sw1 is ON and Sw2 is OFF. In this charging and discharging processes, the voltage of the capacitor V_{c1} is maintained between b_1 and b_2, i.e., $b_1 \leq V_{c1} \leq b_2$.

To implement the QHBG FDI system for real-time monitoring, an experimental system that consists of the electrical hybrid system, MATLAB® SIMULINK®, and MATLAB® Rapid-prototyping Real-Time Window Target (RTWT) is built. Figure 3.46 depicts the experimental system where its layout is shown in Fig. 3.47. In this setup, the PC is interfaced with the hybrid system via a National Instruments 6025E data acquisition card (DAQ). The PC's RTWT application controls the states of the Sw1 and Sw2 via the analog outputs of the DAQ card. The QHBG FDI system is also programmed in the application. Note that the sensor data are interfaced with the RTWT environment via the DAQ card, and the diagnosis algorithms are implemented in real-time at a sampling interval of $t_s = 0.1$ s.

In this experiment, two fast switching Photomos switching relays AQV201 and the MATLAB® RTWT are utilized to realize the two autonomous switches (Sw1 and Sw2). MATLAB® Stateflow chart is also used to implement the control logic (automaton) of the two switches. Two faults are considered in this experiment. In the first run, a faulty resistor (R_1) is simulated where the resistor's parameter decreases from 100 to 62 kΩ. In the second run, a short circuit is implemented on R_2 by reducing its resistive value from 86.2 kΩ to 0 kΩ. These faults are realized by utilizing variable resistors for components R_1 and R_2. The DC voltage V_{in} is supplied by an external DC power supply. The fault-free values of the components' parameters are chosen as follows.

$$\begin{aligned}
&b_1 = 0.2\,\text{V},\ b_2 = 6.5\,\text{V},\ R_1 = 100\,\text{k}\Omega,\\
&V_{in} = 7\,\text{V},\ R_2 = 86.2\,\text{k}\Omega,\ R_3 = 13.8\,\text{k}\Omega,\ C_1 = 10\,\mu\text{F},\\
&\varepsilon_1 = 1.9 \times 10^{-5}, \varepsilon_2 = 0.5 \times 10^{-3}, \varepsilon_3 = 0.5.
\end{aligned}$$

Figure 3.48 depicts the acausal HBG of the considered hybrid system. The three sensors of the hybrid systems are represented as $\{Df, De_1, De_2\}$ on the graph. The HBG has three controlled junctions $\{1_1, 1_3, 0_4\}$ that capture the switching behavior of the system. Since the controlled junctions 1_3 and 0_4 have the same states for $\forall t$, two boolean variables, a_1 and a_2 are defined for the GARR generation. $a_1 \in \{0, 1\}$ represents the state of the junction 1_1, and $a_2 \in \{0, 1\}$ represents the states of the controlled junctions $\{1_3, 0_4\}$. The operating mode of the system is defined as $a = [a_1\ a_2]$. The CSPECs (Control Specification) of the controlled junctions are shown in Figs. 3.49 and 3.50 [31].

To derive the GARRs from the HBG, the integrated procedure (SCAPHD and modeling approximation methods) is applied. Figure 3.51 depicts the derived DHBG of the system. Notice that the model approximation method adds an R_s (with functional form) to the junction 0_2 of the HBG to aid the derivation of the DHBG.

Now with the DHBG, the GARRs can be derived. First, consider the constitutive relation of R_1

$$f_2 - \frac{1}{R_1}(e_2) = 0 \tag{3.94}$$

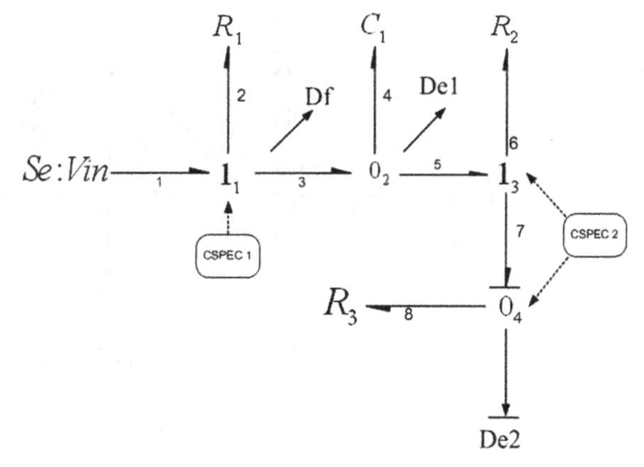

Fig. 3.48 The Acausal HBG of the hybrid system

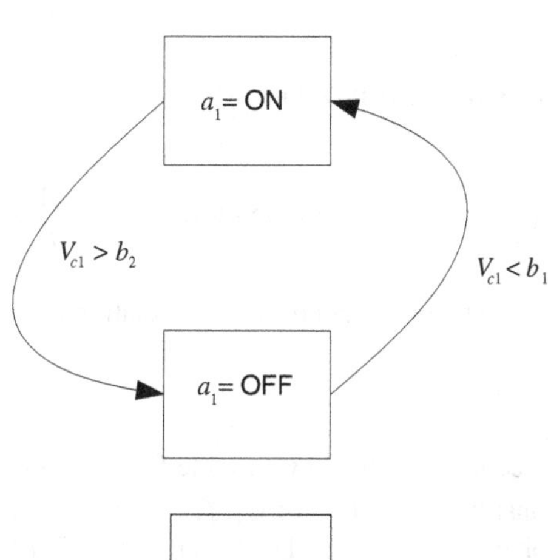

Fig. 3.49 The CSPEC of controlled junction 1_1

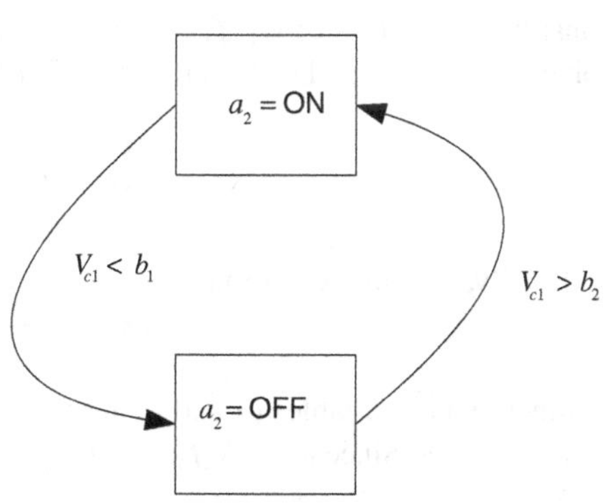

Fig. 3.50 The CSPEC of controlled junctions $\{1_3, 0_4\}$

as a GARR candidate. Since f_2 is measurable, $f_2 = D_f$, and it left unknown variable e_2 to be solved. The causal paths of bonds $\{1, 2, 3\}$ leads e_2 to $e_2 = V_{in} - De_1$. Hence Eq. (3.94) becomes

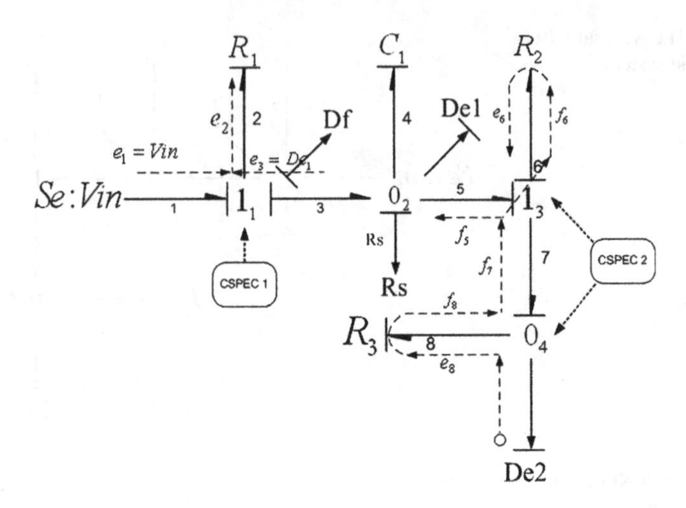

Fig. 3.51 The DHBG of the hybrid system

$$GARR_1 = D_f - \frac{1}{R_1}(V_{in} - De_1) = 0. \tag{3.95}$$

Next, considers constitutive relation of junction 0_2

$$a_1 f_3 - f_4 - a_2 f_5 = 0, \tag{3.96}$$

since $f_s \approx 0$. The flow variable f_3 is measurable by D_f, hence $f_3 = D_f$. By using the constitutive relation of $C_1, f_4 = C_1 \frac{d}{dt}(De_1)$. Now it left unknown variable f_5 to be solved. By causal paths of bonds {5, 6, 7, 8},

$$GARR_2 = a_1 D_f - C_1 \frac{d}{dt}(De_1) - \frac{a_2}{R_3} De_2. \tag{3.97}$$

Finally, consider equation

$$e_5 - e_6 - e_7 = 0 \tag{3.98}$$

of junction 0_4. Variables e_5 and e_7 are measurable by De_1 and De_2, hence $e_5 = De_1$ and $e_7 = De_2$. Since $e_6 = R_2 f_6$, and $f_6 = f_7 = \frac{1}{R_3} De_2$,

$$GARR_3 = De_1 - \frac{R_2}{R_3} De_2 - De_2 = 0. \tag{3.99}$$

Three GARRs are derived from the DHBG based on the generation procedure. Unfortunately, GARRs (3.95) and (3.99) are generated from controlled junctions instead of normal junctions. This means that the equations cannot be used if the respective controlled junction is OFF. Therefore, at mode $a = [1\,0]$, GARR in (3.95)

Table 3.6 Fault signature matrix (FSM) at operating mode $a = [1\ 0]$

$a = [1\ 0]$	r_1	r_2	r_3	D_b	I_b
V_{in}	1	0	×	1	0
R_1	1	0	×	1	0
C_1	0	1	×	1	1
R_2	0	0	×	0	0
R_3	0	0	×	0	0

Table 3.7 Fault signature matrix (FSM) at operating mode $a = [0\ 1]$

$a = [0\ 1]$	r_1	r_2	r_3	D_b	I_b
V_{in}	×	0	0	0	0
R_1	×	0	0	0	0
C_1	×	1	0	1	1
R_2	×	0	1	1	1
R_3	×	1	1	1	1

is disabled when $a_1 = 0$. Similarly GARR (3.99) is disabled when $a = [0\ 1]$. This characteristic implies that there is always two active GARRs that govern the behavior of the hybrid system for $\forall t$. The fault detectability and fault isolability of the components $\{V_{in}, R_1, C_1, R_2, R_3\}$ are presented in the following Tables (Tables 3.6 and 3.7).

The '×' of Tables 3.6 and 3.7 represents 'don't care', which implies that the corresponding residuals are to be ignored at the respective operating modes. For example, Table 3.6 suggests that residual r_3 is ignored at mode $a = [1\ 0]$. From the FSMs, it is clear that C_1 is all-mode detectable and isolable. $\{V_{in}, R_1\}$ are fault detectable, and $\{R_2, R_3\}$ are non-detectable at mode $a = [1\ 0]$. This observation matches with our intuition since Loop 1 (as defined in Fig. 3.45) possesses some energy interaction between the components at this mode. On the other hand, components R_2 and R_3 are not detectable at this mode since there is no energy interaction within these components. The same argument explains the fault detectability and fault isolability of the components at mode $a = [0\ 1]$. At this mode, components $\{R_2, R_3\}$ are fault detectable, and $\{V_{in}, R_1\}$ are non-detectable. With the help of the GARRs, the quantitative bond graph-based FDI design can be applied to monitor hybrid systems. The R_1 fault is injected in the physical system in the first run, whereas the R_2 fault is introduced in the second run. The mode tracer of the diagnosis system is based on the two automatons implemented in the MATLAB® RTWT application. Since the signal $V_{c1}(t)$ that governs the transition between the discrete states of the automatons is measurable, the instantaneous operating state of the hybrid system can be determined based on the states of the automatons. Moreover, since a_1 and a_2 are complementary, i.e., $a_1 = \bar{a}_2$ for $\forall t$, the mode can be deduced based on the state of any one of the automatons. The experimental results of the two runs are presented in the remainder of this section.

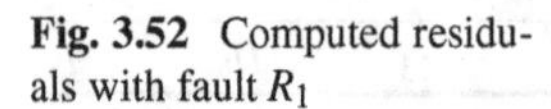

Fig. 3.52 Computed residuals with fault R_1

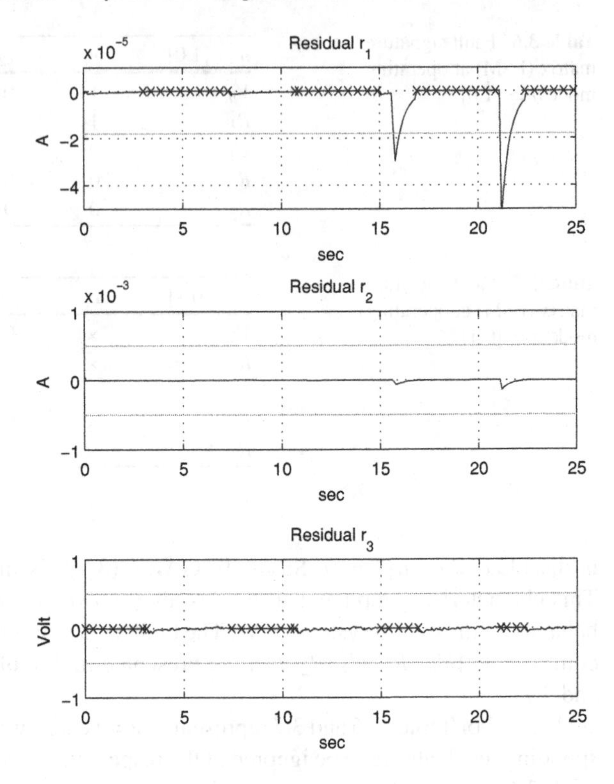

Fig. 3.53 Operating modes: a_1 and a_2 of run 1

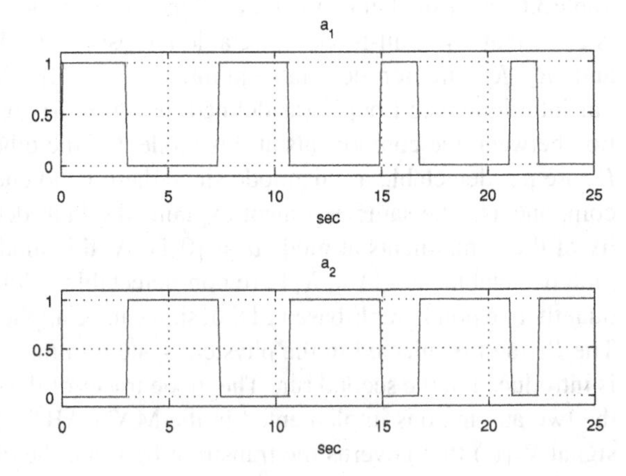

- Faulty R_1 component
  The experimental results of the first run are shown in Figs. 3.52 and 3.53. In this run, the fault on R_1 component is injected at $t = 15.2$ s, by reducing the resistive value of R_1 to 62 kΩ. Figure 3.52 depicts the online computed residuals (r_1, r_2, r_3) based on the GARR$_1$, GARR$_2$, and GARR$_3$, and Fig. 3.53 depicts the operating

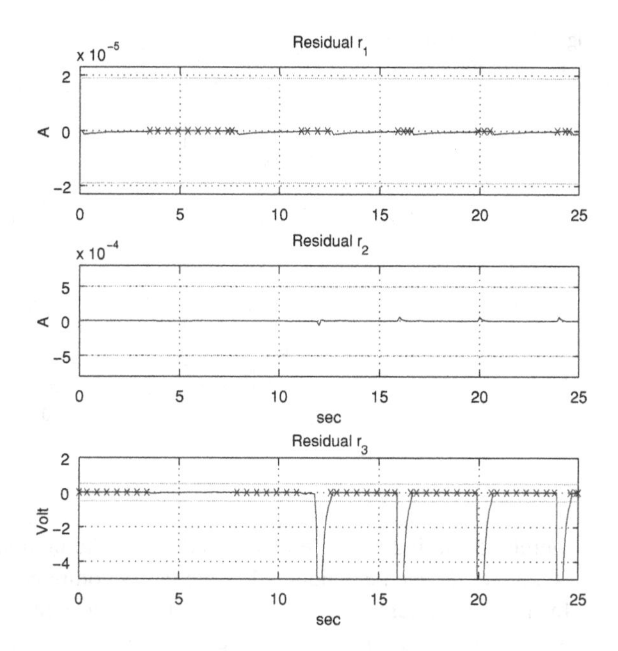

Fig. 3.54 Computed residuals with fault R_2

mode of the system during the run. The marks '×' on the residuals $\{r_1, r_3\}$ represent the time intervals where the respective residuals offer no monitoring information. From the fault detectability and fault isolability analysis presented in this section, it is clear that R_1 component is fault detectable at mode $a = [1\ 0]$ and r_1 is sensitive to the fault. This implies that at this operating mode, r_1 deviates from its small magnitude level should there have a fault occurred in the component. r_1 plot of Fig. 3.52 depicts a deviation occurred and crossed the predefined threshold value $\varepsilon_1 = 1.9 \times 10^{-5}$. On the other hand, the residuals that are insensitive to the fault maintained at low-magnitude levels during the run. The coherence vector at $t = 15.3$ s is $C = [1\ 0]$, which indicates that a fault has occurred in R_1 component. It is worth to note that the minor glitches reside in residual r_2 are due to the numerical computation error of the term $\frac{d}{dt}(De_1)$. This numerical error can be reduced by using a better derivative estimator. These responses confirm our theoretical expectations.

- Faulty R_2 component
  Figures 3.54 and 3.55 depict the online computed residuals and the hybrid system's operating modes during the second run. In this run, the R_2 fault was injected at the time instant $t = 11.7$ s where the system was operating at mode $a = [0\ 1]$. In the same manner, the marks '×' of these figures denote the time intervals where the residuals can be ignored. At time $t = 11.8$ s, the computed residual r_3 deviated from its low-level magnitude and crossed the predefined threshold $\varepsilon_3 = 0.1$ at time $t = 11.9$ s. The coherence vector at this time instant was $C = [0\ 1]$, and

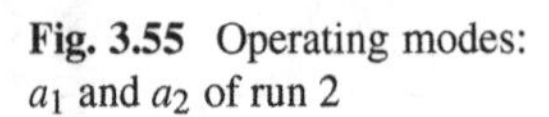
Fig. 3.55 Operating modes: a_1 and a_2 of run 2

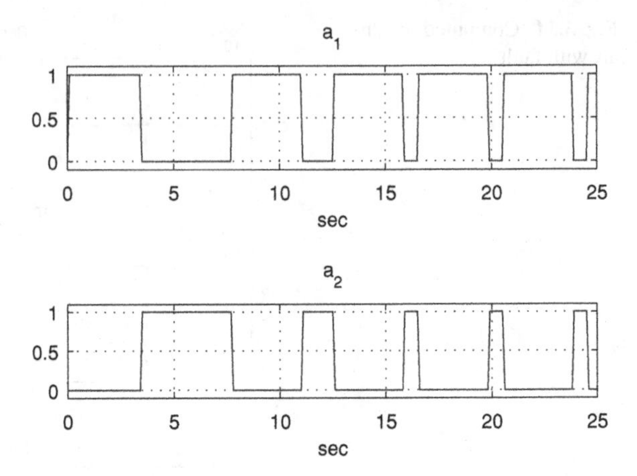

thereby declaring a fault had detected. Since the fault signature of R_2 component is unique, R_2 can be concluded as the faulty component.
In this experiment, the GARRs are utilized to generate informative alarm signals at all operating modes. Based on these alarm signals, a faulty component can be detected and isolated based on its fault detectability and fault isolability. For illustrative purposes, a relative simple hybrid system is chosen in this experiment. Although this experiment is not an exhaustive investigation on all the theory proposed in this chapter, it demonstrates the potential for applying HBG to FDI design for hybrid systems. More importantly, the diagnosis framework provides an efficient and effective method for FDI of hybrid systems.

References

1. D.C. Karnopp, D. Margolis, R.C. Rosenberg, *in System Dynamics: Modeling and Simulation of Mechatronic Systems* (Wiley, Hoboken, 2006)
2. P.J. Mosterman, G. Biswas, Diagnosis of continuous valued systems in transient operating regions. IEEE Trans. Syst. Man Cyber. **29**(6), 554–565 (1999)
3. M. Tagina, J.P. Cassar, G. Dauphin-Tanguy, M. Staroswiecki, Monitoring of systems modelled by bond-graphs, in *Internation Conference on BG Modeling (ICBGM'95)* (Las Vegas, 1995), pp. 275–279
4. B.O. Bouamama, A.K. Samantaray, M. Staroswiecki, G. Dauphin-Tanguy, Derivation of constraint relations from BG models for fault detection and isolation, in *Internation Conference on BG Modeling (ICBGM'03)*, (2003), pp. 104–109
5. A.K. Samantaray, K. Medjaher, B.O. Bouamama, M. Staroswiecki, G. Dauphin-Tanguy, Diagnostic BG for online fault detection and isolation. Simul. Model. Pract. Theor. **14**(3), 237–262 (2006)
6. B.O. Bouamama, K. Medjaher, M. Bayart, A.K. Samantaray, B. Conrard, Fault detection and isolation of smart actuators using BGs and external model. Control Eng. Pract. **13**(2), 159–175 (2005)

7. B.O. Bouamama, K. Medjaher, A.K. Samantaray, M. Staroswiecki, Supervision of an industrial steam generator. Part I: BG Modelling. Control Eng. Pract. **14**(1), 71–83 (2006)
8. K. Medjaher, A.K. Samantaray, B.O. Bouamama, M. Staroswiecki, Supervision of an industrial steam generator. Part II: Online Implementation. Control Eng. Pract. **14**(1), 85–96 (2006)
9. C.H. Low, Y.K. Wong, A.B. Rad, Intelligent system for process supervision and fault diagnosis in dynamic physical systems. IEEE Trans. Ind. Electron. **53**(2), 581–592 (2006)
10. D.A. Linkens, H. Wang, Qualitative BG reasoning in control engineering: fault diagnosis, in *ICBGM'95. In International Conference on BG Modeling* (Las Vegas, 1995), pp. 189–194
11. A.S. Wilsky, E.Y. Chow, X.C. Lou, G.C. Verghese, Analytical redundancy and the design of robust failure detection systems. IEEE Trans. Autom. Control **29**(7), 603–614 (1984)
12. M. Staroswiecki, Quantitative and qualitative models for fault detection and isolation. Mech. Syst. Sig. Process. **14**(3), 301–325 (2000)
13. M. Staroswiecki, G. Comtet-Varga, Analytical redundancy relations for fault detection and isolation in algebraic dynamic systems. Automatica **37**(5), 687–699 (2001)
14. A.K. Samantaray, B. Ould Bouamama, *Model-based Process Supervision: A Bond Graph Approach* (Springer, London, 2008)
15. F. Zhao, X. Koutsoukos, H. Haussecker, J. Reich, P. Cheung, Fault modeling for monitoring and diagnosis of sensor-rich hybrid systems, in *Proceedings of the IEEE Conference on Decision and, Control* (2001), pp. 793–801
16. F. Zhao, X. Koutsoukos, H. Haussecker, J. Reich, P. Cheung, Monitoring and fault diagnosis of hybrid systems. IEEE Trans. Syst. Man Cybern. Part B. Cybern. **35**(6), 1225–1240 (2005)
17. S. Narasimhan, *Model-based Diagnosis of Hybrid Systems*, Ph.D. Thesis, (Vanderbilt University, 2002)
18. S. Narasimhan, G. Biswas, Model-based diagnosis of hybrid systems. IEEE Trans. Syst. Man Cybern. Part A Syst. Hum. **37**(3), 348–360 (2007)
19. V. Cocquempot, T. El Mezyani, M. Staroswiecki, Fault detection and isolation for hybrid systems using structured parity residuals, in *5th Asian Control Conference*, vol. 2, (2004), pp. 1204–1212
20. P.J. Mosterman, G. Biswas, A theory of discontinuities in physical system models. J. Franklin Inst. **335**(B), 401–439 (1998)
21. I. Roychoudhury, M. Daigle, G. Biswas, X. Koutsoukos, P.J. Mosterman, A method for efficient simulation of hybrid bond graph, in *Proceedings of the International Conference Bond Graph Modeling Simulation, ICBGM 2007*, 335(B), (2007), pp. 177–184
22. I. Roychoudhury, M. Daigle, G. Biswas, X. Koutsoukos, Efficient simulation of hybrid systems: an application to electrical power distribution systems, in *Proceedings of the 22nd European Conference on Modeling and Simulation (ECMS 2008)* (2008), pp. 471–477
23. P.J. Mosterman, Diagnosis of physical systems with hybrid models using parameterized causality, in *Hybrid Systems: Computation and Control '01* (Rome, Italy, 2001), pp. 447–458
24. C.B. Low, D. Wang, S. Arogeti, J.B. Zhang, Causality assignment and model approximation for hybrid bond graph: fault diagnosis perspectives. IEEE Trans. Autom. Sci. Eng. **7**(3), 570–580 (2010)
25. A.K. Samantaray, K. Medjaher, B.O. Bouamama, M. Staroswieki, G. Dauphin-Tan, Element-based modelling of thermofuild systems for sensor placement and fault detection. SIMULATION **80**, 381–398 (2004)
26. S.A. Arogeti, D.W. Wang, C.B. Low, Mode tracking and FDI of hybrid systems, in *10th International Conference on ICARCV* (2008), pp. 892–897
27. C.B. Low, D. Wang, S. Arogeti, J.B. Zhang, Monitoring ability and quantitative fault diagnosis using hybrid bond graph, in *International Federation of Automatic Control (IFAC)*, (Seoul, Korea, 2008), pp. 10516–10521
28. C.B. Low, D. Wang, S. Arogeti, M. Luo, Quantitative hybrid bond graph-based fault detection and isolation. IEEE Trans. Autom. Sci. Eng. **7**(3), 558–569 (2010)
29. M.A. Djeziri, R. Merzouki, B. Ould Bouamama, Robust fault diagnosis by using bond graph approach. IEEE/ASME Tranac. Mechatron. **12**(6), 599–611 (2007)

30. M.A. Djeziri, B. Ould Bouamama, R. Merzouki, Modeling and robust FDI of steam generator using uncertain bond graph model. J. Process Control **19**(1), 149–162 (2008)
31. P.J. Mosterman, G. Biswas, A theory of discontinuities in physical system models. J. Franklin Inst. **335B**(3), 401–439 (1998)
32. V. Venkatasubramanian, R. Rengaswamy, K. Yin, S.N. Kavuri, A review of process fault detection and diagnosis. Part I: Quantitative model-based methods. Comput. Chem. Eng. **27**(3), 293–311 (2003)
33. V. Venkatasubramanian, R. Rengaswamy, S.N. Kavuri, A review of process fault detection and diagnosis. Part II: Quanlitative models and search strategies. Comput. Chem. Eng. **27**(3), 313–326 (2003)
34. V. Venkatasubramanian, R. Rengaswamy, S.N. Kavuri, A review of process fault detection and diagnosis. Part III: Process history based methods. Comput. Chem. Eng. **27**(3), 327–346 (2003)
35. J. Cusidó, L. Romeral, J.A. Ortega, J.A. Rosero, A.G. Espinosa, Fault detection in induction machines using power spectral density in wavelet decomposition. IEEE Trans. Ind. Electron. **55**(2), 633–643 (2008)
36. M. Mlödt, P. Granjon, B. Raison, G. Rostaing, Models for bearing damage detection in induction motors using stator current monitoring. IEEE Trans. Ind. Electron. **55**(4), 1813–1822 (2008)

Chapter 4
Fault Identification Techniques

4.1 Nonlinear Least Square Optimization for Fault Identification

In this section, a HBG based fault estimation technique is developed to estimate the size of the parameter change due to system fault. The developed estimation technique estimates fault parameters that can be linearly or nonlinearly parameterized by formulating the estimation problem as a nonlinear least square problem. Through the bond graph based estimation method, one can identify the size of the fault occurrence and refine the isolated fault parameters of a complex hybrid system even when a mode change occurs after the occurrence of faults.

4.1.1 Nonlinear Least Square Method

The system under consideration has a set of n_θ parameters denotes by θ. Here, the set θ is divided into two mutually exclusive sets $\{\theta_t, \theta_k\}$ such that $\theta = \theta_t \bigcup \theta_k$. $\theta_t \subset \theta$ is the set of targeted parameters chosen from the fault isolation module which is $\theta_t = \{\theta_{t,1}, \theta_{t,2}, \ldots, \theta_{t,n_t}\}$ where n_t is a positive constant. The constant $n_t = 1$ implies an isolable parameter, and $n_t > 1$ implies detectable but not isolable parameters. On the other hand, the set $\theta_k \subset \theta$ represents the set of system parameters that does not contain any element belonging to set θ_t and is assumed to be known. Similarly, the set $\theta_k = \{\theta_{k,1}, \theta_{k,2}, \ldots, \theta_{k,n_k}\}$.

In general, a complex hybrid system consists of discrete mode change and also nonlinear behavior due to nonlinear elements; for instance, nonlinear resistor in electrical domain. These attributes of a complex system complicates both FDI and fault estimation.

From FDI perspectives, we may want to access the unknown damage occurs at the component; hence, it is desirable to estimate the value of the faulty parameters once a fault is detected. In this sense, the developed algorithm must be able to estimate the faulty parameters, including those that are nonlinear parameterized. Additionally,

DOI: 10.1007/978-1-4614-7369-5_4,

the samples data of intervals that the hybrid system undergoes mode changes should also be utilized.

An analytical input-output model that describes the system inputs—outputs is required to generate the errors for the cost function that is to be minimized. Analogously, in FDI literatures, ARRs are equations that compute the errors between the analytically computed variables and measured values. These errors are signals due to fault parameters, which are used for fault detection. From previous chapters, algorithms have been developed to derive the GARRs to monitor faults at all operating modes. This application of error signals suggests that these unified equations can be utilized to formulate the fault estimation problem for the complex hybrid systems. In this way, the fault estimation problem becomes a tractable problem, where its solution is computationally efficient.

In this section, we assume that N finite samples data are collected after a fault is detected. Additionally, we assume a set of targeted fault parameters which have been identified by the fault detection and isolation modules. Here, we formulate the estimation problem into the nonlinear least-square problem based on the GARRs to estimate the targeted fault parameters. And also, we utilize the Gauss-Newton iterative solution to solve the nonlinear optimization problem. Other iterative solutions to this nonlinear least-square problem, e.g., steepest-descent and Levenberg-Martquardt algorithms can be also applied to this problem [1].

In this study, we consider a hybrid system with m GARRs which can be expressed as $g_i(\theta, n)$ for $i = 1, \ldots, m$. n denotes the discrete sampling time index. For convenience, we rewrite a GARR as $g_i(\theta_k, \theta_t, n)$.

As we have discussed in the previous chapter, we can detect a fault by evaluating the GARRs $g_i(\theta, n)$ of a hybrid system by $\theta = \bar{\theta}$. $\bar{\theta}$ denotes the nominal fault-free parameters of the hybrid system. If a fault detectable parameter $\theta_i \in \theta$ is changed due to some faults, then the respective GARRs $g_i(\bar{\theta}, n)$ that contains θ_i will be nonzero, and hence a fault is detected and declared. Moreover, if θ_i is fault isolable, then fault signature generated by the $g_i(\bar{\theta}, n)$ for $i = 1, \ldots, m$ will be able to identify the faulty parameter. This means that the targeted parameter θ_t to be estimated is a scalar. On the other hand, if the fault is due to a non-isolable parameter, then we chose the set of fault candidates θ_t that shares the same fault signature. This selection can be chosen from the FSM of the hybrid system.

From parameter identification viewpoint, $g_i(\theta, n) = 0$ for $i = 1, \ldots, m$ if the values of parameters θ are equal to the physical values of parameters of the hybrid system. This property suggests that we can let the evaluated $g_i(\theta, n) = 0$ be the residuals for parameter identification if we assume θ be the unknown variables. Since in general, $g_i(\theta, n)$ is not a linear function of θ; hence without loss of generality, we formulate the parameter estimation problem as a nonlinear least-square problem. In this section, we estimate the targeted parameters $\theta_t \subset \theta$. This greatly simplifies the parameter estimation problem and its computation since we need not have to consider all parameters θ for estimation.

For the convenience of digital implementation, the nonlinear least-square problem is formulated in the discrete-time domain. And since θ_k are known variables, we can conveniently define the cost function for the nonlinear least-square problem based on N samples of data as follows:

$$V(\theta_t) = \frac{1}{2} r^T r, \tag{4.1}$$

where

$$r = \begin{bmatrix} r_1 \\ r_2 \\ \cdot \\ \cdot \\ \cdot \\ r_N, \end{bmatrix}, \text{ and } r_n = \begin{bmatrix} g_1(\theta_t, n) \\ \cdot \\ \cdot \\ \cdot \\ g_m(\theta_t, n), \end{bmatrix} \tag{4.2}$$

for $n = 1, \ldots, N$.

To solve the nonlinear least-square optimization problem using iterative method, we require the gradient of the cost function to be well-defined. Hence, we impose the following assumption.

Assumption 1 *Every g_i for $i = 1, \ldots, m$ is differentiable with respect to $\hat{\theta}_t$.*

Since the cost function $V(\theta_t)$ is linear combination of $g_i(.)^2$, it is clear that V is differentiable with respect to θ_t if Assumption 1 is satisfied.

By Gauss-Newton method, the iterative solution which minimizes the cost function (4.1) is

$$\hat{\theta}_t(k+1) = \hat{\theta}_t(k) - H_V(\theta_t)^{-1} \nabla V(\hat{\theta}_t(k)) \tag{4.3}$$

where $\nabla V(\theta_t)$ is the gradient vector of the cost function V

$$\nabla V(\theta_t) = \frac{\partial r(\theta_t)}{\partial \theta_t}^T r(\theta_t) \tag{4.4}$$

and

$$H_V \approx \frac{\partial r(\theta_t)}{\partial \theta_t}^T \frac{\partial r(\theta_t)}{\partial \theta_t} \tag{4.5}$$

k denotes the iteration index. Note that the $\frac{\partial r(\theta_t)}{\partial \theta_t}$ of Eq. (4.5) can be further rewritten as

$$\frac{\partial r(\theta_t)}{\partial \theta_t} = \left[\frac{\partial r_1(\theta_t)}{\partial \theta_t}, \ldots, \frac{\partial r_N(\theta_t)}{\partial \theta_t} \right]^T \tag{4.6}$$

where

$$\frac{\partial r_n(\theta_t)}{\partial \theta_t} = \begin{bmatrix} \frac{\partial g_1(\theta_t,n)}{\partial \theta_{t,1}} & \cdots & \frac{\partial g_1(\theta_t,n)}{\partial \theta_{t,n_t}} \\ \frac{\partial g_2(\theta_t,n)}{\partial \theta_{t,1}} & \cdots & \frac{\partial g_2(\theta_t,n)}{\partial \theta_{t,n_t}} \\ \cdot & \cdots & \cdot \\ \frac{\partial g_m(\theta_t,n)}{\partial \theta_{t,1}} & \cdots & \frac{\partial g_m(\theta_t,n)}{\partial \theta_{t,n_t}} \end{bmatrix} \tag{4.7}$$

The fault estimation of complex hybrid system becomes complicated when the number of components is large. Therefore, the analytical expression of (4.4) is generally complex and not easily obtained in practice. For large complex systems, we require systematic approach that facilitates implementation. Here, we can exploit the iterative nature of the nonlinear least-square solution to numerically estimate the gradient (4.4) about $\hat{\theta}_t(k)$.

Notice that the cost function is constructed based on information provided by the unified GARRs that describe the hybrid systems at every mode. This generalized formulation allows mode change to occur during the N samples interval.

In this section, we adopt a numerical strategy to estimate ∇V and H_V since the solution to the optimization problem is iterative. The respective gradient information $\partial g_i/\partial \theta_t$ at point $\hat{\theta}_t(k)$ can be approximated using *finite-difference method* [2] shown as in (4.8).

$$\frac{\partial g_i(\hat{\theta}_t(k))}{\partial \theta_{t,j}} \approx \frac{g_i(\hat{\theta}_t(k) + \delta \mathbf{u}_j) - g_i(\hat{\theta}_t(k))}{\delta} \tag{4.8}$$

for $i = 1, \ldots, m$ and $j = 1, \ldots, n_t$. $\mathbf{u}_j \in R^{n_t}$ denotes an unit vector with jth element one and others zero. δ is a small positive constant which is used to approximate the n_t partial derivatives. By using the numerical gradient approximation (4.8), we are able to implement the Gauss-Newton algorithm to estimate the targeted parameters of a complex hybrid system.

Remark 4.1 Since we are utilizing the GARRs to estimate the targeted parameters, we only select GARRs that contain a targeted parameter. In other words, every g_i that does not contain any parameter $\theta_j \in \theta_t$ will not be considered in (4.7).

Remark 4.2 The GARRs describe the behavior of a hybrid system in a compact and efficient way at all modes. Here, we exploit the advantageous attribute of the GARRs to formulate the cost function during the N samples interval. In this way, we can utilize the set of N data to provide a fault estimate on the fault parameter even the system undergoes mode changes during this N samples interval.

In this section, we apply Gauss-Newton to solve the formulated nonlinear least-square problem to estimate the targeted parameters θ_t. The gradient about a point $\hat{\theta}_t(k)$ is computed numerically based on (4.8). If there exists a GARR g_i which has a term where the targeted parameters θ_t are the arguments of a nonlinear relation Φ and a derivative operator $\frac{d}{dt}(.)$ in the form of

$$\frac{d}{dt}(\Phi(\theta_t, s_1(t), \ldots, s_v(t))), \tag{4.9}$$

then we are not able to estimate the partial derivatives based on (4.8). $\{s_1(t), \ldots, s_v(t)\}$ denote the set of time-varying variables.

To overcome such implementation problem, one approach is to rewrite the term of the form of (4.9) using chain rule to extract the target parameters out of the derivative operator. In other words, the term can be written as

$$\frac{d}{dt}(\Phi(\theta_t, s_1(t), \ldots, s_v(t))) = \frac{\partial \Phi(\theta_t, t)}{\partial s_1}\dot{s}_1(t) + \cdots + \frac{\partial \Phi(\theta_t, t)}{\partial s_v}\dot{s}_v(t) \tag{4.10}$$

where $\dot{s}_i$ denotes the time derivative of the variable s_i, and θ_t are constant parameters. By replacing each term that has the form of (4.9) with (4.10), we are able to extract the target parameters θ_t from the derivative operator $\frac{d}{dt}(.)$; and hence, we can apply the numerical method (4.8) to estimate the partial derivatives used in the iterative solution (4.3).

One final point to highlight is that the vector r must contain the parameters θ_t so that the nonlinear least-square problem is meaningful and can be solved. It is clear that we cannot estimate the θ_t if the formulated cost function V of the N data is not a function of the targeted parameters.

4.1.2 Example: A Nonlinear Hybrid Electrical System

Figure 4.1 depicts the electrical hybrid system considered in this case study. The hybrid system has three R elements $\{R_1, R_2, R_3\}$, three C elements $\{C_1, C_2, C_3\}$, a switch Sw, one current (flow) sensor D_f, and two voltage sensors $\{De_1, De_2\}$ connected as shown in the figure.

In this study, resistors $\{R_2, R_3\}$ are linear resistors that follow the ohms law $e = Rf$ faithfully at all range. On the other hand, R_1 is a nonlinear resistor that has a nonlinear constitutive relation as follows

$$e = \Phi_{R_1}(\alpha, \beta, f) = \frac{2\beta}{\pi}\tan^{-1}(\alpha f) \tag{4.11}$$

where $\{\alpha, \beta\}$ are the parameters of the nonlinear resistor. As a result of R_1, the hybrid system is also a nonlinear system.

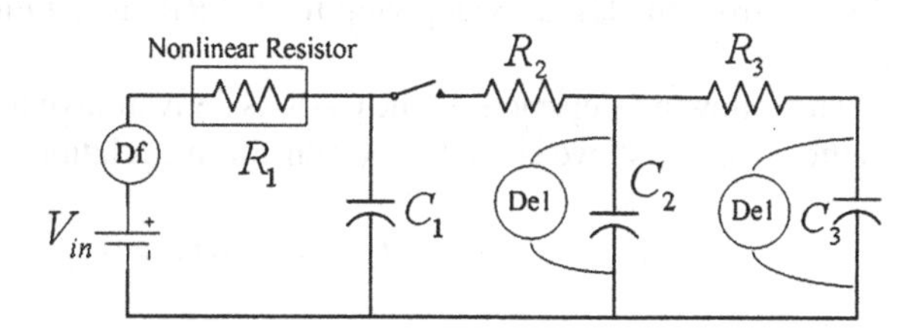

Fig. 4.1 An electrical hybrid system

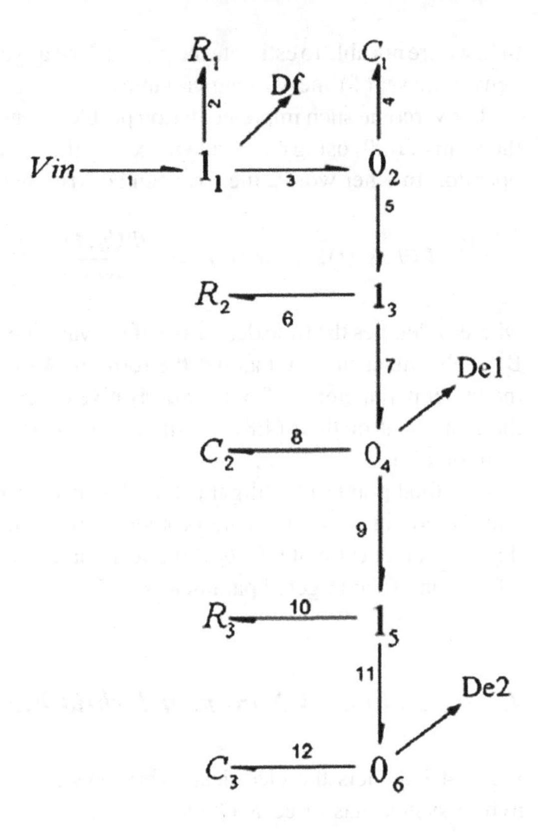

Fig. 4.2 Acausal HBG of the electrical hybrid system

Figure 4.2 depicts the acausal HBG of the nonlinear hybrid system. 1_C denotes the controlled junction that captures the switching behavior of the hybrid system due to the electrical switch.

To demonstrate the fault estimation method, we need to derive the GARRs from the HBG of the system. By applying the integrated procedure developed in Chap. 2, we yield the HBG of the hybrid system as shown in Fig. 4.3.

Notice that an additional R-element R_s appears in the HBG. This element is added via the modeling approximation proposed in Chap. 3 to ease the derivation of HBG. The causal paths of the HBG which are determined by the causality of the HBG are also shown in Fig. 4.3. These causal paths allow systematic derivation of the unified GARRs from the HBG by applying the GARRs generation procedure presented in Chap. 3.

The following steps present the GARRs derivation of the considered system based on the HBG. First, we consider the constitutive relation of junction 0_2

$$f_3 - f_4 - af_5 = 0. \tag{4.12}$$

Fig. 4.3 HBG with causal paths

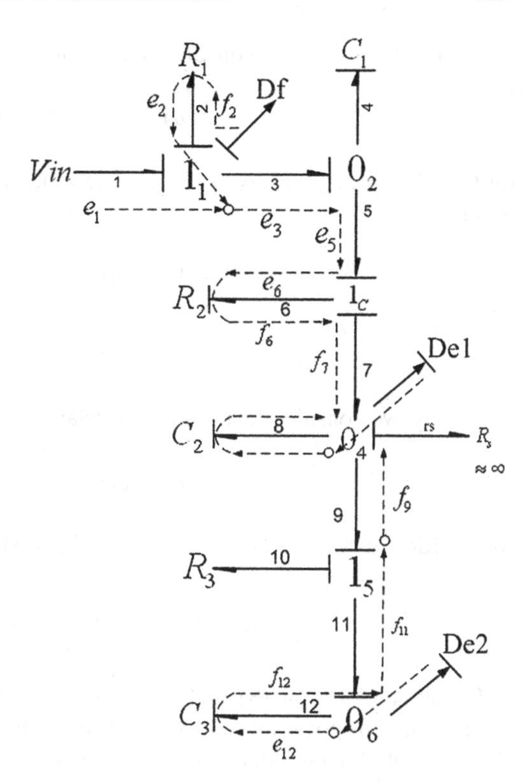

Note that $\{f_2, f_3\}$ are measurable by D_f, and boolean variable a denotes the state of the controlled junction 1_C.

As a result, Eq. (4.12) becomes

$$D_f - f_4 - f_5 = 0. \tag{4.13}$$

By covering the causal paths of bonds 1-2-3-4, and bonds 5-6-7, we have

$$f_4 = C_1 \frac{d}{dt}(V_{in} - \Phi_{R_1}(\alpha, \beta, D_f)) \tag{4.14}$$

$$f_5 = \frac{1}{R_2}(V_{in} - \Phi_{R_1}(\alpha, \beta, D_f) - De_1). \tag{4.15}$$

By substituting Eqs. (4.14) and (4.15) into (4.13), we obtain the GARR

$$\begin{aligned} g_1 = D_f &- C_1 \frac{d}{dt}(V_{in} - \Phi_{R_1}(\alpha, \beta, D_f)) \\ &- \frac{a}{R_2}(V_{in} - \Phi_{R_1}(\alpha, \beta, D_f) - De_1) = 0. \end{aligned} \tag{4.16}$$

Next, we consider the constitutive relation of junction 0_4

$$af_7 - f_8 - f_9 = 0 \tag{4.17}$$

for GARR. Since $f_7 = f_5$, $f_8 = C_2\frac{d}{dt}(De_1)$, $f_9 = f_{11} = C_3\frac{d}{dt}(De_2)$, Eq. (4.17) becomes

$$\begin{aligned} g_2 = &\frac{a}{R_2}(V_{in} - \Phi_{R_1}(\alpha, \beta, D_f) - De_1) \\ &- C_2\frac{d}{dt}(De_1) - C_3\frac{d}{dt}(De_2). \end{aligned} \tag{4.18}$$

Finally, we consider the constitutive relation of junction 1_5

$$e_9 - e_{10} - e_{11} = 0 \tag{4.19}$$

for GARR. Since $e_9 = De_1$, $e_{10} = R_3C_3\frac{d}{dt}(De_2)$, and $e_{11} = De_2$, we obtain

$$g_3 = De_1 - R_3C_3\frac{d}{dt}(De_2) - De_2 \tag{4.20}$$

Equations (4.16), (4.18), and (4.20) are the unified GARRs yield from the HBG. These equations govern the behaviors of the hybrid system and are used to generate alarm signals for fault diagnosis [3]. To apply these equations for fault parameter estimation, we need to rearrange the term $\frac{d}{dt}(\Phi_{R_1}(\alpha, \beta, f_2))$ of Eq. (4.16) using chain rule as it has the form of (4.9). Since α, β are constants, we have

$$\begin{aligned} \frac{d}{dt}(\Phi_{R_1}(\alpha, \beta, f_2)) &= \frac{d\Phi_{R_1}}{df_2}\frac{df_2}{dt} \\ &= \frac{2\beta}{\pi}\frac{\alpha}{1+\alpha^2 f_2^2}\frac{df_2}{dt} \\ &= \frac{2\beta}{\pi}\frac{\alpha}{1+\alpha^2 D_f^2}\frac{dD_f}{dt} \end{aligned} \tag{4.21}$$

Equation (4.21) will be utilized to replace the term $\frac{d}{dt}(\Phi_{R_1}(\alpha, \beta, f_2))$ of GARR g_1 to perform the parameter estimation.

MATLAB® SIMULINK® is adopted to implement the hybrid system to validate the estimation algorithm. In the study, we consider a fault occurred in the nonlinear resistor R_1. For illustration purposes, we assume that the fault causes a change from $\alpha = 2k$ to $\alpha = 4k$ whereas β and V_{in} remain unchanged. The nominal (fault-free) parameters of the hybrid system are chosen as

$$\alpha = 2k,\ \beta = 10,\ R_2 = 10\,\text{k}\Omega,\ R_3 = 10\,\text{k}\Omega,$$

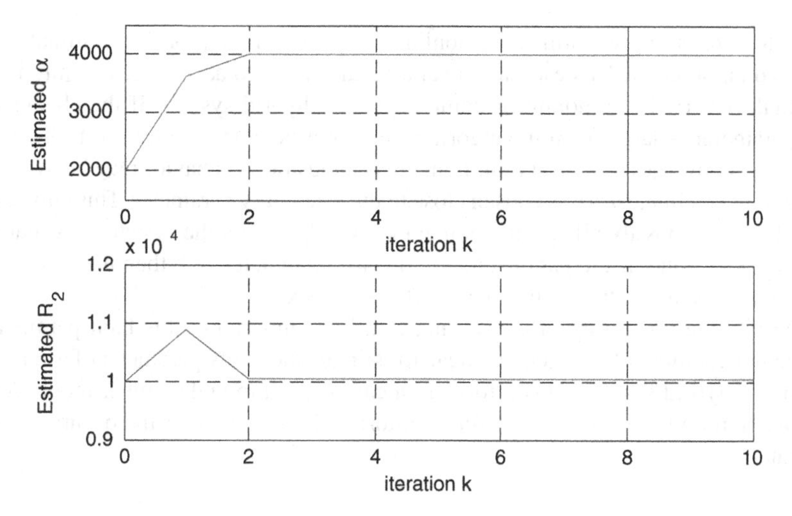

Fig. 4.4 Estimated targeted parameter $\hat{\theta}_t = [\hat{\alpha}\ \hat{R}_2]$

$$C_1 = 100\,\mu\text{F},\ C_2 = 100\,\mu\text{F},\ C_3 = 100\,\mu\text{F},\ V_{in} = 4\,\text{v}.$$

In this simulation, the sampling interval is fixed at $t = 0.1$ s, and we collect the measurements in a 2-s window ($N = 20$ samples) after a fault is detected from the FDI modules. From the GARRs, we learn that parameter α and R_2 share the same fault signature; therefore, we would like to estimate the targeted parameters $\theta_t = [\alpha\ R_2]$. The constant δ chosen for the gradient approximation in (4.8) is $\delta = 1 \times 10^{-5}$. Also note that the derivatives of the known data, e.g., dDf/dt are implemented using numerical derivatives. To make this study more meaningful, the hybrid system undergoes a mode change from mode $a = 1$ to $a = 0$, at $N = 10$ samples.

Figure 4.4 depicts the trajectories of the estimated parameters $\hat{\theta}_t$ for 11 iterations, i.e., $k = 0, \ldots, 11$, based on the 20 samples data collected during the run. The initial estimates are chosen as $\hat{\theta}_t(0) = [\hat{\alpha}(0)\ \hat{R}_2(0)] = [2k\ 10k]$. From the results, we can see that the iterative solution converges to a close estimate point at $\hat{\theta}_t = [3.990k\ 10.046k]$ after 3 iterations. This estimates are close to the actual simulated targeted parameters $\theta_t = [\alpha\ R_2] = [4k\ 10k]$.

Besides providing new information about the changed parameters, the results also suggest that the estimated parameters can be used to refine the faulty candidates where θ_t is not scalar. For instance, in this case study, we have identified that α changes from $2k$ to $4k$, and the resistive value of R_2 remains unchanged from this estimation. Therefore, we can make use of this information to conclude the fault occurred in R_1 and not R_2.

Note that the mild estimation error is due to the numerical computation errors of the derivative of sensors, and the estimates of the partial derivatives about a point (4.8). As we can see, in the presence of these computation errors, the Gauss-Newton algorithm still able to converge to the point that is near the true state. One final point to note is that we utilize the nonlinear least-square approach to estimate the targeted parameters. This estimation solution allows us to access the magnitude of the fault occurred in a nonlinear component of a hybrid system. If the change of the parameter is large, then the algorithm may not be able to converge to the true point and converges to a local minimum. This undesirable behavior may be avoided if the initial estimate $\hat{\theta}_t(0)$ is chosen close to the true faulty parameter. This choice of initial states is possible if specific domain knowledge about the system is available. On the other hand, if the parameters are linearly parameterized, then the algorithm will converges to the true point since $V(\theta_t)$ is convex.

Despite the possibility of converging to a local minimum, this fault parameter estimation method offers a general mean to estimate the faulty parameter of a general nonlinear hybrid system. This information can be used to update other modules of the health monitoring framework for monitoring the system after the occurrence of the fault.

4.2 Simultaneous Fault Parameter and Mode Change Identification

Isolation of fault candidates leads to the identification of fault parameters and evaluation of system health. However, if mode switching is unknown, it is difficult to select the correct model after a fault occurrence. Nonlinear least-square based parameter estimation requires mode information for identification purpose. In order to overcome this problem, the mode changes are parameterized such that they can be represented by the functions of switching time stamps. During the estimation process, the fault parameters together with the switching time stamps are identified by the estimation algorithm to evaluate the size of the fault parameter.

4.2.1 Parametrization of Mode Changes

In this section, a component fault is represented by its parameter value. It is also assumed that there are at most of p mode changes during an observation period.

Definition 4.1 For a hybrid system, during the observation window $[t_0 \; t_f]$, $\alpha = [\alpha_1, \alpha_1, \ldots, \alpha_q]$ is defined as a switching vector whose component α_j, $j = 1, 2, \ldots, q$ takes the form of

$$\alpha_j(t, T_j) = \lambda_j \cdot \alpha_j^1(t, T_j) + (1 - \lambda_j) \cdot \alpha_j^0(t, T_j) \tag{4.22}$$

where λ_j is the Initial Mode Coefficient (IMC), whose value is obtained by the following formula

$$\lambda_j = \begin{cases} 1 & \text{if the state of the jth controlled junction at } t_0 \text{ is } 1 \\ 0 & \text{if the state of the jth controlled junction at } t_0 \text{ is } 1 \end{cases} \tag{4.23}$$

$\alpha_j^1(t, T_j)$ and $\alpha_j^0(t, T_j)$ are defined as

$$\alpha_j^1(t, T_j) = \tau(t) - \tau(t - T_j^1) + \tau(t - T_j^2) + \cdots + (-1)^p \tau(t - T_j^p) \text{ for } \lambda_j = 1 \tag{4.24}$$

$$\alpha_j^0(t, T_j) = \tau(t - T_j^1) - \tau(t - T_j^2) - \cdots - (-1)^p \tau(t - T_j^p) \text{ for } \lambda_j = 0 \tag{4.25}$$

Define function

$$\tau(*) = \begin{cases} 1 & \text{if } * \geq 0 \\ 0 & \text{if } * < 0 \end{cases} \tag{4.26}$$

where $T_j = [T_j^1, T_j^2, \cdots, T_j^p]$ is the switching time stamp vector whose elements are the time stamps for mode switching with $t_0 \leq T_j^1 \leq T_j^2 \leq \cdots \leq T_j^p \leq t_f$.

By definition, if all the switching time stamps T_j^d, $j = 1, 2, \ldots, q$ and $d = 1, 2, \ldots, p$ are equal to t_f, then there are no mode changes during the time interval $[t_0\ t_f]$. If the switching time stamps and IMC are known, the mode switching vector $\alpha = [\alpha_1, \alpha_1, \ldots, \alpha_q]$ can be expressed as functions of time t.

The essence of Definition 4.1 is the choice of $\alpha_j^1(t, T_j)$ and $\alpha_j^0(t, T_j)$, which can be directly observed from Figs. 4.5 and 4.6. It is obvious that the mode changes can be represented as the combination of step functions with different firing time.

4.2.2 Simultaneous Fault Parameter and Mode Switching Identification

The formula in Eq. (4.22) is a discontinuous function, which might be problematic if gradient based methods are used for the switching time identification problem con-

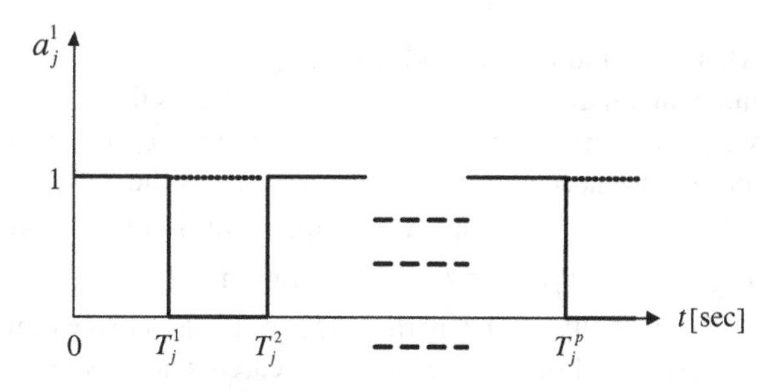

Fig. 4.5 Mode change sequence when the initial state is 1

Fig. 4.6 Mode change sequence when the initial state is 0

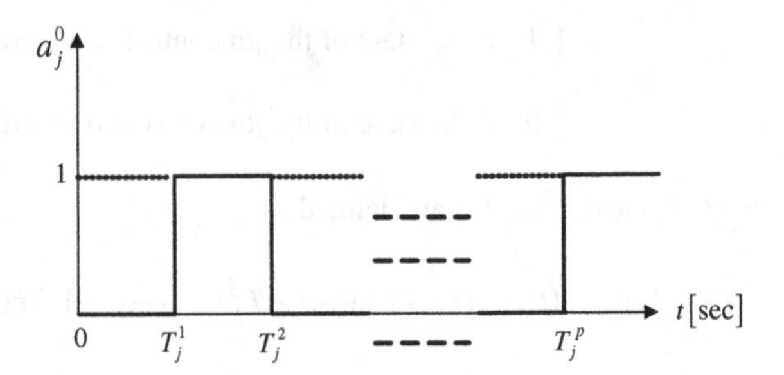

sidering the non-differentiable function. Therefore, heuristics-based Particle Swarm Optimization (PSO) can be an efficient alternative. PSO, developed by Kennedy and Eberhart [4], is a population based stochastic search method inspired by social behavior metaphor.

Unlike genetic algorithm [5, 6], which relies on Darwin's theory of nature selection and the competition between individual chromosomes, the swarm intelligence in the nature is modeled by fundamental Newtonian mechanics in PSO for optimization purposes. This corporative scheme manifests PSO the concise formulation, the ease in implementation and many distinct features in different types of optimizations.

4.2.2.1 Particle Swarm Optimization

The individuals in PSO are called particles, and each particle has its own position and velocity. The possible solutions to the optimization task can be represented by the particles' positions. During each generation, each particle moves according to its own best solution obtained so far, denoted as $P_{best,r}$, as well as the global best solution obtained by all particles of the swarm so far, denoted as G_{best}. If the problem space is D-dimensional and the swarm size is s , the PSO algorithm can be represented as follows:

$$v_{r,g}^{k+1} = w_r v_{r,g}^k + \psi_1 rand_1 (P_{best,r}^k - x_{r,g}^k) + \psi_2 rand_2 (G_{best}^k - x_{r,g}^k) \tag{4.27}$$

$$x_{r,g}^{k+1} = x_{r,g}^k + v_{r,g}^{k+1} \tag{4.28}$$

where r is the particle index in a swarm with $r = 1, 2, \ldots, s$; g is the dimension index in a particle with $r = 1, 2, \ldots, D$; k is the iteration number; w_r is the inertia weight of particle r; $rand_1$ and $rand_2$ are two random values with range [0, 1]; ψ_1 and ψ_2 are acceleration coefficients; $v_{r,g}^k$ is the velocity of particle r at iteration k, $-V_{max} \leq v_{r,g}^k \leq V_{max}$; $x_{r,g}^k$ is the position of particle r at iteration k, $-x_{max} \leq x_{r,g}^k \leq x_{max}$; $P_{best,r}^k$ is $P_{best,r}$ at iteration k; G_{best}^k is G_{best} at iteration k.

The velocity of each particle $v_{r,g}^k$ is clamped to the range $[-V_{max}, V_{max}]$ to reduce the likelihood of particles leaving search space. Note that this does not restrict the

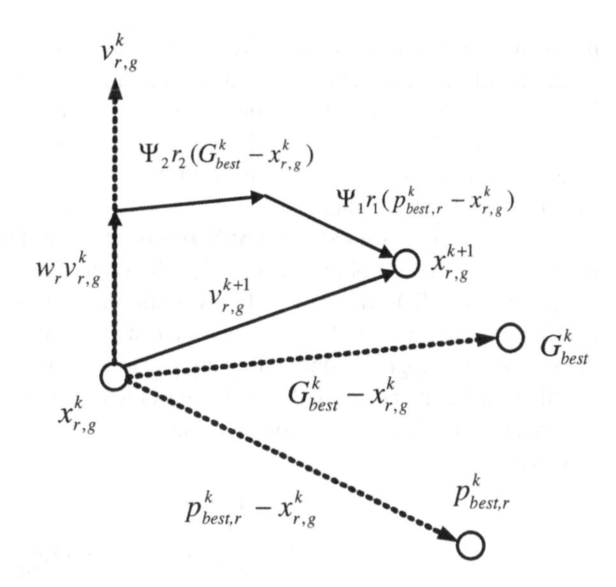

Fig. 4.7 Movement of a particle

position of each particle $v_{r,g}^{k}$ to the range $[-V_{max}, V_{max}]$; it only limits the maximum distance that a particle will move during one iteration. The inertia weight w_r is similar to the momentum term in a gradient descent neural network training algorithm. Acceleration coefficients ψ_1 and ψ_2 also control how far a particle will move in a single iteration. Typically, they are both set to value of 2.0. The movement pattern of a particle is illustrated in Fig. 4.7. The PSO algorithm can be summarized as follows:

1. Initialize the particle swarm. Initialize s particles with random positions and velocities within the search space.
2. Calculate the fitness value of each particle.
3. Compare the fitness values at the present position with its best position $P_{best,r}$ so far for each particle. If the fitness value at the present position is better than that at its best position, set the present position as the best position so far for this particle. Otherwise, keep the best position unchanged for this particle.
4. Compare the fitness value of each particle with the fitness value at the global best position G_{best} within the population. If the fitness value of a particle is better than that at the global best position, set the present position of this particle as the best global position so far for the population. Otherwise, keep the original best global position for the population.
5. Update the velocity and position of each particle by using (4.27) and (4.28).
6. Check the termination criterion. If the criterion is met, then stop. Otherwise, go to step 2.

The PSO mentioned above is Real-valued PSO (RPSO) which is developed to solve the problem where solution elements are continuous real number. Many optimization problems are set in a space featuring discrete, qualitative distinctions between variables and between levels of variables. Some examples like ordering or arranging of discrete elements, as in scheduling and routing problems. Other problems need cast floating-point problems in binary terms, and solve them in a discrete number space. As any problem, discrete or continuous, can be expressed in a binary notation, it is seen that an optimization algorithm which operates on two-valued functions could be advantageous.

A type of Binary-valued PSO (BPSO) was introduced to allow the PSO algorithm to operate in binary problem spaces [7]. The BPSO has a structure almost identical to the standard PSO, where the velocity is still defined to be in the continuous space. In a binary space, a particle moves in a state space restricted to zero and one on each dimension. The logistic function of the particle velocity is used as the probability distribution for the position, that is, the particle position in a dimension is randomly generated using that distribution. The equation that updates the particle position is changed as follows:

$$x_{r,g}^{k+1} = \begin{cases} 0 & \text{if } rand \geq f(v_{r,g}^k) \\ 1 & \text{if } rand < f(v_{r,g}^k) \end{cases} \tag{4.29}$$

with

$$f(v_{r,g}^k) = \frac{1}{1 + exp(-v_{r,g}^k)} \tag{4.30}$$

where *rand* is a random value with range [0,1].

The BPSO is susceptible to sigmoid function saturation, which occurs when velocity values are either too large or too small. In such cases the probability of a change in bit value approaches zero, thereby limiting exploration. For a velocity of 0, the sigmoid function returns a probability of 0.5, implying that there is a 50 % chance for the bit to flip.

In order to ensure that there is always some chance of a bit flipping, the velocity can be limited in the range $[-V_{max}, V_{max}]$ to avoid saturation of the sigmoid function in (4.30). In the BPSO, V_{max} functions similarly to the mutation rate in GA, smaller V_{max} allows a higher mutation rate and vice verse. For example, if $V_{max} = 6.0$, then the probabilities will be limited to $f(v_{r,g}^k)$, between 0.9975 and 0.0025. The result of this is that new vectors will still be tried, even after each bit has attained its best position. Setting V_{max} with higher value, e.g. 10.0, makes new vectors less likely. Note that while high V_{max} in the RPSO increases the range explored by a particle, the opposite occurs in the binary version; smaller V_{max} allows a higher mutation rate. It is also observed that the probability that a bit will be a one is equal to $f(v_{r,g}^k)$, and the probability that it will be a zero is $1 - f(v_{r,g}^k)$. Moreover, if it is a zero already, then the probability that it will change is $f(v_{r,g}^k)$, and if it is a one the probability it will change is $1 - f(v_{r,g}^k)$. Consequently, the probability of the bit changing is given by

$$P_{change} = f(v_{r,g}^k)[1 - f(v_{r,g}^k)] = f(v_{r,g}^k) - f(v_{r,g}^k)^2 \tag{4.31}$$

Equation (4.31) is the absolute (non-directional) rate of change for that bit given a value of $v_{r,g}^k$. Thus change in $v_{r,g}^k$ is still a change in the rate of change, with a new interpretation.

4.2.2.2 Comparison Between Gradient and Gradient Free Methods

Gradient based optimization algorithms are so far the most commonly used techniques in most existing applications due to their effectiveness in solving convex optimization problems and solid mathematical foundation. Nevertheless, new developments in non classical methods are shifting the focus to confidently adapt heuristic tools to similar or more complex problems. The major differences between heuristic and classical optimization methods can be summarized as follows:

1. Most heuristic methods are population based methods that search the solution space by a group of possible solutions. In contrast, gradient based methods use a single path to search for the optimal solution. This difference enhances the chances of locating the near global optima in heuristic methods. It reduces the dependency of successful convergence on the starting search point since most gradient based methods require a good starting point to ensure successful convergence.
2. Heuristic methods introduce randomness into their updating rules, while gradient based methods apply deterministic transition rules. This stochastic nature of the transition rules makes the heuristic methods less likely to get trapped in local optimum points.
3. Heuristic methods are general tools that can suit various optimization problems such as nonlinear, discrete, mixed type and constrained problems. This is due to the fact that they are derivative free and require only a fitness function to measure the quality of a given solution. Moreover, they require more fitness evaluation with less computation efforts compared with the gradient based methods.

In short, the choice of which optimization tools one should use is problem dependent. Careful analysis of the nature of the optimization problem and constraints can narrow the search process to select the most suitable technique.

4.2.2.3 Fault Parameter and Mode Change Estimation

After fault isolation, a set of suspected faults is established. Data is then collected for identification purpose. If there is time difference between FDI and the beginning of data collection, IMC is not equal to the mode that the fault is detected and isolated. Therefore, IMC is also the target of identification task. Since the solution of the identification problem can be split into two parts: real values for targeted fault parameters and switching time stamps and binary values for IMC. The Hybrid PSO (HPSO) algorithm is proposed, where RPSO is used to solve targeted fault parameters and switching time stamps identification problem and binary valued BPSO is utilized

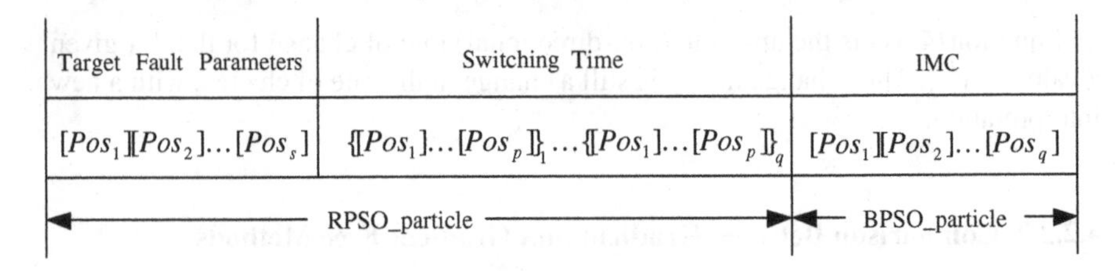

Fig. 4.8 Particle structure of HPSO algorithm

to solve IMC identification problem. The RPSO and BPSO evolve simultaneously and are coupled through the common fitness function.

Figure 4.8 depicts the particle structure of HPSO algorithm. In the figure, s is the number of fault candidates. Since the number of IMC, which determine the initial state of the controlled junctions during the observation window, is equal to the number of controlled junctions in HBG, therefore, there are number of q IMC for BPSO. Both solutions from RPSO and BPSO have to be evaluated to produce a potential solution vector because optimization requires the outcomes from these two parts. A fuzzy system is adopted to adjust the key parameters of the algorithm for better performance.

In this section, it is assumed that no frequent switching occurs in mode changes, and there exists a constant η such that $|T_j^h - T_j^u| > \eta$ for $\forall T_j^h \in (t_0, t_f)$, $\forall T_j^u \in (t_0, t_f)$, $j = 1, 2, \ldots, q$, $h = 1, 2, \ldots, p$, $u = 1, 2, \ldots, p$ and $h \neq u$. Based on this assumption, a frequent switching avoidance strategy is introduced, in which the switching time stamps estimated have to be compared with each other before each particle position is updated, if any of two switching time stamps satisfy $|T_j^h - T_j^u| \leq \eta$, all the switching time stamps will be reset to the following values

$$T_j^d = \frac{d[(t_0 - t_f) - 2\eta]}{p} \tag{4.32}$$

The underlying meaning in the above strategy is that when two switching time stamps are close, all the switching time stamps will be distributed evenly on the search space for next iteration search.

In the undertaken problem, the objective is to search for fault parameter values and switching time stamps with the minimum residual values in certain operating mode. To evaluate the fitness, N samples are collected after a fault is detected. If the parameter values estimated are the true physical parameters and mode switching vector is accurately identified, all GARRs, g_l, for $l = 1, 2, \ldots, m$, are zero. These lead to the definition of the following fitness function

$$F_{fitness} = \frac{1}{\sum_{l=1}^{m} \sum_{n=1}^{N} |g_l^n| + \varepsilon} \tag{4.33}$$

with n represents the sampling time index, and ε is a small positive constant to avoid zero division during the search process. Since the GARRs are used to identify the targeted fault parameters, only those GARRs containing targeted parameters are selected.

In PSO, it is known that the inertia weight is used to control the influence of the previous velocities on the current velocity, thereby influencing the tradeoff between global and local search abilities of RPSO. Since the maximum permissible velocity affects global exploration indirectly and the inertia weight affects it directly, it will generally be better to control global exploration through inertia weight only. Because of its clear physical meaning and its effectiveness, inertia weight has become a very important part of standard particle swarm optimization. The inertial weight has greatly influence on both global and local search; so many attempts have been tried by using various strategies to make its performance better. Shi and Eberhart [8] proposed a linear decreasing strategy in which

$$w^k = w_{max} - k(w_{max} - w_{min})/k_{max} \tag{4.34}$$

where w_{max} represents the maximal value of w, w_{max} denotes the minimal value of w, k is the iteration number, k_{max} denotes the maximal iteration in a run.

In this method, the particle in the beginning of the search could move fast, so that a sufficiently optimal region could be quickly accessed, and the inertial weight changes to be smaller with the iteration of the method later, then the particle slow down to search in the local region. Some other similar techniques are proposed to tune the inertia weight for better performance [9, 10]. However, these algorithms are on the population level, which means that every particle shares the same inertial weight in each iteration. Therefore, the inherent difference of particles' search state is ignored.

As for BPSO, it is clear that using small maximum velocities will make the sigmoid function sensitive to the noise introduced by the random number multiplication. This will result in the bit values easily switching between 0 and 1, in turn increases global search ability. On the contrary, bigger maximum velocity will provide finer search. In this work, a novel individual level adaptive scheme is introduced, in which a fuzzy system is employed to dynamically adjust the inertial weight of RPSO and $V_{max,r}$ of BPSO according to the improvement of each particle.

The improvement index for *rth* particle is defined by

$$\zeta_r = \frac{1/F_{fitness}(P^k_{best,r})}{1/F_{fitness}(P^{k-1}_{best,r})} = \frac{F_{fitness}(P^{k-1}_{best,r})}{F_{fitness}(P^k_{best,r})} \tag{4.35}$$

According to Eq. (4.35), when ζ_r is close to 1, it means that the rth particle has a deficient improvement, then it is desirable to increase its search space, so the inertial weight of RPSO should be increased and $V_{max,r}$ of BPSO should be decreased. On the other hand, when ζ_r is close to 0, it means that the *rth* particle has a significant improvement, then the area this particle needs to explore should be

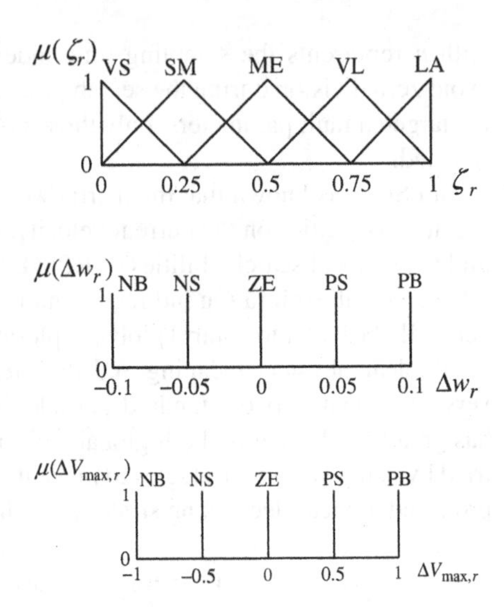

Fig. 4.9 Membership function of the improvement index ζ_r

Fig. 4.10 Membership function of the change of inertia weight Δw_r

Fig. 4.11 Membership function of the change of maximum velocity $\Delta V_{max,r}$

smaller, so the inertial weight of RPSO should be decreased and $V_{max,r}$ of BPSO should be increased accordingly. Based on the qualitative analysis, a fuzzy system is adopted to dynamically adjust the inertial weight of RPSO and $V_{max,r}$ of BPSO. The improvement index ζ_r is employed as the fuzzy input variable, and the change of the inertial weight Δw_r of RPSO and the change of the maximum velocity $\Delta V_{max,r}$ of BPSO are the fuzzy output variables. To get a balance of global search and local search ability of RPSO, w_r cannot be too large or too small, typical w_r is between 0.4 and 0.9. Similarly, $V_{max,r}$ of BPSO is often set at 4, so that there is always a good chance that a bit will change state. Both positive and negative corrections are needed for the inertial weight of RPSO and $V_{max,r}$ of BPSO. Therefore, a range of -0.1 to $+0.1$ is chosen for Δw_r and a range of -1 to $+1$ is selected for $\Delta V_{max,r}$.

The input fuzzy variable is defined in five fuzzy sets: very small (VS), small (SM), medium (ME), large (LA) and very large (VL). Each fuzzy set has its own triangular membership function as shown in Fig. 4.9. Triangular membership function is a proper choice here due to its simplicity. Two output fuzzy variables consist of five fuzzy sets: negative big (NB), negative small (NS), zero (ZE), positive small (PS) and positive big (PB). Each fuzzy set has its membership function with singleton shape as shown in Figs. 4.10 and 4.11. The singleton output membership functions are utilized here because of the efficient combination with the weighted average defuzzification process, which is easy for implementation. The fuzzy "IF-THEN" inference rules are established in Table 4.1 based on the aforementioned qualitative analysis. The output Δw_r and $\Delta V_{max,r}$ are derived from the fuzzy inference decision and defuzzification operation.

The block diagram of AHPSO algorithm is shown in Fig. 4.12, and it is seen that after the updating of RPSO solutions and BPSO solutions, their combination will be evaluated by the fitness function for next iteration. Moreover, when new candidate

Table 4.1 Rules tables for AHPSO

Antecedent	Consequent	
ζ_r	Δw_r	$\Delta V_{max,r}$
VS	NB	PB
SM	NS	PS
ME	ZE	ZE
LA	PS	NS
VL	PB	NB

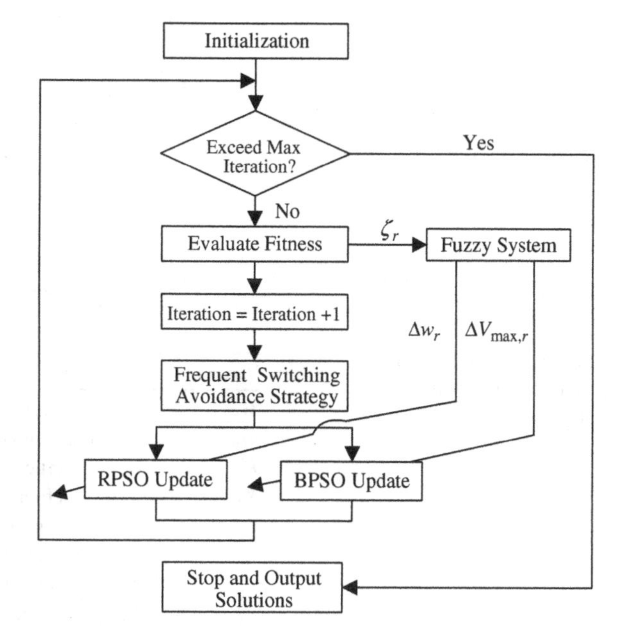

Fig. 4.12 Block diagram of AHPSO algorithm

solutions are generated in the next iteration, they should be checked before they enter new updating using frequent switching avoidance strategy. Finally, when the number of iteration exceeds the predefined maximal iteration, the algorithm will stop and then output the final solutions which contain fault parameter values, switching time stamps and IMC. The fault parameter values will be compared with their corresponding nominal values, and those parameters whose values are distinct from their nominal values are considered as true fault parameters.

4.2.3 Example I: An Electro-Hydraulic Suspension

The model of the quarter-car active suspension system, shown in Fig. 4.13, is used in this investigation. In the figure, some physical variables are defined in Table 4.2.

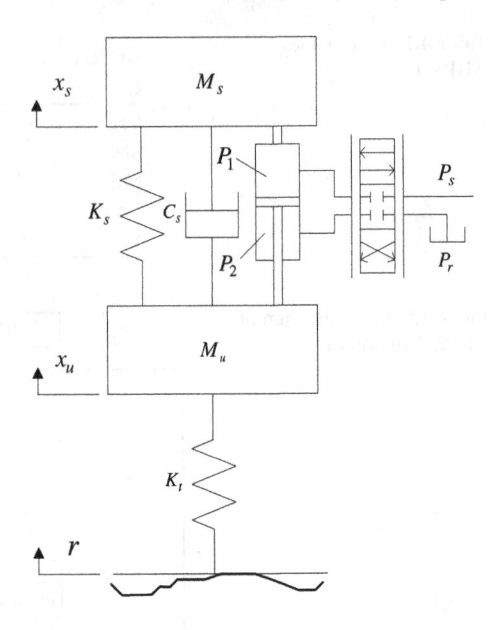

Fig. 4.13 Model of a quarter car

Table 4.2 Definition of physical variables in the quarter car

Variable	Definition
M_s	Sprung mass of the quarter car body
M_u	Unsprung mass of the wheel
K_s	Spring coefficients of the suspension
K_t	Spring coefficients of the wheel
x_s	Displacement of the quarter car body
x_u	Displacement of the wheel
x_u	Displacement of the wheel
C_s	Damper coefficient
P_1	Pressure inside the top chamber of the cylinder
P_2	Pressure inside the bottom chamber of the cylinder
P_s	Supply pressure of hydraulic actuator
P_r	Return pressure of hydraulic actuator
P_r	Road disturbance due to road roughness

Two types of discrete transitions for this system are considered in the study. The first one is autonomous mode change in the four way spool valve. The spool valve flow rate dynamic equation has two modes, depending on the valve position as shown in Fig. 4.14, where V_1, V_2 are the original total control volumes of the two cylinder chambers, respectively (including the volume of the servo valve, pipelines, and cylinder chambers), A_1 and A_2 are the ram area of the two chambers, respectively,

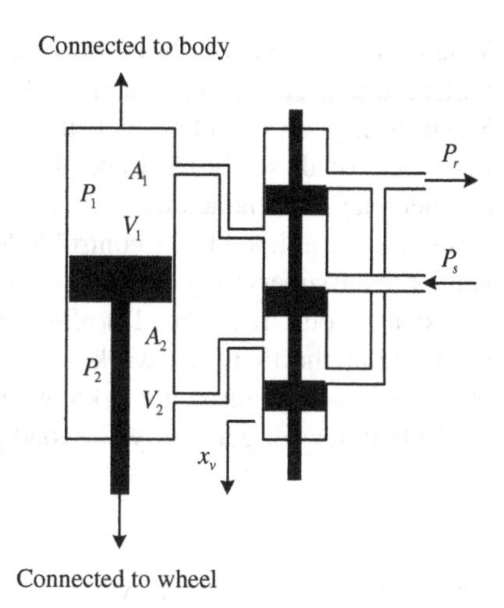

Fig. 4.14 Actuator with four way spool valve

x_v is the displacement of spool valve. The mode changes are governed by the values of internal state variables.

Using the equation for hydraulic fluid flow through an orifice, the relationship between spool valve displacement x_v, and the supply flow rate to the forward chamber (or cylinder end) Q_1, the return flow rate of the return chamber (or rod-end) Q_2, is given as [11]

$$Q_1 = C_d w x_v \left[s(x_v)\sqrt{\frac{2}{\rho}(P_s - P_1)} + s(-x_v)\sqrt{\frac{2}{\rho}(P_1 - P_r)} \right] \tag{4.36}$$

$$Q_2 = C_d w x_v \left[s(x_v)\sqrt{\frac{2}{\rho}(P_2 - P_r)} + s(-x_v)\sqrt{\frac{2}{\rho}(P_s - P_2)} \right] \tag{4.37}$$

with

$$s(*) = \begin{cases} 1 & \text{if } * \geq 0 \\ 0 & \text{if } * < 0 \end{cases} \tag{4.38}$$

where C_d is the discharge coefficient; ρ is the fluid density and w is the spool valve area gradient.

The second type of discrete transition is the road profile which has multiple modes [12]. Since the transition from one mode to another mode is not determined by the system internal state, and is also not decided by the driver, this type of mode transition can be considered as exogenous mode changes which are unknown for the designer, the proposed method can be used to identify these mode changes. Note that in hybrid systems, some mode-changes are known, because they are initiated by a supervisory controller which brings different control inputs at different modes.

When consider road profile, it may be characterized as the bumpy road, and the bump will be interpreted as equivalent velocity as required for bond graph flow input. Since during the vehicle travel, it will undergo several different road conditions, like plane road, sandy road and the like. The transition from one road profile to another may be characterized as mode change which will bring different inputs (excitations), although not controllable, at different modes. This interpretation is similar to the supervisory controller of hybrid systems and assumes that the transition is instantaneous. If the road profile does not match any of the given profiles, the proposed method will not work. As a result, assume that all types of road profiles the vehicle will go through are known in advance.

Without loss of generality, the road profile can be formulated as

$$r = \alpha^T SF \tag{4.39}$$

where SF is the road input vector with $SF = [Sf_1\, Sf_2\, \ldots\, Sf_z]^T$, where z is the total number of road profiles that the vehicle goes through; α is the state vector of controlled junctions with $\alpha = [\alpha_1\, \alpha_2\, \cdots\, \alpha_z]^T$.

It is assumed that the following relationship be satisfied, meaning that at each time moment the vehicle can only go through one road profile

$$\alpha_1 + \alpha_2 + \cdots + \alpha_z = 1 \tag{4.40}$$

For simplicity, only two road profiles are considered. One is random road, and the other is bumpy road. The road profile under random road can be represented as [13]

$$\dot{x}(t) = 2\pi q_0 \sqrt{G_0 V} w_n(t) \tag{4.41}$$

where q_0 is the reference spatial frequency; G_0 is the road roughness coefficient; V is the vehicle forward velocity; $w_n(t)$ is the zero-mean white noise.

The road displacement due to a bump is represented by [14]

$$x_2(t) = \begin{cases} \frac{A}{2} - \frac{A}{2}\cos(\frac{2\pi V t}{L}) & 0 \le t \le \frac{L}{V} \\ 0 & t > \frac{L}{V} \end{cases} \tag{4.42}$$

where A is the height of the bump; V is the length of the bump.

HBG model of electro-hydraulic suspension is developed in Fig. 4.15. Here four controlled junctions, i.e., 0_{a1}, 0_{a2}, 1_{a3} and 1_{a4}, are used and Boolean variables a_1, a_2, a_3 and a_4 are used to represent the state (ON/OFF) of these junctions.

The following relationships are valid

$$a_1 = 1 - a_2, \quad a_3 = 1 - a_4 \tag{4.43}$$

where a_3 depends on the state of the displacement of valve x_v

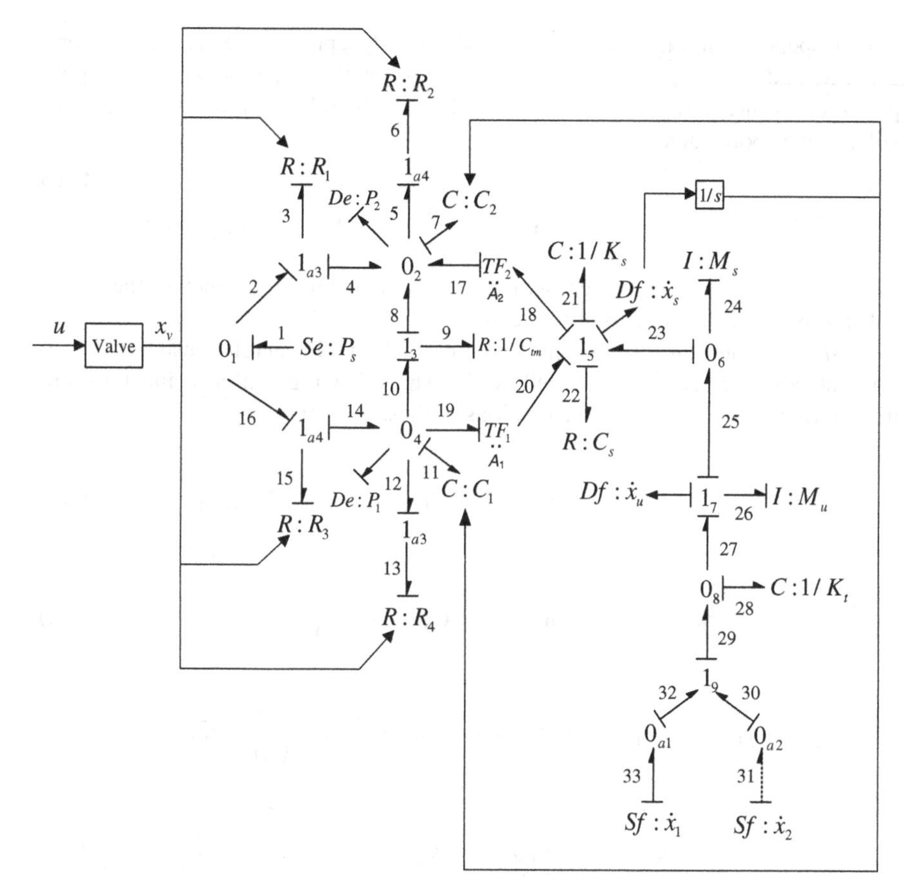

Fig. 4.15 HBG of quarter-car electro-hydraulic suspension system

$$a_3 = \begin{cases} 1 & \text{if } x_v \geq 0 \\ 0 & \text{if } x_v < 0 \end{cases} \tag{4.44}$$

Then the road input can be represented as follows

$$r = a_1\dot{x}_1 + a_2\dot{x}_2 \tag{4.45}$$

where $\dot{x}_1$ is the random road profile and $\dot{x}_2$ is the bumpy road profile.

Note that at junction 1_9, there are two bonds are flow-deciding bonds, i.e., two bonds do not have causal bars at the junction. Traditionally, this is an unacceptable 1-junction model. However, based on Eq. (4.43), it is clear that the junction flow is chosen from either of the two flow-deciding bonds at mutually exclusive time instants, there should be no conflict with the BG modeling principle. If the flow-deciding bond connected to $\dot{x}_2$ is given in dotted line as shown in Fig. 4.15, this clarification can also be specified. In Fig. 4.15, $De : P_1$ and $De : P_2$ represent the pressure sensors; $Df : \dot{x}_s$ and $Df : \dot{x}_u$ denote the velocity sensors; $Se : P_s$ represents the supply pressure. Furthermore, the block "Valve" denotes the relationship between control voltage and spool valve

$$\frac{x_V}{u} = \frac{K_V}{\left(\frac{s}{w_V}\right)^2 + 2\xi \frac{s}{w_V} + 1} \tag{4.46}$$

where K_V is the gain of the servo-valve; w_V is the natural frequency of the servo-valve; ξ is the damping ratio of the servo-valve.

Resistor elements R_1, R_2, R_3 and R_4 represent the nonlinear relationships between flow rate and pressure in (4.36) and (4.37). The following nonlinear functions are used to represent the constitutive relations of these elements

$$f_3 = g_1(e_3) = -R_1\sqrt{e_3} = -C_d \cdot w \cdot x_v \cdot \sqrt{\frac{2}{\rho}} \cdot \sqrt{e_3} \tag{4.47}$$

$$f_6 = g_2(e_6) = R_2\sqrt{e_6} = C_d \cdot w \cdot x_v \cdot \sqrt{\frac{2}{\rho}} \cdot \sqrt{e_6} \tag{4.48}$$

$$f_{13} = g_4(e_{13}) = -R_3\sqrt{e_{13}} = -C_d \cdot w \cdot x_v \cdot \sqrt{\frac{2}{\rho}} \cdot \sqrt{e_{13}} \tag{4.49}$$

$$f_{15} = g_3(e_{15}) = R_4\sqrt{e_{15}} = C_d \cdot w \cdot x_v \cdot \sqrt{\frac{2}{\rho}} \cdot \sqrt{e_{15}} \tag{4.50}$$

From the above equations, it is observed that the coefficient for these resistor elements, i.e., $C_d \cdot w \cdot x_v \cdot \sqrt{\frac{2}{\rho}}$, is not constant because it is modulated by the displacement of spool valve x_v, thus R_1, R_2, R_3 and R_4 are modulated resistors. In Fig. 4.15, these modulated resistors are represented by common R-elements with additional information bonds from the signal x_v.

External leakage flow is negligible in this study. Resistive element $R : 1/C_{tm}$ represents the internal leakage of the cylinder. TF_1 and TF_2 represent the relationship between velocity of the single-rod of the cylinder and the flow rate changes of two chambers. C_1 and C_2 represent the fluid compliance of the two chambers of the cylinder. The constitutive relations are nonlinear functions as follows

$$f_{11} = g_{C_1}(e_{11}) = C_1\frac{d}{dt}(e_{11}) = \frac{V_1 - A_1\int \dot{x}_s dt}{\beta} \cdot \frac{d}{dt}(e_{11}) \tag{4.51}$$

$$f_7 = g_{C_2}(e_7) = C_2\frac{d}{dt}(e_7) = \frac{V_2 + A_2\int \dot{x}_s dt}{\beta} \cdot \frac{d}{dt}(e_7) \tag{4.52}$$

where β represents the effective bulk modulus of the hydraulic fluid.

By inspection of (4.51), it is observed that the parameter representing C-element, i.e., $\frac{V_1 - A_1 \int \dot{x}_s dt}{\beta}$, is not a constant and is modulated by the flow sensor measuring the velocity of the quarter car body $\dot{x}_s$, so it is a modulated C-element. Thus, in the model of Fig. 4.15, it is modeled by using standard C-element with additional information bond from the time integral of signal $\dot{x}_s$. The same solution also applies to C_2.

Elements $C : 1/K_t$ and $C : 1/K_s$ in Fig. 4.15 represent the compliance of two springs, $R : C_s$ denotes the damper coefficient, $I : M_s$ and $I : M_u$ stand for the inertias of the unsprung mass and sprung mass, respectively.

The GARRs can be derived from the Diagnostic Hybrid Bond Graph (DHBG) of the system. DHBG is a HBG with suitable causalities such that every active BG component remains valid at all operating modes. This feature of the DHBG allows consistent causal description and derivation of GARRs of a hybrid system from its DHBG. By applying the procedure developed in [15], the DHBG of the hybrid system is obtained as shown in Fig. 4.16.

The constitutive relations of junctions $0_1, 0_2, 1_3, 0_4, TF_1, TF_2, 1_5, 0_6, 1_7, 0_8$ and 1_9 are given by the following equations

$$a_3 f_2 + a_4 f_6 - f_1 = 0 \tag{4.53}$$

$$a_3 f_4 + f_8 + f_{17} - f_7 - a_4 f_5 = 0 \tag{4.54}$$

$$e_{10} - e_8 - e_9 = 0 \tag{4.55}$$

$$a_4 f_{14} - f_{10} - f_{11} - a_3 f_{12} - f_{19} = 0 \tag{4.56}$$

$$f_{19} - f_{20}/TF_1 = 0 \tag{4.57}$$

$$e_{20} - f_{19}/TF_1 = 0 \tag{4.58}$$

$$e_{18} - TF_2 \cdot e_{17} = 0 \tag{4.59}$$

$$f_{17} - TF_2 \cdot f_{18} = 0 \tag{4.60}$$

$$e_{20} + e_{23} - e_{18} - e_{21} - e_{22} = 0 \tag{4.61}$$

$$f_{25} - f_{23} - f_{24} = 0 \tag{4.62}$$

$$e_{27} - e_{25} - e_{26} = 0 \tag{4.63}$$

$$f_{29} - f_{27} - f_{28} = 0 \tag{4.64}$$

$$a_1 e_{32} - a_2 e_{30} - e_{29} = 0 \tag{4.65}$$

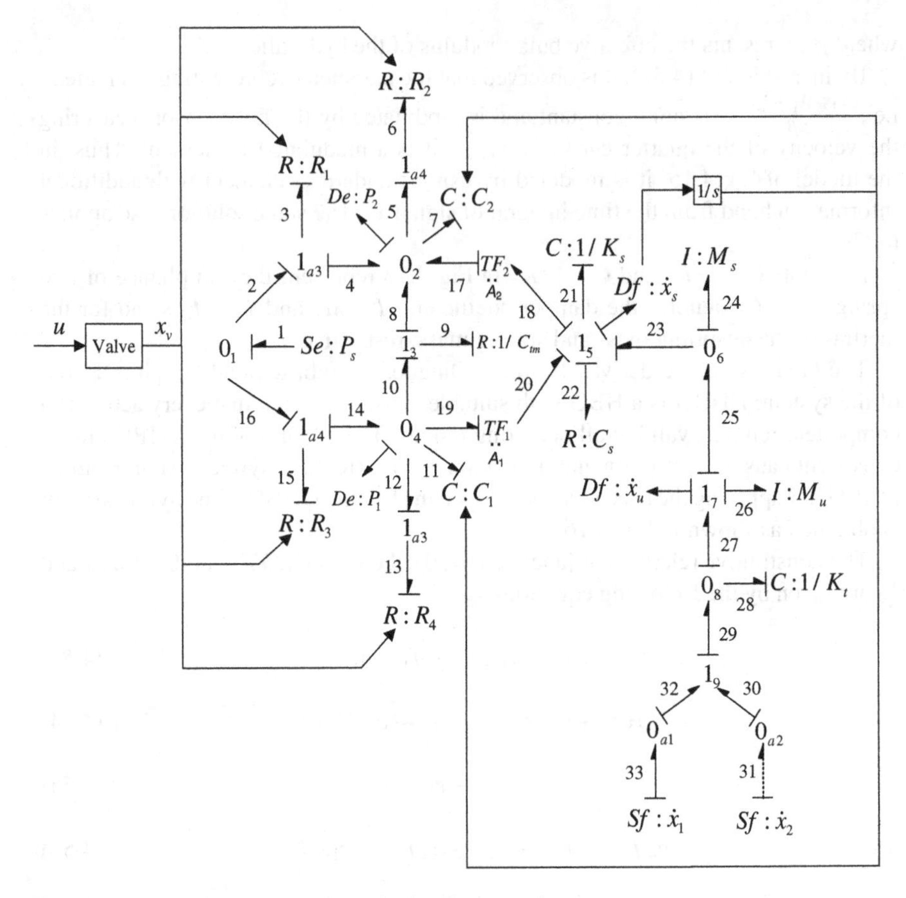

Fig. 4.16 DHBG of quarter-car electro-hydraulic suspension system

Four structurally independent GARR equations can be generated from Eqs. (4.54), (4.56), (4.61) and (4.64) after eliminating the unknown variables. The unknown variables elimination process is achieved by following the causal path, from known to unknown, on the DHBG model. For Eq. (4.54), the unknown variables f_4, f_5, f_7, f_8 and f_{17} can be calculated as follows

$$f_4 = f_3 = g_1(e_3) = g_1(e_2 - e_4) = -C_d \cdot w \cdot x_v \cdot \sqrt{\frac{2}{\rho}} \cdot \sqrt{P_s - P_2} \tag{4.66}$$

$$f_5 = f_6 = g_2(e_6) = g_1(e_5) = C_d \cdot w \cdot x_v \cdot \sqrt{\frac{2}{\rho}} \cdot \sqrt{P_2} \tag{4.67}$$

$$f_7 = g_{C_2}(e_7) = C_2 \frac{d}{dt}(e_7) = \frac{V_2 + A_2 \int \dot{x}_s dt}{\beta} \cdot \frac{d}{dt}(P_2) \tag{4.68}$$

$$f_8 = f_9 = C_{tm}(P_1 - P_2) \tag{4.69}$$

$$f_{17} = TF_2 \cdot f_{18} = -A_2 \dot{x}_s \tag{4.70}$$

Then the first GARR is obtained as follows

$$\begin{aligned} GARR_1 = & -a_3 C_d \cdot w \cdot x_v \cdot \sqrt{\frac{2}{\rho}} \cdot \sqrt{P_s - P_2} + C_{tm}(P_1 - P_2) \\ & - A_2 \dot{x}_s - \frac{V_2 + A_2 \int \dot{x}_s dt}{\beta} \cdot \frac{d}{dt}(P_2) \\ & - a_4 C_d \cdot w \cdot x_v \cdot \sqrt{\frac{2}{\rho}} \cdot \sqrt{P_2} \end{aligned} \tag{4.71}$$

For Eq. (4.56), the unknown variables f_{10}, f_{11}, f_{12}, f_{14} and f_{19} can be calculated as follows

$$f_{10} = f_9 = C_{tm}(P_1 - P_2) \tag{4.72}$$

$$f_{11} = g_{C_1}(e_{11}) = \frac{V_1 - A_1 \int \dot{x}_s dt}{\beta} \cdot \frac{d}{dt}(P_1) \tag{4.73}$$

$$f_{12} = f_{13} = g_4(e_{13}) = g_4(e_{12}) = -C_d \cdot w \cdot x_v \cdot \sqrt{\frac{2}{\rho}} \cdot \sqrt{P_1} \tag{4.74}$$

$$f_{14} = f_{15} = g_3(e_{15}) = g_3(e_{16} - e_{14}) = C_d \cdot w \cdot x_v \cdot \sqrt{\frac{2}{\rho}} \cdot \sqrt{P_s - P_1} \tag{4.75}$$

$$f_{19} = f_{20}/TF_1 = -A_1 \dot{x}_s \tag{4.76}$$

Then the second GARR is obtained

$$\begin{aligned} GARR_2 = & a_4 C_d \cdot w \cdot x_v \cdot \sqrt{\frac{2}{\rho}} \cdot \sqrt{P_s - P_1} - C_{tm}(P_1 - P_2) \\ & + A_1 \dot{x}_s - \frac{V_1 - A_1 \int \dot{x}_s dt}{\beta} \cdot \frac{d}{dt}(P_1) \\ & + a_3 C_d \cdot w \cdot x_v \cdot \sqrt{\frac{2}{\rho}} \cdot \sqrt{P_1} \end{aligned} \tag{4.77}$$

For Eq. (4.61), the unknown variables e_{18}, e_{20}, e_{22} and e_{23} can be calculated as follows

$$e_{18} = TF_2 e_{17} = -A_2 P_2 \tag{4.78}$$

$$e_{20} = e_{19}/TF_1 = -A_1 P_1 \tag{4.79}$$

$$e_{22} = C_s \dot{x}_s \tag{4.80}$$

$$e_{23} = e_{24} = M_s \frac{d}{dt}(f_{24}) = M_s \frac{d}{dt}(\dot{x}_u - \dot{x}_s) \tag{4.81}$$

The constitutive relation of the component $C : 1/K_s$

$$\frac{1}{K_s} \cdot \frac{d}{dt}(e_{21}) = \dot{x}_s \tag{4.82}$$

Then the third GARR is obtained from the component $C : 1/K_s$ after combining (4.78)–(4.82)

$$\begin{aligned} GARR_3 = \; & -A_1 \frac{1}{K_s} \frac{d}{dt}(P_1) + M_s \frac{1}{K_s} \frac{d^2}{dt^2}(\dot{x}_u - \dot{x}_s) \\ & + A_2 \frac{1}{K_s} \frac{d}{dt}(P_2) - C_s \frac{1}{K_s} \frac{d}{dt}(\dot{x}_s) - \dot{x}_s \end{aligned} \tag{4.83}$$

For Eq. (4.64), the unknown variables f_{27}, f_{28} and f_{29} can be calculated as follows

$$f_{27} = \dot{x}_u \tag{4.84}$$

$$f_{28} = \frac{1}{K_t} \frac{d}{dt}(e_{25} + e_{26}) = \frac{1}{K_t} \frac{d}{dt}\left[M_s \frac{d}{dt}(\dot{x}_u - \dot{x}_s) + M_u \frac{d}{dt}(\dot{x}_u)\right] \tag{4.85}$$

$$f_{29} = a_1 f_{33} + a_2 f_{31} = a_1 \dot{x}_1 + a_2 \dot{x}_2 \tag{4.86}$$

Then the fourth GARR is obtained as follows

$$GARR_4 = a_1 \dot{x}_1 + a_2 \dot{x}_2 - \frac{1}{K_t} \frac{d}{dt}\left[M_s \frac{d}{dt}(\dot{x}_u - \dot{x}_s) + M_u \frac{d}{dt}(\dot{x}_u)\right] - \dot{x}_u \tag{4.87}$$

The MD-FSM tables (Tables 4.3, 4.4, 4.5, 4.6) are obtained.

MATLAB® Simulink® is used to build the HBG model of the electro-hydraulic suspension system. In MATLAB® Simulink®, BG basic elements are built using the "subsystem" block in the Simulink library. For example, a R-element is established by set icon drawing command of the block as "disp(['R'])". The constitutive relation of the R-element is built under the mask of the block. For the example of $R : Cs$ in Fig. 4.16, its constitutive relation described in (4.80) is modeled as shown in Fig. 4.17.

Table 4.3 MD-FSM at $a_1 = 1, a_3 = 1$

	r_1	r_2	r_3	r_4	D_b	I_b
C_d	1	1	0	0	1	0
w	1	1	0	0	1	0
C_{tm}	1	1	0	0	1	0
β	1	1	0	0	1	0
V_1	0	1	0	0	1	0
V_2	1	0	0	0	1	0
A_1	0	1	0	0	1	0
A_2	1	0	0	0	1	0
K_s	0	0	1	0	1	0
K_t	0	0	0	1	1	0
C_s	0	0	1	0	1	0
M_s	0	0	1	1	1	1
M_u	0	0	0	1	1	0
S_e	1	1	0	0	1	0
S_{f1}	0	0	0	1	1	0
S_{f2}	0	0	0	0	0	0

Table 4.4 MD-FSM at $a_1 = 0, a_3 = 1$

	r_1	r_2	r_3	r_4	D_b	I_b
C_d	1	1	0	0	1	0
w	1	1	0	0	1	0
C_{tm}	1	1	0	0	1	0
β	1	1	0	0	1	0
V_1	0	1	0	0	1	0
V_2	1	0	0	0	1	0
A_1	0	1	0	0	1	0
A_2	1	0	0	0	1	0
K_s	0	0	1	0	1	0
K_t	0	0	0	1	1	0
C_s	0	0	1	0	1	0
M_s	0	0	1	1	1	1
M_u	0	0	0	1	1	0
S_e	1	1	0	0	1	0
S_{f1}	0	0	0	0	0	0
S_{f2}	0	0	0	1	1	0

Similarly, as for other elements, such as transformer, gyrator and junctions, they are modeled in Simulink using the "subsystem" blocks.

The half arrow bond is also modeled using the "subsystem" block, the half arrow line and the stroke are drawn through the icon drawing command. For controlled junctions, they are modeled by common junctions with Boolean variables. For the example of controlled junction 0_{a1} in Fig. 4.16, the constitutive relation in Simulink

Table 4.5 MD-FSM at $a_1 = 1, a_3 = 0$

	r_1	r_2	r_3	r_4	D_b	I_b
C_d	1	1	0	0	1	0
w	1	1	0	0	1	0
C_{tm}	1	1	0	0	1	0
β	1	1	0	0	1	0
V_1	0	1	0	0	1	0
V_2	1	0	0	0	1	0
A_1	0	1	0	0	1	0
A_2	1	0	0	0	1	0
K_s	0	0	1	0	1	0
K_t	0	0	0	1	1	0
C_s	0	0	1	0	1	0
M_s	0	0	1	1	1	1
M_u	0	0	0	1	1	0
S_e	1	1	0	0	1	0
S_{f1}	0	0	0	1	1	0
S_{f2}	0	0	0	0	0	0

Table 4.6 MD-FSM at $a_1 = 0, a_3 = 0$

	r_1	r_2	r_3	r_4	D_b	I_b
C_d	1	1	0	0	1	0
w	1	1	0	0	1	0
C_{tm}	1	1	0	0	1	0
β	1	1	0	0	1	0
V_1	0	1	0	0	1	0
V_2	1	0	0	0	1	0
A_1	0	1	0	0	1	0
A_2	1	0	0	0	1	0
K_s	0	0	1	0	1	0
K_t	0	0	0	1	1	0
C_s	0	0	1	0	1	0
M_s	0	0	1	1	1	1
M_u	0	0	0	1	1	0
S_e	1	1	0	0	1	0
S_{f1}	0	0	0	0	0	0
S_{f2}	0	0	0	1	1	0

is depicted in Fig. 4.18. Finally, all components are connected together by bonds to form a complete HBG as shown in Fig. 4.19.

The nominal parameters of the electro-hydraulic suspension are listed in Table 4.7. The input is $u = 2\sin(\pi t_s)$, sampling time t_s is 0.005 s. A fault occurrence is considered in unsprung spring K_t, with parameter value changes from 190000 N/m to 95000 N/m, which can also be taken as the fault of flat tire. Two fault scenarios have

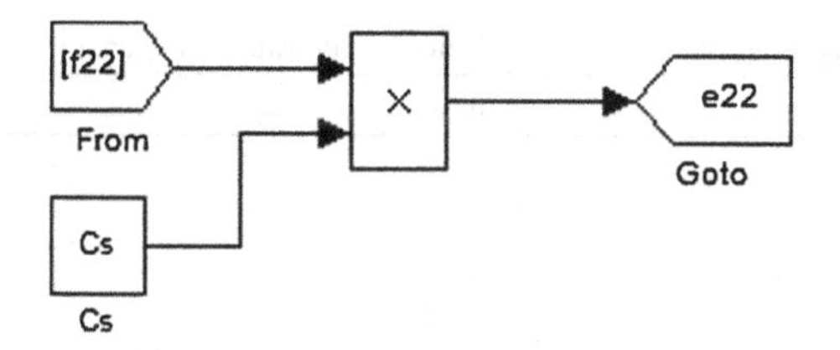

Fig. 4.17 Simulink model representing constitutive relation of R-element

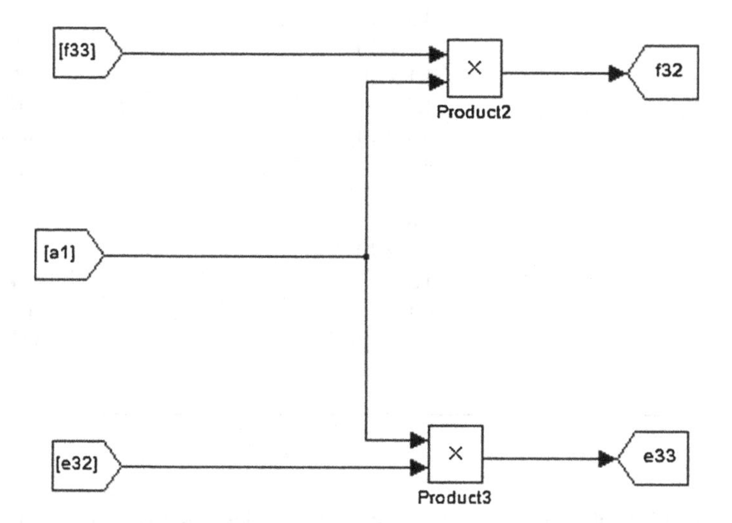

Fig. 4.18 Simulink model of constitutive relation of 0-junction

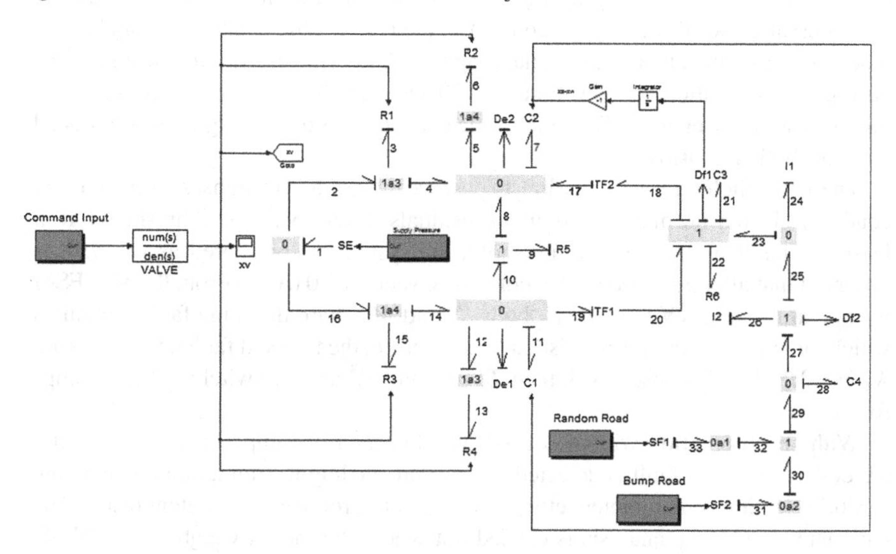

Fig. 4.19 HBG model of the quarter car in Simulink environment

Table 4.7 Nominal physical parameters of the electro-hydraulic suspension

Parameter	Value	Unit
M_s	290	kg
M_u	59	kg
K_s	16800	N/m
K_t	190000	N/m
C_s	1000	N s/m
A_1	3.8×10^{-4}	m^2
A_2	3.5×10^{-4}	m^2
V_1	1.2×10^{-4}	m^3
V_2	1.1×10^{-4}	m^3
P_s	1.5×10^{7}	Pa
C_{tm}	0.3×10^{-12}	$m^3\ s^{-1}/Pa$
ρ	900	$kg\ m^{-3}$
β	1.2×10^{9}	Pa
C_d	0.6	/
w	1.5×10^{-3}	m
V	12.5	m/s
q_0	0.1	m^{-1}
G_0	512×10^{-6}	m^3
A	0.05	m
L	5	m

been simulated. In the first case, the road profile undergoes from random road to bumpy road, featuring changing from 1 to 0, the transition time is set at 7 s during the observation window. In the second case, the road profile undergoes from bumpy road to random road, featuring a_1 changing from 0 to 1, the transition time is set at 8 s during the observation window. Figures 4.20 and 4.21 show the road profiles according to these two scenarios. The fault occurrence in unsprung spring K_t is introduced at 5 s for both scenarios.

These thresholds are set according to observe the residual responses during normal condition. Under normal operation, all residuals should below the thresholds. If at least one residual exceeds the threshold, the system is considered as faulty. It is observed that after 5 s, the detected coherence vector is [0 0 0 1]. From the MD-FSM for operating modes $\forall a_1 \in \{0, 1\}$, both K_t and M_u share the same fault signatures which indicates the fault is non-isolable. Therefore, the targeted fault parameters are K_t and M_u. If p is chosen as 3, then T_1^1, T_1^2 and T_1^3 are the switching time stamps for a_1.

With an observation window of 5 s, one thousand of sample data ($N = 1000$) are collected after a fault is detected. Before proceeding to simulation comparison, careful selection of parameter settings is crucial to produce a competent result. The parameter sensitivity analysis is carried out where the inertia weight w_r of RPSO decreases from 0.8 to 0.5 with decrement of 0.1, and the $V_{max,r}$ of BPSO increases from 4 to 6 with increment 1. In order to eliminate the influence of randomness, each

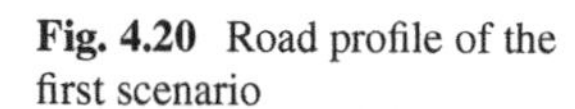
Fig. 4.20 Road profile of the first scenario

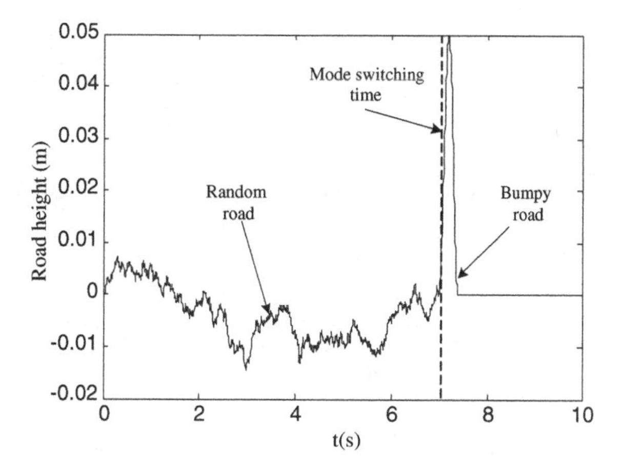

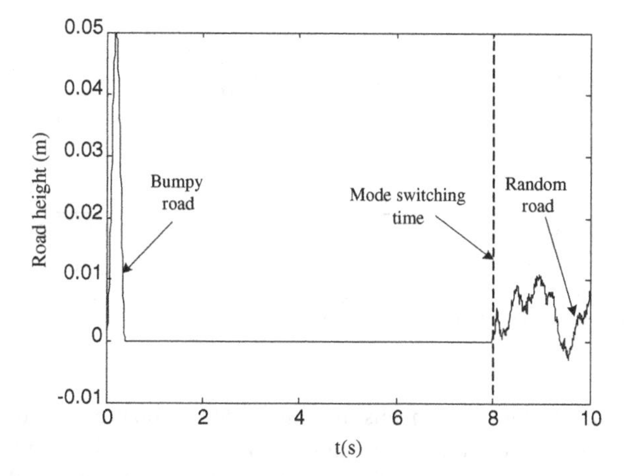

Fig. 4.21 Road profile of the second scenario

parameter setting of AHPSO is carried out 50 times from different initial seeds of the random number to ensure the repetitiveness of convergence. It has been observed that the final solutions obtained with different parameter settings do not differ much (standard deviation are less than 1 % of the mean value). Therefore, results presented are the mean of the 50 times. All the other parameters associated with AHPSO are taken as: Particle number = 50, Maximum iterations = 150, $\psi_1 = 2$, $\psi_2 = 2$, $\eta = 0.5$.

Figures 4.22 and 4.23 present the residual responses for these two scenarios. Note that the residual theoretically should be zero under normal condition. In practice, however, the residual is within certain error bounds as long as no faults occur during system operation. The value is not exactly zero over some time interval due to numerical inaccuracies caused by derivative approximation in Simulink® and noise signal in road profile. Usually, the threshold is set by observation of system responses under

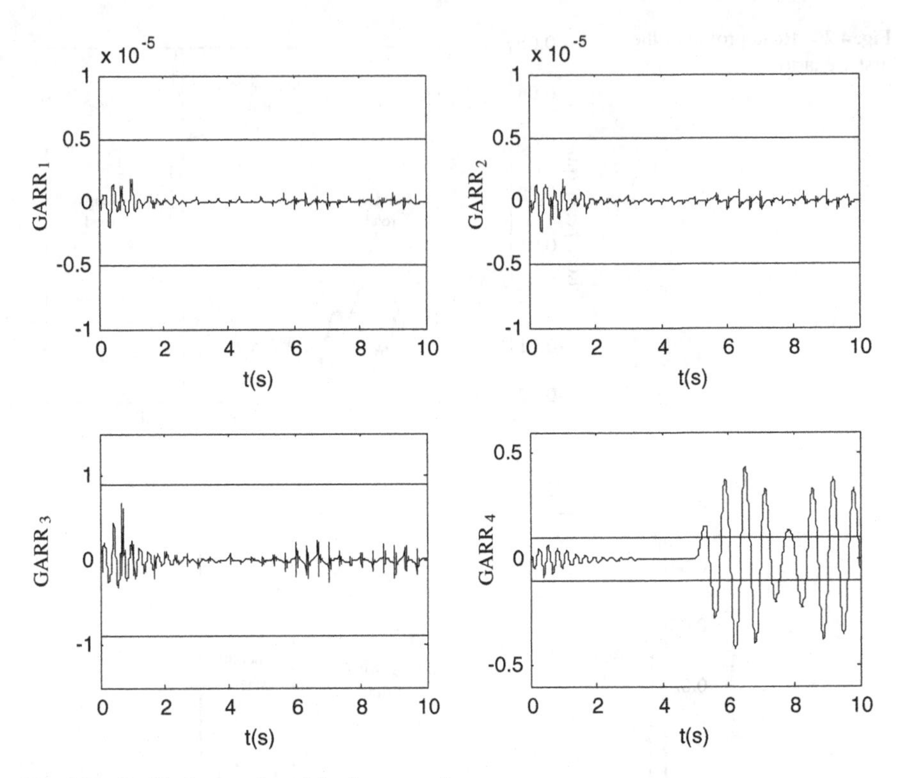

Fig. 4.22 Residual responses of the first scenario

normal condition. Misdetection means missing to detect the presence of an actually occurred fault and false alarm refers to an indication of fault which in fact does not happen. They are two contradictory factors when selecting the proper threshold for FDI purpose. Consideration of tradeoff is then made depending on experience and trial and error experiment. For the abrupt fault in this example, its effect on residual is larger than other factors and thus a simple fixed threshold is sufficient to detect the fault. The thresholds for these two scenarios are shown in Table 4.8.

During the optimization process, the parameters identified fall into the bounded ranges, including $K_t \in [50000, 300000]$N/m, $M_u \in [10, 150]$kg, $T_1^1 \in [0, 5]$s, $T_1^2 \in [0, 5]$s and $T_1^3 \in [0, 5]$s [16]. The optimum solution is denoted as $F_{fitness}$ in (4.33), so the bigger $F_{fitness}$ is, the better the global search ability of the method is. Tables 4.9 and 4.10 list the results of parameter sensitivity analysis, and the best $F_{fitness}$ is shown in boldface for both cases. According to these results, $w_r = 0.5$ for RPSO and $V_{max,r} = 5$ for BPSO are chosen for the AHPSO algorithm of the first scenario and $w_r = 0.7$ for RPSO and $V_{max,r} = 6$ for BPSO are chosen for the AHPSO algorithm of the second scenario. The different optimal parameter settings of AHPSO for different scenarios reveal that the parameter settings of AHPSO are actually problem dependent.

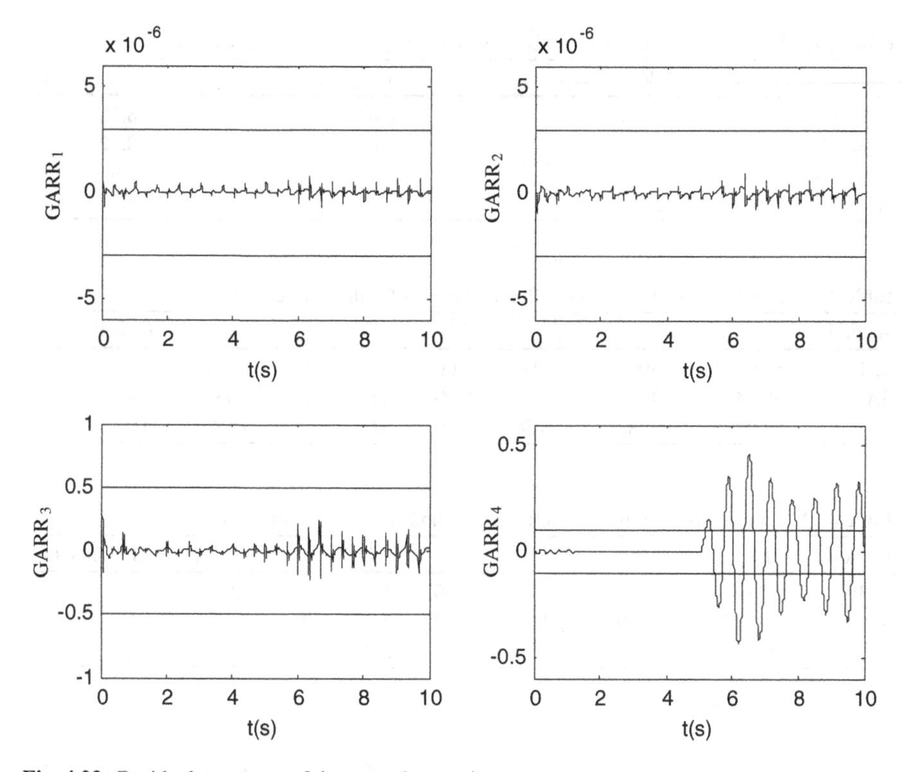

Fig. 4.23 Residual responses of the second scenario

Table 4.8 Thresholds for the two scenarios

	Th_1	Th_2	Th_3	Th_4
First scenario	$0.5e^{-5}$	$0.5e^{-5}$	0.9	0.15
Second scenario	$0.3e^{-6}$	$0.3e^{-6}$	0.5	0.12

Table 4.9 Parameter sensitivity analysis for identification for the first scenario

	$V_{max,r} = 4$	$V_{max,r} = 5$	$V_{max,r} = 6$
$w_r = 0.8$	7.32	6.27	6.85
$w_r = 0.7$	6.98	6.34	7.31
$w_r = 0.6$	7.49	7.25	6.85
$w_r = 0.5$	7.22	**7.81**	7.43

For comparison, AHPSO, GA and AGA [17] approaches have been tested on the same data collected after the fault occurring. The parameters of GA are taken as: Maximum iterations = 150, Population = 50, $P_c = 0.6$, $P_m = 0.05$.

Table 4.10 Parameter sensitivity analysis for identification for the second scenario

	$V_{max,r} = 4$	$V_{max,r} = 5$	$V_{max,r} = 6$
$w_r = 0.8$	8.71	8.63	8.70
$w_r = 0.7$	8.25	7.68	**8.92**
$w_r = 0.6$	8.51	8.46	7.98
$w_r = 0.5$	7.87	8.14	8.42

Table 4.11 Comparison results using different methods for the first scenario

Method	M_u	K_t	T_1^1	T_1^2	T_1^3	λ_1	$F_{fitness}$	CPU time (s)
AHPSO	59.0052	95004	1.9990	5.00	5.00	1	7.81	27
GA	58.7443	94946	1.9424	5.00	5.00	1	6.18	39
AGA	58.9261	94982	1.9758	5.00	5.00	1	7.28	41

Table 4.12 Comparison results using different methods for the second scenario

Method	M_u	K_t	T_1^1	T_1^2	T_1^3	λ_1	$F_{fitness}$	CPU time (s)
AHPSO	59.1024	94956	3.0002	5.00	5.00	0	8.92	25
GA	58.6773	94839	2.9832	5.00	5.00	0	8.17	38
AGA	59.0782	94877	2.9990	5.00	5.00	0	8.48	43

The parameters of AGA are the same as those of GA, the adaptive scheme of AGA involves some rules to adjust adaptively the crossover and mutation rates according to the performance of the current genetic operators. It increases the probability of the genetic operator if it consistently produces a better offspring during the search process; on contrast, it reduces the probability of the genetic operator if it always produces a poorer offspring. This method can adaptively regulate the balance between the exploration and exploitation of the solution space [17].

Comparison results of all methods are shown in Tables 4.11 and 4.12, it is seen that AHPSO is superior to the other two methods because AHPSO has better information sharing and conveying mechanism than GA and AGA. AHPSO also has good dynamics of balance between global and local search abilities. The results show the consistency between $F_{fitness}$ and accuracy of identified parameters which means the highest $F_{fitness}$ can provide the most accurate estimated parameters, except for M_u value in the second scenario, whereby the difference is so small that it can be neglected. Average execution CPU time for AHPSO approach is minimum, 27 s for the first scenario and 25 s for the second scenario, because it can easily adjust inertia weight of RPSO and maximum velocity of BPSO using simple "IF-THEN" rules. These results illustrate that the proposed method can identify the fault of K_t, the flat tire fault, and rule out M_u, the unsprung mass, as a fault; and that the proposed method can estimate the parameters accurately. The estimated switching time stamps also show a good matching with the implemented values.

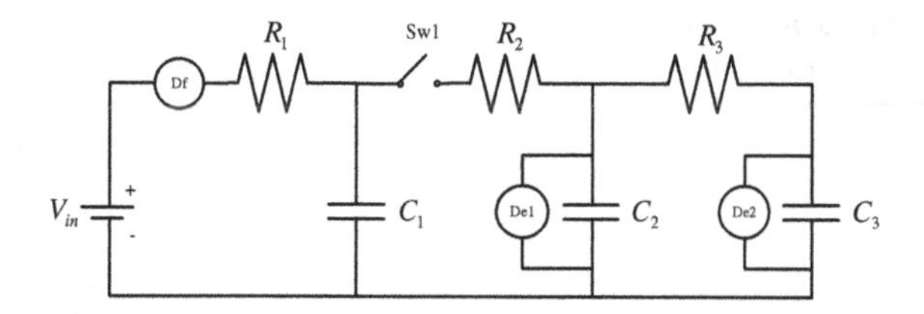

Fig. 4.24 A hybrid system: an electric circuit

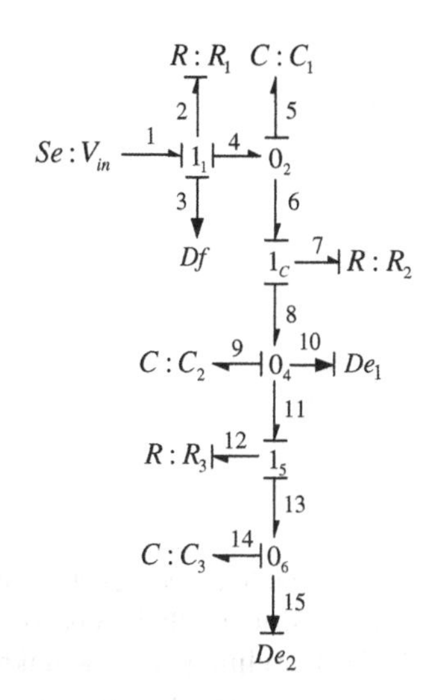

Fig. 4.25 HBG of the electric circuit

4.2.4 Example II: A Hybrid Electrical System

Figure 4.24 shows an example of hybrid system in electrical domain whose HBG is depicted in Fig. 4.25. The system consists of three R elements $\{R_1, R_2, R_3\}$, three C elements $\{C_1, C_2, C_3\}$, one switch Sw_1, one current (flow) sensor Df, and two voltage sensors $\{De_1, De_2\}$. The HBG has a 1-type controlled junction denoted by 1_c that represents the physical switch in electrical domain. The DHBG of the hybrid system is shown in Fig. 4.26.

The constitutive relations of junctions 1_1, 0_2, 1_c, 0_4, 1_5 and 0_6 are given by the following equations

$$e_1 - e_2 - e_3 - e_4 = 0 \tag{4.88}$$

$$f_4 - f_5 - af_6 = 0 \tag{4.89}$$

$$a(e_6 - e_7 - e_8) = 0 \tag{4.90}$$

$$af_8 - f_9 - f_{10} - f_{11} = 0 \tag{4.91}$$

$$e_{11} - e_{12} - e_{13} = 0 \tag{4.92}$$

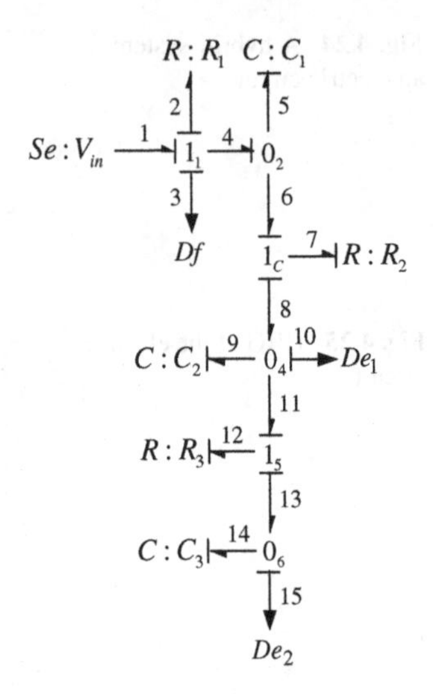

Fig. 4.26 DHBG of the electric circuit

$$f_{13} - f_{14} - f_{15} = 0 \tag{4.93}$$

where variable a denotes the state of switch Sw_1.

Three structurally independent GARRs can be generated from (4.89), (4.91) and (4.93) after eliminating the unknown variables. The unknown variables elimination process is achieved by following the causal path, from known to unknown, on the DHBG model.

In (4.89) the unknown variables f_4, f_5 and f_6 can be calculated as follows

$$f_4 = Df \tag{4.94}$$

$$f_5 = C_1 \frac{d}{dt} e_5 = C_1 \frac{d}{dt} (V_{in} - R_1 Df) \tag{4.95}$$

$$f_5 = f_6 = \frac{1}{R_2} (V_{in} - R_1 Df - De_1) \tag{4.96}$$

The first GARR is obtained as

$$GARR_1 = Df - C_1 \frac{d}{dt} (V_{in} - R_1 Df) - \frac{a_2}{R_2} (V_{in} - R_1 Df - De_1) \tag{4.97}$$

The unknown variables f_8, f_9, f_{10} and f_{11} in (4.91) can be calculated as

$$f_8 = f_7 = \frac{1}{R_2} (V_{in} - R_1 Df - De_1) \tag{4.98}$$

$$f_9 = C_2 \frac{d}{dt} De_1 \tag{4.99}$$

$$f_{10} = 0 \tag{4.100}$$

$$f_{11} = f_{12} = \frac{1}{R_3}(De_1 - De_2) \tag{4.101}$$

The second GARR is obtained as

$$GARR_2 = \frac{a_2}{R_2}(V_{in} - R_1 Df - De_1) - C_2 \frac{d}{dt} De_1 - \frac{1}{R_3}(De_1 - De_2) \tag{4.102}$$

For (4.93) the unknown variables f_{13}, f_{14} and f_{15} can be rewritten as

$$f_{13} = f_{12} = \frac{1}{R_3}(De_1 - De_2) \tag{4.103}$$

$$f_{14} = C_3 \frac{d}{dt} De_2 \tag{4.104}$$

$$f_{15} = 0 \tag{4.105}$$

The third GARR can be represented as

$$GARR_3 = \frac{1}{R_3}(De_1 - De_2) - C_3 \frac{d}{dt} De_2 \tag{4.106}$$

With these three GARRs, the MD-FSM tables for the circuit are shown (Tables 4.13, 4.14).

To implement the hybrid electrical system for verification of proposed fault identification algorithm, an experimental system that consists of the electrical hybrid system, MATLAB® Simulink®, and MATLAB® Real-Time Window Target is built. Figure 4.27 depicts the experimental system where the high speed photoMOS switching relay (AQV251) in Fig. 4.28 is used to act as a switch. The nominal parameters

Table 4.13 MD-FSM at $a = 1$

	r_1	r_2	r_3	D_b	I_b
R_1	1	1	0	1	0
R_2	1	1	0	1	0
R_3	0	1	1	1	1
C_1	1	0	0	1	1
C_2	0	1	0	1	1
C_3	0	0	1	1	1

Table 4.14 MD-FSM at $a = 0$

	r_1	r_2	r_3	D_b	I_b
R_1	1	0	0	1	0
R_2	0	0	0	0	0
R_3	0	1	1	1	1
C_1	1	0	0	1	0
C_2	0	1	0	1	1
C_3	0	0	1	1	1

Fig. 4.27 Experiment setup

of the hybrid system are: $R_1 = 135\,\Omega$, $R_2 = 135\,\Omega$, $R_3 = 135\,\Omega$, $C_1 = 10000\,\mu\text{F}$, $C_2 = 10000\,\mu\text{F}$, $C_3 = 10000\,\mu\text{F}$.

Fig. 4.28 High speed photo-MOS switching relay

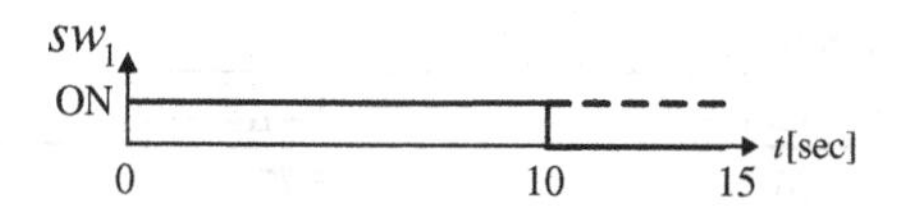

Fig. 4.29 Switching signal of the electric circuit

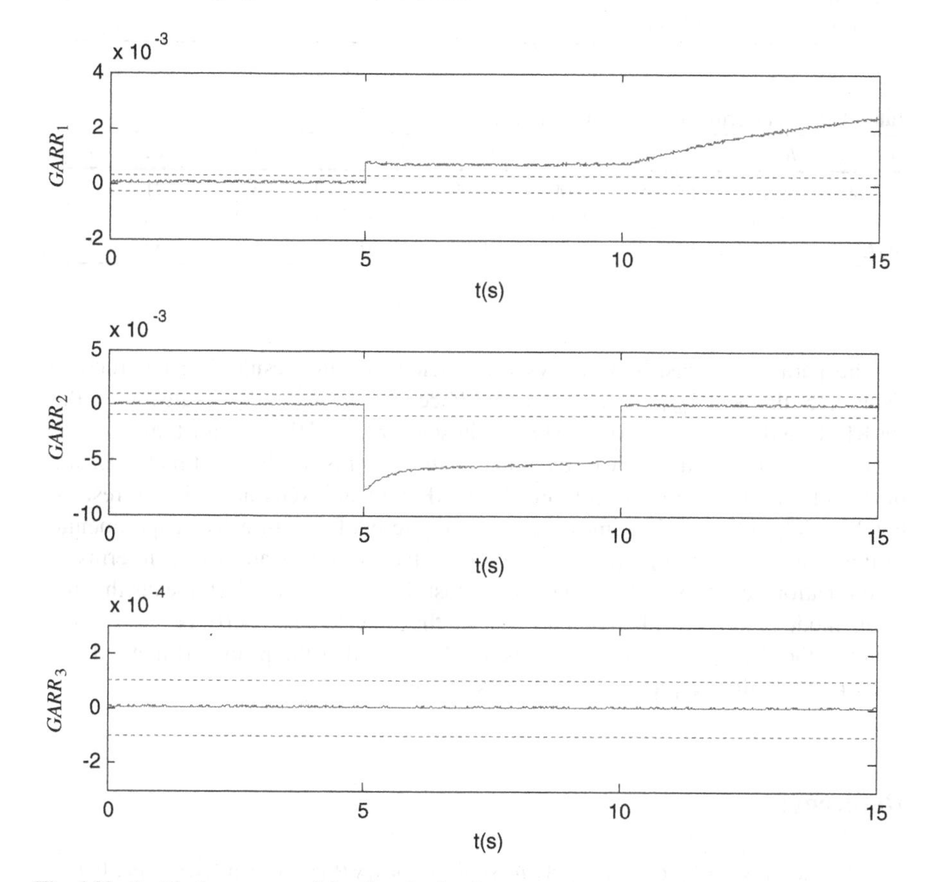

Fig. 4.30 Residual responses of the electric circuit

The input $V_{in} = 5\sin(0.1t_s)$, sampling time t_s is 0.01 s. A fault occurred in resistor R_1 from 135 Ω to 67.5 Ω at 5 s is considered in the system, and the switching signal is depicted in Fig. 4.29. The thresholds are set as: $\text{Th}_1 = 3\text{e}^{-4}$, $\text{Th}_2 = 1\text{e}^{-3}$, $\text{Th}_3 = 1\text{e}^{-4}$.

Figure 4.30 shows the residual responses of the system. It is observed that after 5 s, the detected coherence vector is [1 1 0]. From the Table 3.10, both R_1 and R_2 share the same fault signatures which indicates the fault is non-isolable. Therefore, the targeted fault parameters are R_1 and R_2. If p is set as 3, then T_1^1, T_1^2 and T_1^2 are the switching time stamps for a. The measurement in a 10 s window is collected, and number of data $N = 1000$ is gathered after a fault is detected.

Table 4.15 MD-FSM at $a = 0$

	$V_{max,r} = 4$	$V_{max,r} = 5$	$V_{max,r} = 6$
$w_r = 0.8$	4.76	**5.77**	4.74
$w_r = 0.7$	3.21	5.60	5.05
$w_r = 0.6$	3.98	4.92	5.32
$w_r = 0.5$	3.64	4.86	5.37

Table 4.16 Comparison results using different methods

Method	R_1	R_2	T_1^1	T_1^2	T_1^3	λ_1	$F_{fitness}$	CPU time (s)
AHPSO	66.721	133.887	5.001	10.00	10.00	1	5.77	21
GA	65.315	132.257	5.176	10.00	10.00	1	4.23	33
AGA	66.572	133.163	5.087	10.00	10.00	1	5.15	36

The parameter sensitivity analysis is carried out and results are recorded in Table 4.15, the best $F_{fitness}$ is shown in boldface. According to the results, $w_r = 0.8$ for RPSO and $V_{max,r} = 5$ for BPSO are chosen for the AHPSO algorithm.

Table 4.16 is the summary of the identification results for different methods, and the results are the average of 50 runs. Both AHPSO and AGA have similar results, but AHPSO uses less CPU time due to the simplicity of structure. The experimental results confirm that the proposed AHPSO is superior to GA and AGA in terms of identification accuracy and computational cost. $\lambda_1 = 1$ in Table 4.16 means that the initial mode of the controlled junction during the collected data is 1(ON state) which matches the designed value. These results illustrate that the proposed method can identify the fault of R_1 and rule out R_2 as fault.

References

1. M.T. Heath, *Scientific Computing: An Introductory Survey* (McGraw-Hill, New York, 1996)
2. R. Fletcher, *Practical Methods of Optimization*, 2nd edn. (Wiley, New York, 2000)
3. C.B. Low, D. Wang, S. Arogeti, M. Luo, Quantitative hybrid bond graph-based fault detection and isolation. IEEE Trans. Autom. Sci. Eng. **7**(3), 558–569 (2010)
4. J. Kennedy, R. Eberhart, Particle swarm optimization, in *IEEE Proceeding of International Conference on Neural Networks*, vol. 4, pp. 1942–1948 (1995)
5. D.E. Goldberg, *Genetic Algorithm in Search, Optimization, and Machine Learning* (Addison-Wesley, Boston, 1989)
6. M. Gen, R. Cheng, *Genetic Algorithm & Engineering Optimization* (Wiley, New York, 2000)
7. J. Kennedy, R. Eberhart, A discrete binary version of the particle swarm optimisation algorithm, in *IEEE Proceeding of International Conference on Neural Networks*, vol. 4, pp. 4104–4108 (1997)
8. Y. Shi, R. Eberhart, Empirical study of particle swarm optimization, in *Proceedings of the IEEE Congress on Evolutionary Computation*, pp. 1945–1950 (1999)
9. A. Chatterjee, P. Siarry, Nonlinear inertial weight variation for dynamic adaption in particle swarm optimization. Comput. Oper. Res. **33**(3), 859–871 (2004)

10. G. Chen, X. Huang, J. Jia, Z. Min, Natural exponential inertia weight strategy in particle swarm optimization, in *Proceedings of the 6th World Congress on Intelligent Control and Automation*, pp. 3672–3675 (2006)
11. H.E. Merritt, *Hydraulic Control Systems* (Wiley, New York, 1967)
12. J.H. Luo, K.R. Pattipati, L. Qiao, S. Chigusa, Model-based prognostic techniques applied to a suspension system. IEEE Trans. Syst. Man Cybern.: Part A: Syst. Humans **38**(5), 1156–1168 (2008)
13. H. Chen, P.Y. Sun, K.H. Guo, A multi-objective control design for active suspensions with hard constraints, in *Proceedings of the American Control Conference*, vol. 5, pp. 4371–4376 (2003)
14. H. Chen, K.H. Guo, Constrained H∞ control of active suspensions: an LMI approach. IEEE Trans. Control Syst. Technol. **13**(3), 412–421 (2005)
15. C.B. Low, D. Wang, S. Arogeti, J.B. Zhang, Causality assignment and model approximation for hybrid bond graph: fault diagnosis perspectives. IEEE Trans. Autom. Sci. Eng. **7**(3), 570–580 (2010)
16. J. Wang, J.D. Wang, N.Q. Daw, H. Wu, Identification of pneumatic cylinder friction parameters using genetic algorithms. IEEE/ASME Trans. Mechatron. **9**(1), 100–107 (2004)
17. L.K. Mak, Y.S. Wong, X.X. Wang, An adaptive genetic algorithm for manufacturing cell formation. Int. J. Adv. Manuf. Technol. **16**(7), 491–497 (2000)

Chapter 5
Mode Tracking Techniques

5.1 Mode Tracking of Hybrid Systems in FDI Framework

Model based health monitoring of dynamical systems is based on residuals that measure the distinction between the actual system and the model. The model represents the system normal condition and as long as the residuals' absolute-value is below a certain threshold, it is concluded that the system is normal. When the monitored system is hybrid and in the case of an unpredicted mode-change, the prevailing continuous model (used for system monitoring) is no longer valid for the monitored system. As a result, residuals used for model-based health monitoring exhibit abnormal behavior which can be interpreted as two different phenomena, namely as a component-fault (i.e., a change of component parameter to an unknown value, this fault is not represented as a mode of the hybrid system), or as a change of mode. In health monitoring of hybrid systems, it is essential to distinguish between these two scenarios. If a mode change is detected, the health monitoring system is required to identify the new mode, and if a components-fault is detected, a fault isolation process is necessary.

5.1.1 Mode Change Signatures of a Hybrid System

In order to study the influence of mode-changes on these residuals, a table, named the Mode-Change Signature Matrix (MCSM) is established. The MCSM is utilized for an efficient mode tracking method, and an efficient quantitative FDI framework for hybrid systems is presented in this section.

Let us consider an example of a two-tank system, as given in Fig. 5.1.

The two-tank system consists of two tanks, regulated centrifugal pump (which is modeled as a source of pressure p_{in}) and four valves represented by R_1, R_2, R_3 and R_4. A_1 and A_2 are the two tanks cross-section areas. The system is equipped with two pressure sensors, namely p_1 and p_2 which measure the pressure at the bottom of

D. Wang et al., *Model-based Health Monitoring of Hybrid Systems*,
DOI: 10.1007/978-1-4614-7369-5_5,

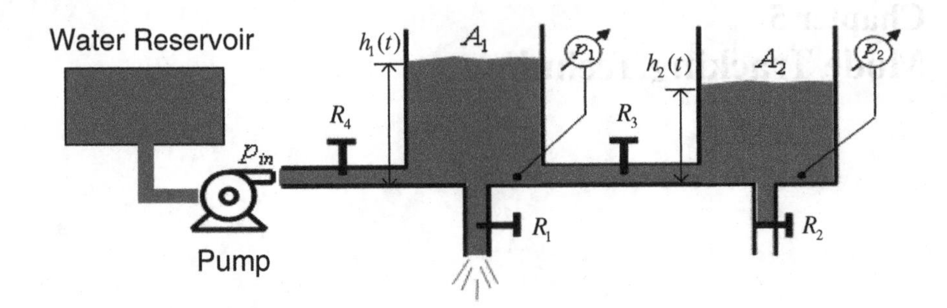

Fig. 5.1 A two-tank system

the two tanks; this pressure is proportional to the liquid level, according to:

$$p_i(t) = \rho g h_i(t), \qquad i = 1, 2. \tag{5.1}$$

where ρ is liquid density; g is the acceleration due to gravity, $h_i(t)$ is the liquid height in the tank.

Each valve can be in one of two discrete states, open and closed. These valves are operated by users and their state is unknown to the FDI system. Without loss of generality, and because of the analogy with an electrical resistor (which is used later), a valve linear dynamics is assumed (to describe the valve open state), and a negligible switching-time between open and closed states. The valves dynamics is given by:

$$f_j(t) = 0, \quad j = 1, 2, 3, 4 \text{ when the valve is closed} \tag{5.2}$$

$$f_j(t) = \frac{\Delta p_j(t)}{R_j}, \qquad \text{when the valve is open} \tag{5.3}$$

where f_j is the liquid-flow through the valve; $\Delta p_j(t)$ is the pressure difference across the valve and R_j is the valve's resistance to the liquid.

The DHBG of the two-tank plant is presented in Fig. 5.2. In this bond graph, the flow variable is the liquid volumetric flow (i.e., m^3/s), and the effort variable is pressure (i.e., [Pa]). The two tanks are modeled by the two storage components with coefficient $C_i = \frac{A_i}{g}$ for $i = 1, 2$. Each one of the four valves is modeled by a set of resistor with parameter R_j for $j = 1, 2, 3, 4$ and a controlled-junction.

In order to derive GARRs, only junctions with attached sensor are considered. Two structurally independent GARR equations can be generated after eliminating the unknown variables

$$GARR_1 = a_4 \frac{p_{in} - p_1}{R_4} - C_1 \dot{p}_1 - a_1 \frac{p_1}{R_1} - a_3 \frac{p_1 - p_2}{R_3} \tag{5.4}$$

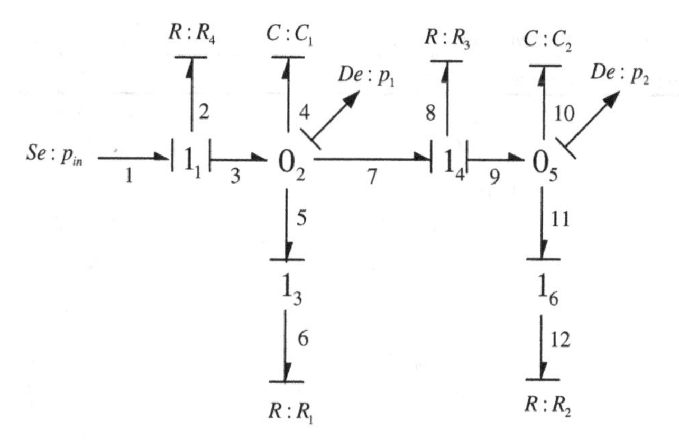

Fig. 5.2 The two-tank system DHBG

$$GARR_2 = a_3 \frac{p_1 - p_2}{R_3} - C_2 \dot{p}_2 - a_2 \frac{p_2}{R_2} \tag{5.5}$$

The global relations in (5.4) and (5.5) are mode-dependent, using these relations for real time fault detection and isolation, requires knowledge of the system current mode and this information comes from a mode-tracker. GARRs can also be used for off-line monitoring ability analysis using FSM. In the case of a hybrid system and the GARR concept, a FSM is derived for any mode; the result is the MD-FSM. At the first stage a Mode-GARR Table (MGT) is generated from the GARRs, to present the ARRs at each systems mode. The MGT of the two-tank system is listed in Table 5.1.

As an example of a MD-FSM, Table 5.2 presents the MD-FSM of the two-tank system in mode 14 (i.e., $[a_4\ a_3\ a_2\ a_1] = [1\ 1\ 1\ 0]$). It is important to note that although modes residuals (i.e., the MGT) are required for online health monitoring in the proposed framework, ARRs of all modes are not required to be stored in memory. Only the GARRs can be saved for the online process. The GARRs represent the ARRs of all system modes, and the ARRs of a particular mode are easily generated from the GARRs in real-time.

5.1.2 ARR-Based Mode Change Identification

MGT is an intermediate stage in the development of the MD-FSM. In the MGT each row represents a single mode and a set of ARRs. From Table 5.1, it is clear that for the two-tank system each system mode is characterized by a unique set of ARRs. This suggests that modes can be identified by ARRs and an ARR-based mode tracker can be developed.

Definition 5.1 A hybrid system is said to be *mode identifiable* if each one of the system modes is consistent with a unique set of ARRs.

Table 5.1 The MGT of the two-tank system

Mode $[a_4\ a_3\ a_2\ a_1]$	$GARR_1$	$GARR_2$
[0 0 0 0]	$-C_1\dot{p}_1 = 0$	$-C_2\dot{p}_2 = 0$
[0 0 0 1]	$-\frac{p_1}{R_1} - C_1\dot{p}_1 = 0$	$-C_2\dot{p}_2 = 0$
[0 0 1 0]	$-C_1\dot{p}_1 = 0$	$-\frac{p_2}{R_2} - C_2\dot{p}_2 = 0$
[0 0 1 1]	$-\frac{p_1}{R_1} - C_1\dot{p}_1 = 0$	$-\frac{p_2}{R_2} - C_2\dot{p}_2 = 0$
[0 1 0 0]	$-C_1\dot{p}_1 - \frac{p_1-p_2}{R_3} = 0$	$\frac{p_1-p_2}{R_3} - C_2\dot{p}_2 = 0$
[0 1 0 1]	$-\frac{p_1}{R_1} - C_1\dot{p}_1 - \frac{p_1-p_2}{R_3} = 0$	$\frac{p_1-p_2}{R_3} - C_2\dot{p}_2 = 0$
[0 1 1 0]	$-C_1\dot{p}_1 - \frac{p_1-p_2}{R_3} = 0$	$-\frac{p_2}{R_2} + \frac{p_1-p_2}{R_3} - C_2\dot{p}_2 = 0$
[0 1 1 1]	$-C_1\dot{p}_1 - \frac{p_1}{R_1} - \frac{p_1-p_2}{R_3} = 0$	$-\frac{p_2}{R_2} + \frac{p_1-p_2}{R_3} - C_2\dot{p}_2 = 0$
[1 0 0 0]	$\frac{p_{in}-p_1}{R_4} - C_1\dot{p}_1 = 0$	$-C_2\dot{p}_2 = 0$
[1 0 0 1]	$-\frac{p_1}{R_1} + \frac{p_{in}-p_1}{R_4} - C_1\dot{p}_1 = 0$	$-C_2\dot{p}_2 = 0$
[1 0 1 0]	$\frac{p_{in}-p_1}{R_4} - C_1\dot{p}_1 = 0$	$-\frac{p_2}{R_2} - C_2\dot{p}_2 = 0$
[1 0 1 1]	$-\frac{p_1}{R_1} + \frac{p_{in}-p_1}{R_4} - C_1\dot{p}_1 = 0$	$-\frac{p_2}{R_2} - C_2\dot{p}_2 = 0$
[1 1 0 0]	$-C_1\dot{p}_1 + \frac{p_{in}-p_1}{R_4} - \frac{p_1-p_2}{R_3} = 0$	$\frac{p_1-p_2}{R_3} - C_2\dot{p}_2 = 0$
[1 1 0 1]	$\frac{p_{in}-p_1}{R_4} - \frac{p_1}{R_1} - C_1\dot{p}_1 - \frac{p_1-p_2}{R_3} = 0$	$\frac{p_1-p_2}{R_3} - C_2\dot{p}_2 = 0$
[1 1 1 0]	$-C_1\dot{p}_1 + \frac{p_{in}-p_1}{R_4} - \frac{p_1-p_2}{R_3} = 0$	$-\frac{p_2}{R_2} + \frac{p_1-p_2}{R_3} - C_2\dot{p}_2 = 0$
[1 1 1 1]	$\frac{p_{in}-p_1}{R_4} - \frac{p_1}{R_1} - C_1\dot{p}_1 - \frac{p_1-p_2}{R_3} = 0$	$-\frac{p_2}{R_2} + \frac{p_1-p_2}{R_3} - C_2\dot{p}_2 = 0$

Table 5.2 The MD-FSM of the two-tank system in mode 14

Mode [1 1 1 0]	$GARR_1$	$GARR_2$	D_b	I_b
C_1	1	0	1	0
C_2	0	1	1	0
R_1	0	0	0	0
R_2	0	1	1	0
R_3	1	1	1	1
R_4	1	0	1	0
P_{in}	1	0	1	0

If the system is mode identifiable then the ARR-based mode-tracker observes the complete set of ARRs of all modes, simultaneously. In FDI framework, each ARR forms a residual. If the residual absolute value is below a threshold, then the ARR is said consistent with the monitored system. If only a set of ARRs, which describes only one of the system modes, is consistent, then it is deduced that the system operates in that mode (Note that as long as the monitored system is normal, the complete set of ARRs of its operating (current) mode, is consistent). A simple configuration is given as in Fig. 5.3. The simple configuration suffers from several drawbacks. The system mode is identified by ARRs and the mode information is fed to the GARRs. The GARRs at the FDI module with the mode information is equivalent to the set

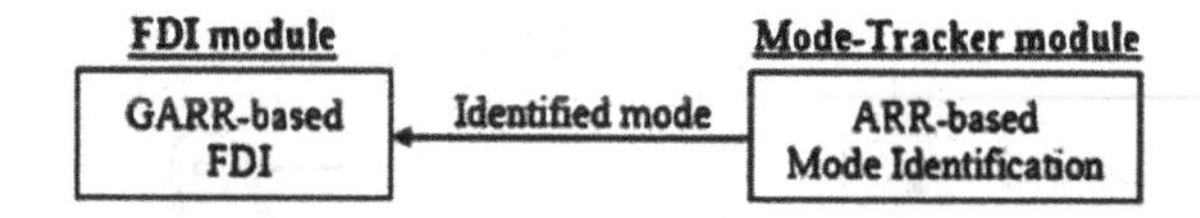

Fig. 5.3 A simple configuration for health monitoring of hybrid system

of ARRs which is used for the mode identification (at the mode-tracker module). Consequently, identical set of ARRs is used in both modules. Moreover, as long as the GARRs at the FDI module are consistent with the monitored system it is clear that the system is normal and its current mode is known; therefore mode identification is not required, at that stage. Mode identification is necessary only when GARRs at the FDI module show inconsistency. The notation $GARR \uparrow$ is used to describe an event in which $GARR_i$ at the FDI module is crossing a threshold. The event $GARR \uparrow$ is an indication of discrepancy between the model and the system. Two different causes may explain this discrepancy, one is a components-fault and the other is a change-of-mode.

The ARR-based mode identification is useful and has an important role in the proposed health monitoring strategy, but its continuous running is inefficient. Important information for mode-tracking is hidden in the $GARR \uparrow$ events. In a HBG model, system modes are represented by the states of controlled-junctions. The GARRs are mode dependent, and the controlled-junctions state-variables are part of the GARR expression. This suggests that the GARR structural-properties can be explored, to deduce on the GARR response to a change-of-mode. Consider for example $GARR_1$ in (5.4), the variables a_1, a_3 and a_4 represent controlled-junctions state. Any inconsistency between the actual controlled-junctions state and the values of a_1, a_3 and a_4 (that are used in the GARR) is reflected by the GARR. If a sudden change of controlled-junction state causes inconsistency, an event $GARR_1 \uparrow$ is expected. The $GARR_1$ is said *sensitive* to the change-of-state of controlled-junctions 1_1, 1_3 and 1_4. While monitoring the hybrid system, these GARR-events are observed and utilized for mode tracking; this process is based on the MCSM.

Definition 5.2 A MCSM is a matrix that represents cause-effect relations between mode-changes and GARRs.

The MCSM is analogous to the FSM which is used for FDI of continuous system, but instead of representing cause-effect relations between component-faults and ARRs, it represents cause-effect relations between mode changes and GARRs. In the HBG modeling framework, the mode-change is represented by the change of state of controlled-junctions; two controlled-junctions configurations are defined:

Definition 5.3 An *independent controlled-junction* is a controlled-junction that its state is independent of other controlled junctions states. Any *independent controlled-junction* is represented by a single state variable $a_i \in \{0, 1\}$.

Definition 5.4 A *synchronized set of controlled-junctions* is a set of controlled-junctions such that the complete set can be described by only two discrete states.

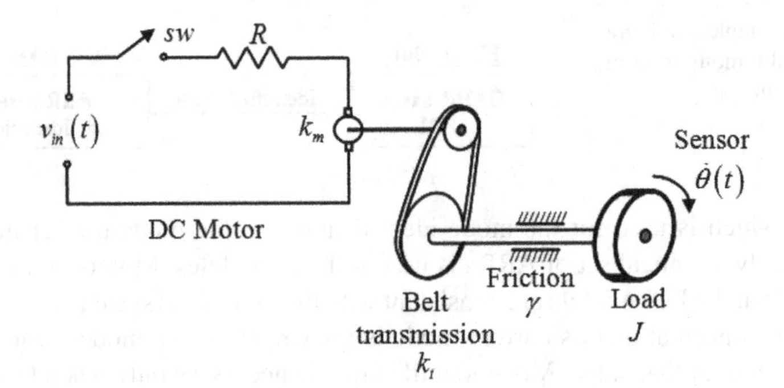

Fig. 5.4 A motor-belt-load system

Any *synchronized set of controlled-junctions* is represented by a single state variable $a_i \in \{0, 1\}$.

To exemplify the concept of synchronized set of controlled-junctions, a Motor-Belt-Load system, shown in Fig. 5.4, is utilized. The DC motor is represented by its electric resistance R and the motor gain k_m (where k_m represents the ratio between motor-current and torque). The motor inductance and inertia are negligible, compared with the load. The input signal to the DC motor is the voltage V_{in}, and the load velocity is measured by the sensor $\dot{\theta}(t)$. The switch *sw* is used to turn the motor ON and OFF, and may also represent a brunt motor fault. A belt mechanism with transmission-ratio k_t delivers the power from the motor to the load. A common failure of such transmission is when the belt is broken. These structural changes (such as a brunt motor and a broken belt) are modeled by controlled junctions. The DHBG of the Motor-Belt-Load system is presented in Fig. 5.5. When a broken belt fault happens,

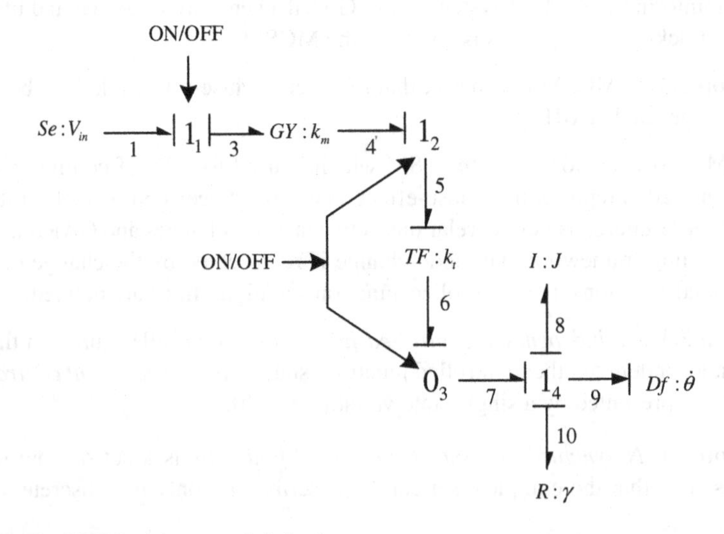

Fig. 5.5 The DHBG of the motor-belt-load system

a_2 is forced to zero, then both junctions 1_2 and 0_3 are deactivated. Therefore, $\{1_2, 0_3\}$ is a *synchronized set of controlled-junctions*.

In the MCSM, each row represents a single discrete state-variable a_i, and each GARR is represented by a column. The term '1' in the matrix, e.g., $MCSM_{ij} = 1$, indicates that the $GARR_j$ is sensitive to the mode-change represented by the change of state of a_i. Assumes that any change of mode is due to a change of a single state variable a_i (this assumption is analogues to the single fault assumption, in the context of parametric-fault isolation). Using this assumption, any row of the MCSM is a mode-change signature and all possible mode-changes are represented by the matrix (some of the signatures may be null). The mode change signatures represent cause-effect relations between mode-changes and GARRs. Each controlled junction has a corresponding signature and its mode-change is isolable if its signature is unique in the matrix. Similar to FSM based continuous system FDI, mode-change detectability (D_b) and mode-change isolability (I_b) are listed at the last two columns of the MCSM. As an example, Table 5.2 presents the MCSM of the two-tank system:

From Table 5.3, it is clear that all mode-changes of the two-tank system are detectable by GARRs, but not all of the mode changes are isolable. Nevertheless, this MCSM is still very useful for mode tracking; the principles are described as follows:

The mode evolution of hybrid system is represented by a series of controlled-junctions states, and is noted as follows

$$[a_4^0\ a_3^0\ a_2^0\ a_1^0] \rightarrow [a_4^1\ a_3^1\ a_2^1\ a_1^1] \rightarrow \cdots [a_4^k\ a_3^k\ a_2^k\ a_1^k] \tag{5.6}$$

where the upper index k represents mode evolution.

Assume the initial mode of the system is known and the mode information is fed to the two GARRs. As long as the two GARRs are consistent with the monitored system, the system mode is $[a_4^0\ a_3^0\ a_2^0\ a_1^0]$ and hence, mode-tracking is not required. In the case of a $GARR \uparrow$ event, a coherence vector given by $C = [c_1\ c_2]$ is generated, where $c_i, i = 1, 2$, is a binary variable representing the consistency of $GARR_i$ (If $GARR_i$ is consistent, then $c_i = 0$, otherwise $c_i = 1$). Next, the coherence vector is compared with the rows of the MCSM to generate mode-hypotheses; this is the mode-change isolation process. For example, if the coherence vector is $C = [c_1\ c_2] = [1\ 0]$, then, based on the MCSM in Table 5.2, the conclusion is that one of the two discrete states a_1 or a_4 is not consistent with the monitored system. Consequently, the next mode is one of two mode-hypotheses, as follows

$$\text{mode-hypothesis 1} : [\overline{a_4^0}\ a_3^0\ a_2^0\ a_1^0] \rightarrow [a_4^1\ a_3^1\ a_2^1\ a_1^1] \tag{5.7}$$

Table 5.3 The MCSM of the two-tank system

CJ	$GARR_1$	$GARR_2$	D_b	I_b
a_1	1	0	1	0
a_2	0	1	1	1
a_3	1	1	1	1
a_4	1	0	1	0

$$\text{mode-hypothesis 1} : [a_4^0\ a_3^0\ a_2^0\ \overline{a_1^0}] \rightarrow [a_4^1\ a_3^1\ a_2^1\ a_1^1] \tag{5.8}$$

where the upper bar represents a binary complement.

Using this information and the MGT in Table 5.1, the ARRs of the two mode-hypotheses are checked simultaneously to identify the hybrid system new mode (e.g., if the initial mode is $[a_4\ a_3\ a_2\ a_1] = [1101]$, and the coherence vector is $C = [c_1\ c_2] = [1\ 0]$, then the ARRs of modes [0 1 0 1] and [1 1 0 0] will be checked simultaneously). If a new mode is identified, then it is fed to the GARRs and the role of the mode tracker is terminated until the next *GARR* ↑ event. If the initial mode of hybrid system is unknown then the first step is to check all ARRs of all modes simultaneously to identify the system initial mode.

Definition 5.5 A process of ARR-based mode identification that is based on the complete set of ARRs (of all modes) is named the *full ARR-based mode identification*.

Definition 5.6 A process of ARR-based mode identification that is based on a subset of ARRs due to mode-hypotheses (i.e., the mode-change isolation result) is named the *narrowed ARR-based mode identification*.

Initial mode identification is a full-ARR-based mode identification. Another reason for a full-ARR-based mode identification is a loss of mode tracks. ARR-based mode identification is a process, and some time is required to complete the mode identification. If two successive mode-changes are taking place too quickly and the mode identification process after the first change is not completed, then the MCSM-based mode-hypotheses are no longer relevant, and the narrowed ARR-based mode identification will fail. If the narrowed ARR-based mode identification has failed then a full ARR-based mode identification is applied to identify the mode. If a new mode is identified, it is fed to the GARRs and the mode tracking process continues as proposed in this session. If a new mode is not identified by the ARR based mode identification (i.e., narrowed and full), then the conclusion is that the GARR event (which had triggered the mode identification process) is due to a parametric-fault, and the MD-FSM of the last know mode is utilized for fault isolation (under assumption of persistent mode, from the time of fault-occurrence to the time of fault-detection).

The GARR-based mode tracking and FDI principles of the two-tank system example are generalized to a health monitoring method for hybrid systems and its scheme is presented in Fig. 5.6.

For some hybrid systems all mode changes are isolable. Consider for example the two-tank system, when the valve R_4 is controlled by a switching-controller as shown in Fig. 5.7. The switching-controller goal is to maintain the liquid level in tank 1, between two fixed predetermined levels h_{min} and h_{max} (which are equivalent to two pressure values, p_{min} and p_{max}), and its discrete dynamics is based on the automaton presented in Fig. 5.8 (where it is assumed that the two tanks are initially empty). Since a_4 is known the MCSM of the two-tank system is narrowed, and the result is given in Table 5.3.

From Table 5.4, it is clear that all mode-changes are isolable. The ability to isolate any change of mode is significant to the mode tracking process. In such systems,

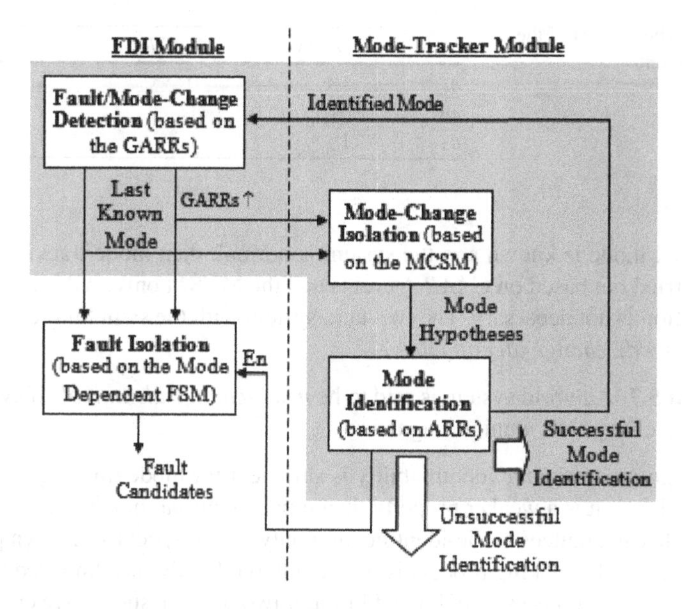

Fig. 5.6 A GARR-based Framework for FDI of hybrid system

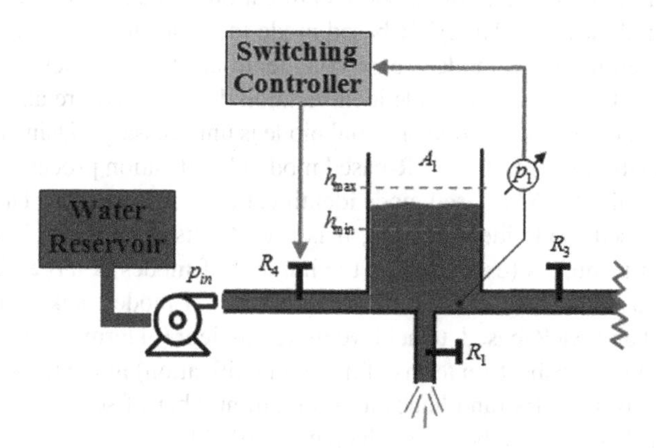

Fig. 5.7 The two-tank system with level control

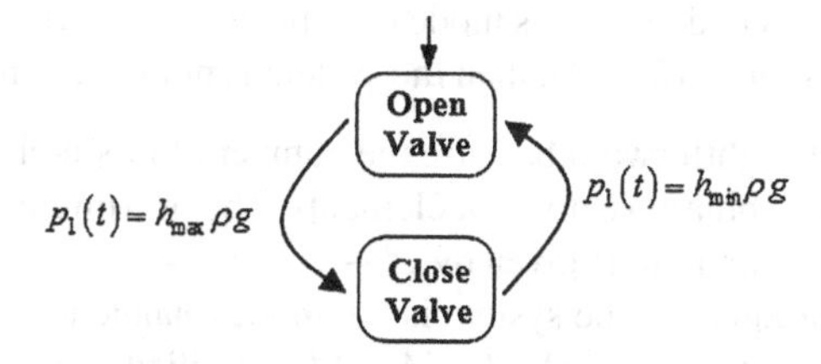

Fig. 5.8 Switching controller operation

Table 5.4 The MCSM of the two-tank system

CJ	$GARR_1$	$GARR_2$	D_b	I_b
a_1	1	0	1	1
a_2	0	1	1	1
a_3	1	1	1	1

if the initial mode is known and the system is normal, then mode tracking process can be carried out based on $GAAR \uparrow$ events and the MCSM only; ARR-based mode identification is not necessary. The two-tank system with the switching-controller is said to be *mode-change identifiable.*

Definition 5.7 A hybrid system is said to be *mode-change identifiable* if every possible mode-change has a unique signature.

Although mode-change identifiability is sufficient for mode tracking (under certain conditions), it is a weaker property than mode identifiability. The mode-change identifiability guarantees mode-identification only with respect to a known previous mode. If the mode tracking process is based only on the MCSM, knowledge of discrete initial condition is essential. In addition, if two or more successive changes are taking place too quickly, and the mode identification process is not quick enough, then the method may fail. The ARR-based mode identification is stronger, because its ability to identify a new mode is based on identification of the new mode continuous dynamics. Consequently, mode identification does not require any knowledge of previous mode (and knowledge of initial mode is unnecessary). Using the analogy of encoder position-sensor, the ARR-based mode identification process is said to be absolute, while the MCSM-based mode identification is incremental. The drawback of the ARR-based mode identification, if is used by itself, is its high demand for computational resources (especially, if the number of modes is large). To summarize, the proposed method utilizes the benefits of both mode tracking techniques, MCSM-based and ARR-based, to achieve more efficient (in terms of computational resources) and more robust (in terms of mode identification) mode tracking. A minimum number of GARRs (and hence a minimum number of sensors) is necessary, so that a hybrid system can be mode-change identifiable.

Property 5.1 Given a hybrid system with number of n independent GARRs, and m discrete state variables to describe its mode (i.e., the mode is given by $[a_m \cdots a_2\, a_1]$ where $a_i \in \{0, 1\}$). If $m > (2^n - 1)$, then the system is not mode-change identifiable.

Property 5.1 is straightforward because the number of mode-change signatures is m ; each signature is a binary vector of n elements. The maximum number of unique signatures, which are not null, is given by $2^n - 1$. If $m > (2^n - 1)$, then some of the signatures are not unique and the system is not mode-change identifiable.

In the two-tank system example the MCSM is unified (i.e., a unified-MCSM) and the structural monitoring properties (i.e., mode-change detection-ability and mode-change isolation-ability) are identical at all modes. For other hybrid systems the MCSM could be mode-dependent (i.e., a MD-MCSM) and its representation is analogues to the MD-FSM. One example of such conditions, which yield a MD-MCSM, is when some of the GARRs are composed of products of controlled-junctions state-variables. In that case, an OFF state of one controlled-junction may block the mode-change signature of other controlled-junctions.

To illustrate the concept of MD-MCSM, the GARR of the Motor-Belt-Load system in Fig. 5.4 is developed. The DHBG is given in Fig. 5.5, and the GARR derivation is based on the 1-junction constitutive-relation (i.e., the junction attached to the sensor $\dot{\theta}$):

$$f_9 = f_8 = f_7 = f_{10} = \dot{\theta} \tag{5.9}$$

$$e_9 = e_7 - e_8 - e_{10} = 0 \tag{5.10}$$

The unknowns variables e_7, e_8 and e_{10} in (5.10) can be obtained as follows

$$\begin{aligned} e_7 &= a_2 e_6 = a_2 k_t e_5 = a_2 k_t e_4 \\ &= a_2 k_t k_m f_3 = a_1 a_2 k_t k_m f_1 \\ &= a_1 a_2 k_t k_m \frac{e_1}{R} \\ &= a_1 a_2 k_t k_m \frac{e_2 - e_3}{R} \\ &= a_1 a_2 k_t k_m \frac{V_{in} - k_m f_4}{R} \\ &= a_1 a_2 k_t k_m \frac{V_{in} - k_m a_2 f_5}{R} \\ &= a_1 a_2 k_t k_m \frac{V_{in} - k_t k_m a_2 f_6}{R} \\ &= a_1 a_2 k_t k_m \frac{V_{in} - k_t k_m a_2 \dot{\theta}}{R} \end{aligned} \tag{5.11}$$

$$e_8 = J \dot{f_8} = J \ddot{\theta} \tag{5.12}$$

$$e_{10} = \gamma f_{10} = \gamma \dot{\theta} \tag{5.13}$$

By combining (5.10)–(5.13), the GARR of the Motor-Belt-Load system is achieved as

$$GARR : a_1 a_2 k_t k_m \frac{V_{in} - k_t k_m \dot{\theta}}{R} - J\ddot{\theta} - \gamma\dot{\theta} = 0 \tag{5.14}$$

The mode is defined by the discrete-state $[a_2\ a_1]$. It is known that if the system mode is $[0\ a_1]$, then the GARR is no longer sensitive to any mode-change that is modeled by the changes of a_1. Although the GARR is insensitive to a_1, it is still sensitive to a_2. When the mode is [0 0], then the GARR is not sensitive to any change of mode (due to the assumption that any change of mode is due to a single change of a single state variable a_i), and if the current mode is [1 1] then all mode changes are

Table 5.5 The MD-MCSM of the Motor-Belt-Load system

Mode [0 0]	*GARR*	D_b	I_b	Mode [0 1]	*GARR*	D_b	I_b
a_1	0	0	0	a_1	0	0	0
a_2	0	0	0	a_2	1	1	1
Mode [1 0]	*GARR*	D_b	I_b	Mode [1 1]	*GARR*	D_b	I_b
a_1	1	1	1	a_1	1	1	0
a_2	0	0	0	a_2	1	1	0

detectable. Base on these insights, The MD-MCSM of the Motor-Belt-Load system is given in Table 5.5.

Using the MD-MCSM for mode tracking is based on principles, similar to those presented previously. The only difference is that the previous mode information is used, in order to determine the relevant MCSM, for mode-change isolation. Since the mode-change signatures are mode dependent, each mode is analyzed individually to derive a MD-MCSM. A unified MCSM is preferred and its derivation is more straightforward (from the GARRs). The following property states a sufficient condition on the DHBG to form a unified MCSM.

Property 5.2 Assume a DHBG and a set of GARRs such that each GARR is based on a constitutive relation of a sensor junction. The set of GARRs forms a unified MCSM if none of the sensor-junctions is a controlled-junction, and if any causal-path that starts at a controlled-junction and ends at a sensor or at a known input does not pass through any controlled-junction other than the controlled-junction where the path stats at.

It is known that a GARR is derived from the constitutive relation of a sensor-junction; such relations do not include products of effort and flow. Any unknown variable such as effort or flow is solved using the bond graph causal paths (from known to unknown variables). These causal paths start at sensors or at known inputs. Using the sufficient condition of Property 5.2 any of these causal paths do not pass through more than a single controlled-junction, and a product of two or more controlled-junctions state-variables is not possible.

Based on the sufficient condition of Property 5.2, one can deduce whether the MCSM is unified, just by checking the DHBG (and before the GARRs are derived). Consider the DHBG of the Motor-Belt-Load system, it is clear that for the causal-path:

$$a_1 \rightarrow f_3 \rightarrow e_4 \rightarrow e_5 \rightarrow a_2 e_6 \rightarrow e_7 \rightarrow e_9 \rightarrow \dot{\theta} \tag{5.15}$$

From (5.15) it is observed that the condition of Property 5.2 does not hold, and the MCSM is not unified. On the other hand, in the two-tank system example in Fig. 5.2, the complete set of causal paths from controlled-junctions to known variables is given as follows:

$$\text{Path 1}: a_1 \rightarrow f_5 \rightarrow p_1$$

$$\begin{aligned}
&\text{Path 2}: a_2 \rightarrow f_{11} \rightarrow p_2\\
&\text{Path 3}: a_3 \rightarrow f_7 \rightarrow p_1\\
&\text{Path 4}: a_3 \rightarrow f_9 \rightarrow p_2\\
&\text{Path 5}: a_4 \rightarrow f_3 \rightarrow p_1
\end{aligned} \tag{5.16}$$

From (5.16) it is concluded that the condition of Property 5.2 holds and the MCSM is unified. If a sensor-junction is also a controlled-junction and the constitutive relation of that junction is utilized for GARR generation, then the MCSM is definitely mode dependent. A constitutive relation of a controlled-junction in its ON state is not valid for the OFF state. Consequently, a GARR that is derived form the constitutive relation of a controlled-junction is useful for mode tracking and FDI only in modes when the junction is ON (and its signatures are not valid when the junction is OFF).

5.1.3 Illustrative Example

The hybrid system health monitoring method is demonstrated on a test-bed. The test-bed is the electric circuit presented in Fig. 5.9. This circuit shares the same bond graph model with the two-tank system in Fig. 5.1 and therefore its health monitoring (i.e., FDI) is based on the GARRs and the method that are presented in the previous section (this equivalency stands behind the design of the circuit). In the circuit, the two tanks are replaced by two electric capacitors and the valves are represented by electric-switches and resistors. The only differences to the DHBG are the two power-variables which are now voltage and current (instead of pressure and liquid-flow), and the two voltage-sensors v_1 and v_2 (instead of the two pressure-sensors p_1 and p_2).

The physical parameters values are: $E = 12\,\text{V}$, $R_1 = 100\,\text{K}\Omega$, $R_2 = 30\,\text{K}\Omega$ $R_3 = 60\,\text{K}\Omega$, $R_4 = 50\,\text{K}\Omega$, $C_1 = 120\,\mu\text{F}$, $C_1 = 240\,\mu\text{F}$. The real-time implementation of the FDI framework is based on the MATLAB® Real-Time Windows Target, and two data acquisition cards (NI PCI-6025E and NI PCI-6713). Such configuration

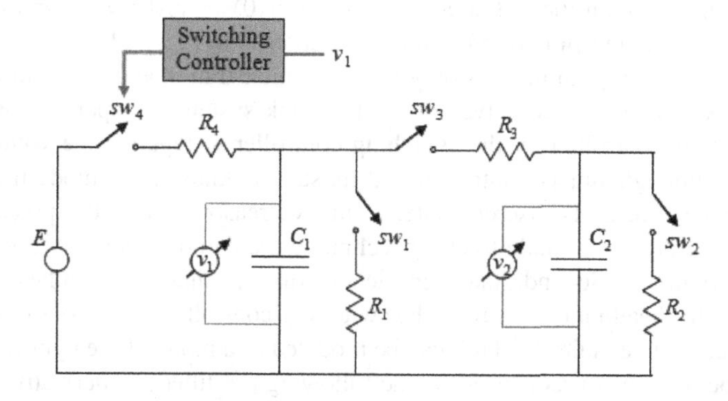

Fig. 5.9 An electric circuit with hybrid dynamics

Fig. 5.10 A test-bed for the FDI framework

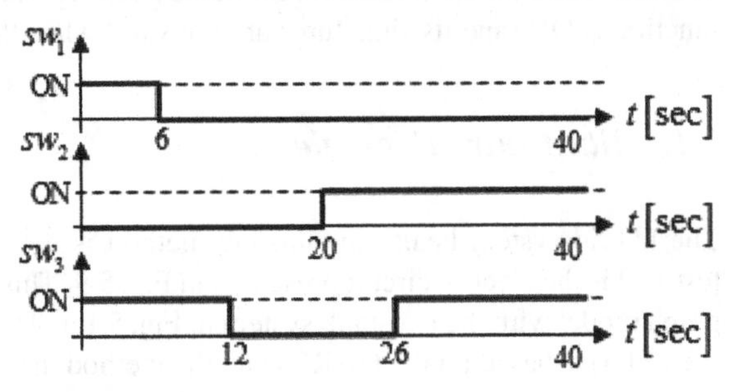

Fig. 5.11 Switches programming

enables design in MATLAB®/Simulink®/Stateflow and automatic code-generation for real-time execution in Windows environment.

The switching-controller is implemented by the MATLAB®/Stateflow according to the automaton in Fig. 5.8 (except that pressure is replaced with voltage). The controller goal is to maintain the voltage over C_1 between two levels: $v_{min} = 7\,\mathrm{V}$ and $v_{max} = 9\,\mathrm{V}$. The switches are implemented by three PhotoMOS relays (AQV251) and controlled electronically by the real-time software. The real-time program sampling-time is 0.01 s. A snapshot of the test-bed is presented in Fig. 5.10. Two fault scenarios are presented; in both the initial conditions are $v_1(0) = v_2(0) = 0$, and the three switches sw_1, $sw_2$and sw_3 are programmed according to Fig. 5.11.

The switches programming is only for the demonstration and is unknown to the mode-tracker (as the three valves of the two-tank system are operated by users). Switch sw_4 is controlled by the switching-controller to regulate the voltage level over C_1. Although sw_4 is controlled and its state is known, the mode tracker (of this example) tracks its discrete state, from two reasons. First, the potential and power of the proposed mode-tracking technique are demonstrated, by tracking the complete 16 modes. Second, state identification of sw_4 enhances redundancy and can be utilized for health monitoring of the switching-controller or the controlled-valve (e.g., a stuck-valve fault; this fault can be modeled as a mode of the hybrid system).

To cope with measurement-noise, the following pre-filter and derivative approximation are implemented in the real time program:

$$\text{Pre-filter:}\frac{10}{s+10},\quad \text{Derivative Approximation:}\frac{10s}{s+10} \tag{5.17}$$

Two scenarios are presented and in each scenario a different parametric-fault is demonstrated. In the first scenario the value of R_3 is abruptly changed from 60 to 100 KΩ, at about $t = 35$ s. This change represents a partial blockage at the central tube of the two-tank system (in practice, the potentiometer resistance is manually changed). In the second scenario the value of the voltage-supplier is abruptly dropped to 0 V, at about $t = 35$ s; this change represents a pump breakdown.

The experimental results are presented in Figs. 5.12 and 5.13. Each figure corresponds to a fault scenario. The first two graphs (of each figure) show the two GARRs (The threshold is presented by a dashed line and its value is $1e^{-5}$), the third graph presents the actual mode (dashed) versus the mode tracker result, and the system response is presented in the bottom graph.

As long as the system is normal, the mode tracking process of the two scenarios is identical. First, a process of initial mode identification is carried out. In this stage all ARRs of all modes (Table 5.1) are running simultaneously and mode 13 (i.e., [1 1 0 1]) is identified. Note that the system starts from $v_1(0) = v_2(0) = 0$, hence at the first few seconds, modes are rather alike and mode identification takes longer time. At $t = 6$ s, an event $GARR \uparrow$ triggers the mode tracker. Based on the coherence vector $C = [1\ 0]$ and the MCSM in Table 5.3, the two mode-hypotheses are: mode 5 and mode 12 (i.e., a change of a_1 or a change of a_4). The mode tracker runs the ARRs of these two suspected modes and mode 12 is identified. At $t = 12$ s the coherence vector is $C = [1\ 1]$, since the last known mode is mode 12, then mode 8 is suspected (a change of a_3) and identified by the mode tracker. At approximately $t = 15$ s, the voltage level over C_1 reaches the maximum limit (9 V) and the switching-controller turns off the switch sw_4; this change is unknown to the mode tracker. Consequently, the coherence vector is $C = [1\ 0]$ and mode 0 is identified. At $t = 20$ s the coherence vector is $C = [0\ 1]$ and the mode-hypothesis is mode 2. At $t = 26$ s the coherence vector is $C = [1\ 1]$ and mode 6 is identified. At approximately $t = 28$ s, the switching-controller turns sw_4 on, the coherence vector is $C = [1\ 0]$ and mode 14 is identified; this ends the mode tracking process of the system under normal condition (before fault occurrence).

The first fault scenario is presented in Fig. 5.12. At approximately $t = 35$ s the coherence vector $C = [1\ 1]$ is detected, but a new mode is not identified by the mode tracker. Consequently, a parametric fault is detected and isolated. Since the last know mode is mode 14, the MD-FSM of that mode (Table 5.2) is utilized for fault isolation. From the MD-FSM, it is concluded that the faulty component is R_3.

Consider the second fault-scenario presented in Fig. 5.13; at approximately $t = 35$ s the coherence vector is $C = [1\ 0]$ and a new mode is not identified. Based on the MD-FSM in Table 5.2 the three suspected parametric-faults are $\{C_1, R_4, P_{in}\}$. A further fault identification process can be utilized to refine the fault isolation result and to reveal the true fault in P_{in}.

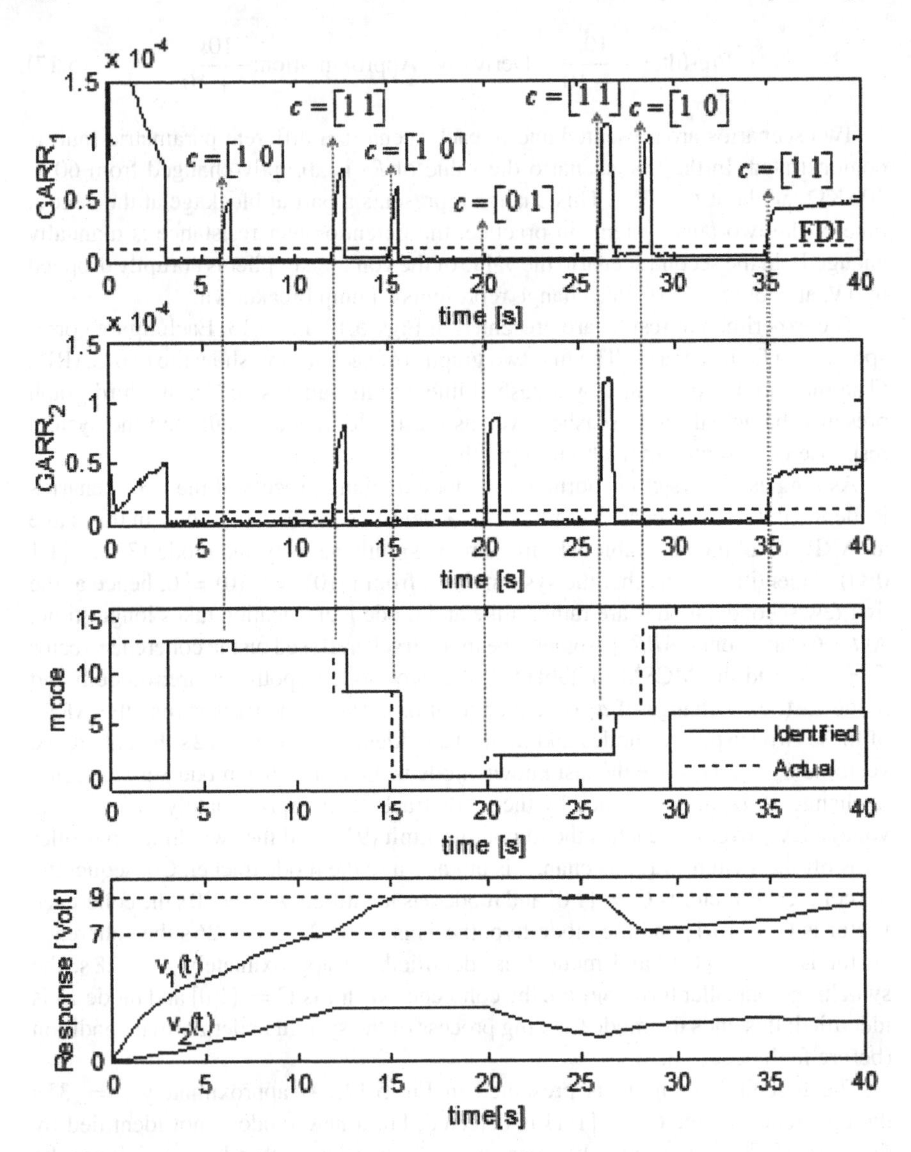

Fig. 5.12 Experiment results: first fault scenario

5.2 Mode Identification of Hybrid Systems in the Presence of Fault

After the fault is detected and isolated, a set of fault candidates is generated. The set of fault candidates contains the faulty components, together with some other non-isolable normal components. The fault candidates are the targeted parameters

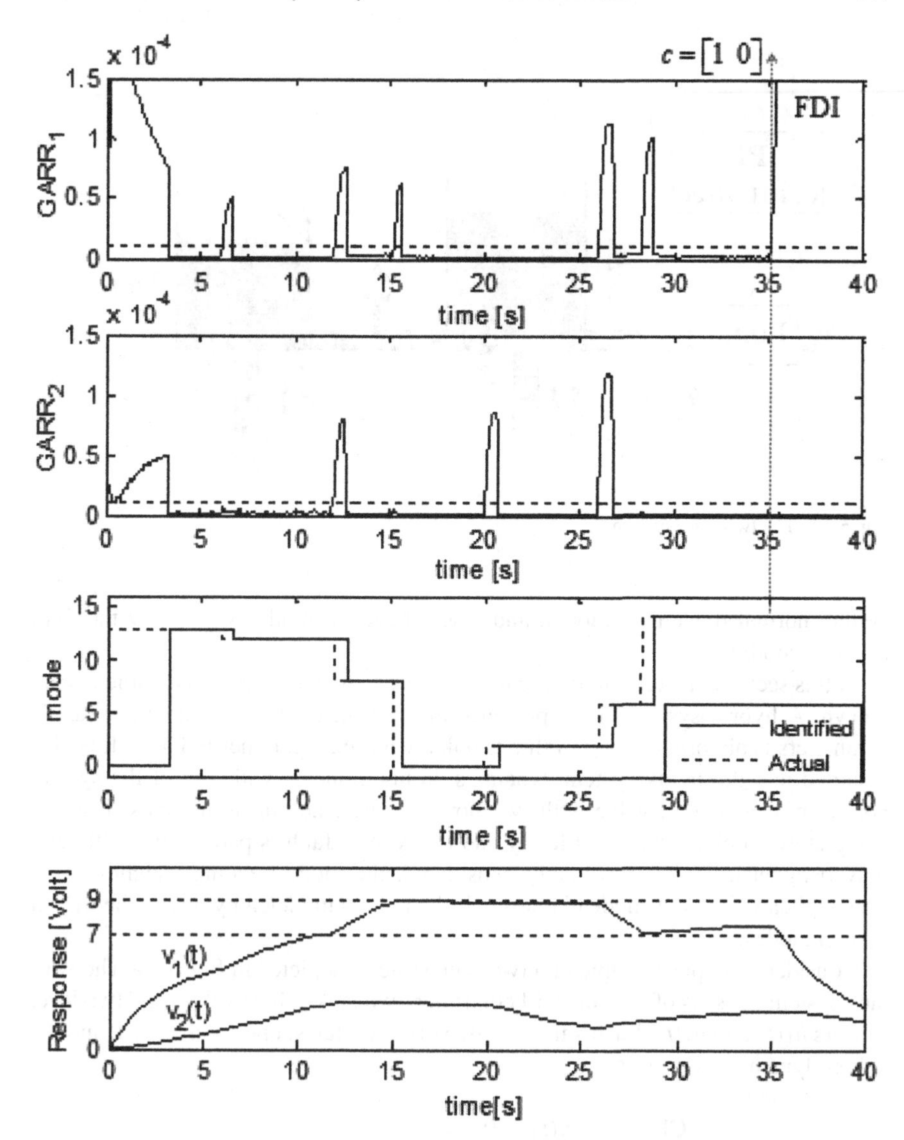

Fig. 5.13 Experiment results second fault scenario

for the fault parameter estimation module. The fault parameter estimation process is based on data collected from the system while the system is faulty, and for some parameters, the actual value is unknown. Fault parameter estimation also requires information of the system mode history. Since the system is faulty, standard model-based mode-identification techniques are not useful and could even mislead. To solve the problem, this section presents a novel mode identification technique for hybrid system in the presence of a parametric fault. The only known information is the

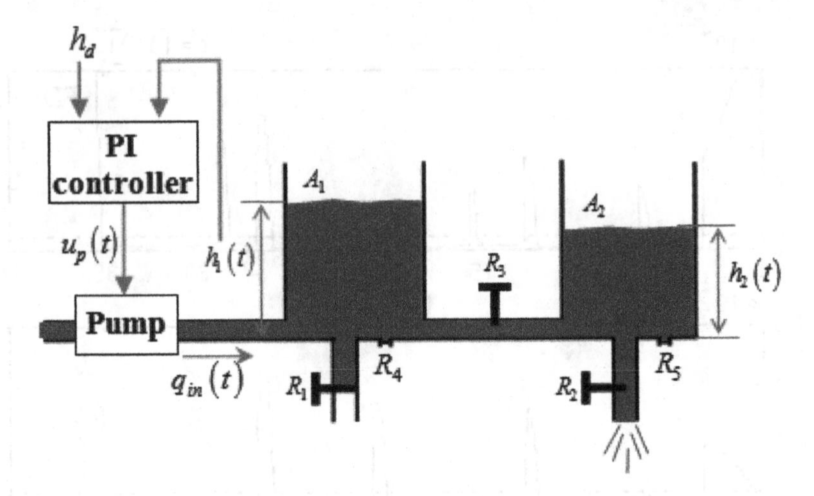

Fig. 5.14 The two-tank system

system normal dynamical model, and a set of fault candidates given by the fault isolation module.

In this section, the term normal condition means that all system's parameters are known. A hybrid system in the presence of fault means that one of the system's parameters is abnormal and its value is unknown (i.e., a parametric fault). It is also assumed that the faulty system dynamics do not coincide with any of the system normal modes, e.g., a valve with two normal states, i.e., open and closed, is not completely blocked when it is faulty and the type of fault is parametric. Although only parametric faults are directly considered, the mode tracking techniques are also applicable to fault-mode diagnosis, if the fault is modeled by one of the hybrid system modes.

Consider a simple example of a two-tank system, depicted in Fig. 5.14. The two-tank system consists of a pump, a PI controller, two tanks, three valves and two level sensors $h_1(t)$ and $h_2(t)$. Each valve has two discrete states: open and closed, and the valves dynamics is given as

$$\begin{aligned} \text{Closed} &\Rightarrow f_j(t) = 0 \\ \text{Open} &\Rightarrow f_j(t) = \frac{1}{R_i}\operatorname{sign}(\Delta h_j(t))\sqrt{|\Delta h_j(t)|} \end{aligned} \tag{5.18}$$

where $\{i, j\} = \{1, 13\}, \{2, 14\}, \{3, 3\}$, $f_j(t)$ is the volumetric flow of the liquid through the valve and $\Delta h_j(t)$ is the liquid level difference across the valve ($\Delta h_j(t)$ is proportional to the pressure difference across the valve). One hole at the bottom of each tank represents a leakage fault. The flow through the hole is given by $f_j(t) = \frac{\sqrt{h_j(t)}}{R_i}$, $\{i, j\} = \{4, 10\}, \{5, 12\}$, where R_4 and R_5 represent leakages in tanks A_1 and A_2, respectively, and $h_j(t)$ is the liquid level in the corresponding tank. When the system is normal, the value of R_4 and R_5 is very high (i.e., $\rightarrow \infty$) and the flow through the holes is zero. In case of leakage in one of the two tanks, the size of the leakage coefficient (i.e., R_4 or R_5) is unknown. For simplicity, the notation $f_j = g(R_i, \Delta h_j(t))$ is adopted to describe the nonlinear dynamics of an Open-valve or a leakage (note that for $i = 1, 2, 4, 5$, $\Delta h_j(t)$ is $h_j(t)$). The pump dynamics is assumed to be negligible, and the pump is modeled as a source of flow modulated by a PI controller, as given in (5.19) (note that $u_p(t)$ is the controller output and the input to the pump, whereas $q_{in}(t)$ is the pump output and has the physical meaning of liquid flow). The controller's goal is to maintain the liquid level in tank A_1 close to h_d, i.e.,

$$q_{in}(t) = u_p(t) = K_p(h_d - h_1(t)) + K_I \int_0^t (h_d - h_1(t))dt \tag{5.19}$$

The DHBG of the two-tank system is presented in Fig. 5.15. In this graph the flow variable is the liquid volumetric flow, and the effort is liquid level. Controlled-junctions 1_1, 1_2 and 1_3 model the Open-Closed nature of the three valves R_1, R_2 and R_3 respectively (an ON junction represents an Open valve, and vice versa). The hybrid system mode is determined by the state of all controlled-junctions. In the two-tank system the mode is given by the discrete state vector $[a_3\ a_2\ a_1]$. For example, $[a_3\ a_2\ a_1] = [110]$ represents mode 6.

The two GARRs of the two-tank system are given as follows

$$GARR_1 = q_{in} - a_1 g(R_1, h_1) - g(R_4, h_1) - a_3 g(R_3, h_1 - h_2) - A_1 \dot{h}_1 = 0 \tag{5.20}$$

$$GARR_2 = a_3 g(R_3, h_1 - h_2) - a_2 g(R_2, h_2) - g(R_5, h_2) - A_2 \dot{h}_2 = 0 \tag{5.21}$$

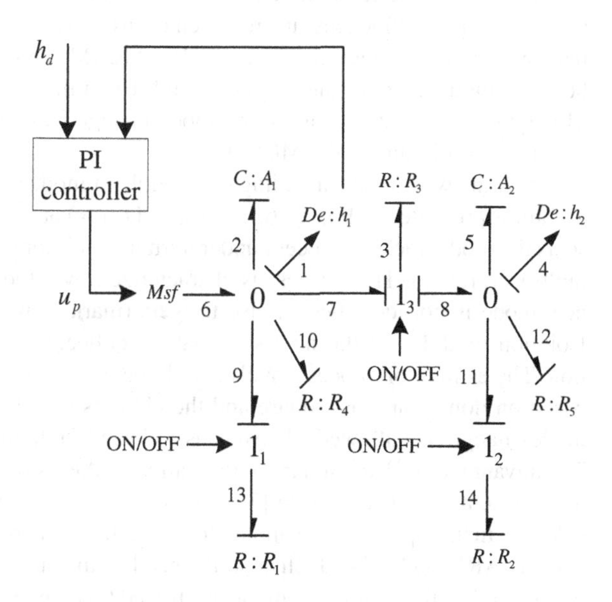

Fig. 5.15 The two-tank system's DHBG

Table 5.6 GFSM of the two-tank system

Mode $[a_3\ a_2\ a_1]$	$GARR_1$	$GARR_2$	D_b	I_b
q_{in}	1	0	1	0
R_1	a_1	0	a_1	0
R_2	1	a_2	a_2	0
R_3	a_3	a_3	a_3	a_3
R_4	1	0	1	0
R_5	0	1	1	$\bar{a}_2$

Table 5.7 MCSM of the two-tank system

CJ	$GARR_1$	$GARR_2$	D_b	I_b
a_1	1	0	1	1
a_2	0	1	1	1
a_3	1	1	1	1

For a hybrid system, a MD-FSM can be developed for each operating mode. A more compact representation is given by the Global-FSM (GFSM). The GFSM is generated directly from the GARRs, to present fault signatures and monitoring abilities at all operating modes. This representation is new and unique to the GARR concept. The GFSM of the two-tank system is presented in Table 5.6, where the meaning of the over bar is logical NOT (i.e., the fault is isolatable only when $a_2 = 0$, under the single fault hypothesis). In [1], the duality between parametric faults and mode changes is investigated. For a hybrid system, residuals inconsistency is not necessarily an indication of a parametric fault, but may also indicate an unexpected change of mode. Different mode-changes may have different influences on system's residuals; these influences are represented by a MCSM. The MCSM of the two-tank system is presented in Table 5.7. The MCSM represents cause-effect relations between mode-changes and residuals and it is utilized for mode-change isolation. Mode-change detection ability and mode-change isolation ability are presented in the last two columns of the MCSM.

The framework chart including the module of mode identification in the presence of fault is presented in Fig. 5.16. The shaded (gray) area, marked on the chart, represents the mode tracking process under normal conditions (i.e., before any parametric fault occurrence). This process is characterized by a loop, and the loop breaks if a new mode is not identified; it also triggers (marked as En, on the chart) the Fault Isolation module. At the next step, data is collected for Fault Parameter Estimation. The estimation process involves only Fault Candidates (FC). The estimation is based on nonlinear least square and the GARRs are utilized for the cost function; mode-changes are allowed when data is collected for fault parameters estimation [2]. The advantages of least square optimization of ARRs for fault estimation in continuous systems are discussed in [3], and are relevant also to the hybrid case, and the GARRs. In this approach, the model for the estimation process is reduced and considers only ARRs/GARRs which are influenced by the fault, there is no need to estimate initial conditions and a single model is utilized for detection, isolation and estimation.

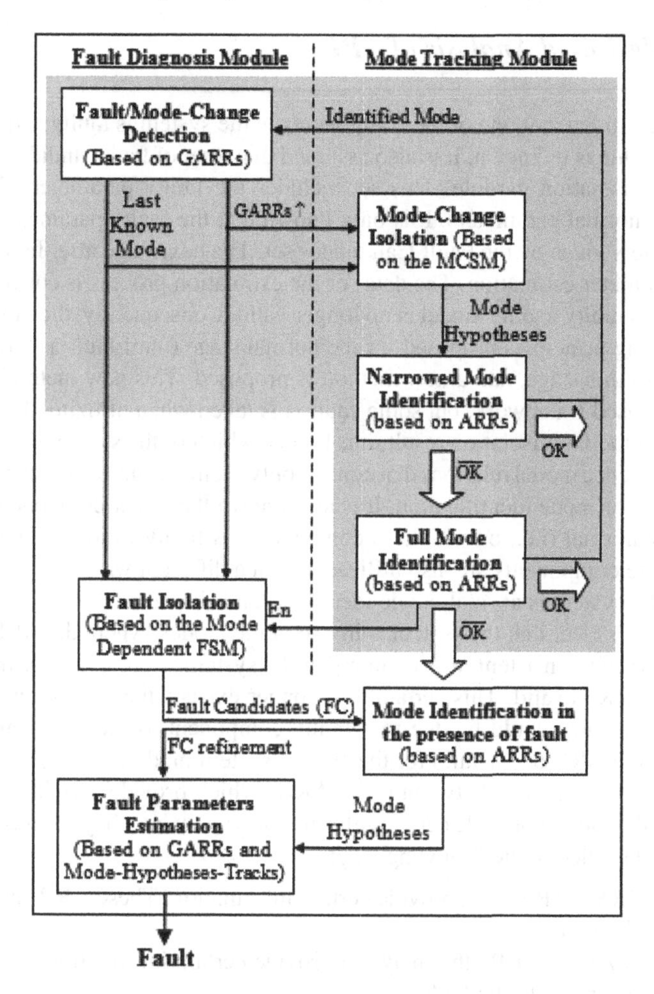

Fig. 5.16 HBG based quantitative health monitoring framework

Systematic methods which generate the residual sensitivity to parametric changes are proposed in [3]. These are required for the numerical solution of the ARRs/GARRs least-square optimization problem.

Under parametric fault conditions, the MCSM-based mode tracking method is not useful, and new strategy is required. This new strategy is the Mode-Identification in the presence of fault, and it is presented in the following sections. The new mode identification method has the potential to refine the fault candidates set (given by FC refinement, on the chart), and the Fault Parameters Estimation is carried out only on those faults given by the refined set.

5.2.1 Rule-Based Analysis of ARRs

Here, it is assumed that one of the components in the system is faulty and the faulty parameter value is unknown. It is also assumed that a set of fault candidates is given by the fault isolation module. This set includes the faulty parameter, along with some other normal parameters. It is only known that the faulty parameter is one of the parameters given by the Fault-Candidates set. The next stage after fault isolation is fault parameter estimation. The data for the estimation process is collected while the system is faulty and its model is no longer valid. Consequently, the model-based mode tracking principles proposed for the normal stage (until fault occurrence) are not useful in this stage, and a new method is proposed. This new method is still in the model-based framework, but some caution is taken when utilizing the model.

Based on the fault isolation result, it is known which of the system's components are normal; hence model relations that contain only normal components can be trusted and utilized for mode identification. It is also known that other components are not necessarily normal (i.e., one of these components is faulty); model equations that contain these components are also utilized, but in a different way.

When the system is normal, mode identification is based on ARRs. If an ARR is consistent, it is clear that the system is in one of the modes, where the ARR is valid. If an ARR is not consistent, it is certain that the system is not in any of the modes, where the ARR is valid. This certainty no longer exists when the system is faulty. ARR Inconsistency could be due to a faulty component, or due to a mismatched mode (i.e., the ARR is not valid for the current system mode).

In order to distinguish between those ARRs which provide certain information and others that do not provide certain information (e.g., when they are inconsistent), ARRs are classified in the following way:

1. *Certain ARRs*: ARRs that provide certain information. These ARRs are not sensitive to the fault.
2. *Uncertain ARRs*: ARRs that may not provide certain information. These ARRs could be sensitive to the fault.

A simple mode identification approach of hybrid system in the presence of fault is to utilize only Certain ARRs. This approach assumes that no useful information is provided by Uncertain ARRs. This approach is wrong, because Uncertain ARRs are not necessarily incorrect. The Fault-Candidate set may include normal components due to a structurally non-isolable fault (Note that non-isolatable faults may be isolated by using a bank of fault models and transient information [4]; however, this study is out of the scope of this study). These normal components can cause ARRs with only normal parameters to be classified as uncertain. If these Uncertain ARRs are not consistent with the monitored system, no useful information is derived. However if these ARRs are found consistent, then useful information is achievable. Hence, the proposed algorithm considers Certain and Uncertain ARRs.

Certain ARRs are more important for mode identification and, in general, provide more useful information, e.g., a Certain ARR always provides some useful information for mode identification, even if is found inconsistent, whereas Uncertain ARR provides useful information only if found consistent. The ARR set includes all different ARRs, which are derived from the GARRs, to describe system's modes. It is clear that better mode identification, i.e., less mode hypotheses, is achieved if more ARRs are certain. The following definitions are presented:

Definition 5.8 A set of ARRs is said to be *comprehensive* if for any possible mode at least one ARR is valid (under normal conditions).

Definition 5.9 A GARR is said to be *comprehensive* if for any possible mode the GARR is not null (note that a comprehensive GARR generates a comprehensive set of ARRs).

Definition 5.10 A set of ARRs is said to be *all-fault certain* if for any possible fault the subset of certain-ARRs is not null.

Definition 5.11 A set of ARRs is said to be *all-fault all-mode certain* if for any possible fault the subset of certain-ARRs is comprehensive.

An *all-fault all-mode certain* set of ARRs is attractive because it guarantees that under any conditions of fault and mode at least one ARR is valid, and can be found consistent by proper evaluation. The set of ARRs is derived from the set of GARRs; a sufficient condition on the set of GARRs is given, to yield an *all-fault all-mode certain* set of ARRs.

Property 5.3 A set of comprehensive GARRs yields an *all-fault all-mode certain* set of ARRs if any component, which is not fault-free, is not found in at least one GARR.

Proof: Under the condition of Property 5.3 and a single fault assumption, in the case of any possible fault, at least one GARR is valid (and can be found consistent by proper evaluation). If a GARR is consistent, then none of the components in the GARR expression is a fault candidate. Any comprehensive GARR generates a comprehensive set of ARRs. Any consistent GARR generates a set of only Certain ARRs. Hence, if for any possible fault at least one GARR is comprehensive and consistent, then the set of GARRs generates an *all-fault all-mode certain* set of ARRs.

A set of GARRs that yields an *all-fault all-mode certain* set of ARRs is preferred. However, the set of independent GARRs is fixed by the number of sensors. For example, for the two-tank system, two level sensors are given, and these sensors generate only two independent GARRs, i.e., $GARR_1$ and $GARR_2$. For mode tracking under normal conditions and FDI, dependent GARRs may not be useful. However, for mode identification in the presence of fault, dependent GARRs can help, e.g., in order to satisfy Property 5.3. Hence, existing GARRs are manipulated to generate more (dependent) GARRs, such that Property 5.3 is achieved.

Observing the two-tank example, the only component that is not fault-free and appears in all independent GARRs is R_3. In order to satisfy Property 5.3, a dependent

GARR is generated, in which the component R_3 is not found. From the two GARRs in (5.20) and (5.21):

$$GARR_3 = q_{in} - a_1 g(R_1, h_1) - g(R_4, h_1) - a_2 g(R_2, h_2) \\ - g(R_5, h_2) - A_2 \dot{h}_2 - A_1 \dot{h}_1 = 0 \quad (5.22)$$

The global relation GARR_3 is dependent, and under the single fault assumption it is not useful for fault isolation, in the FSM sense (i.e., fault isolation ability is not improved). Hence GARR_3 is not utilized for FDI, or mode tracking in normal conditions. The role of GARR_3 begins when mode identification in the presence of fault is required.

Each GARR of the two-tank system generates eight ARRs (as the number of modes), but not all of these ARRs are unique. For the two-tank example, 12 different ARRs are given by the three GARRs. All these ARRs are grouped into a set called the ARR-Set, and each ARR is attached with two codes, which represent its relevant information to the mode identification process. The ARR-codes facilitate online implementation of the algorithm, and are defined as follows:

1. *Mode code*: a binary code that represents the modes where the ARR is valid, under normal conditions.
2. *Component code*: a binary code that represents the faults to which the ARR is sensitive.

The ARR-Set of the two-tank system and the ARR codes are given in Table 5.8. The codes structure is presented in the first row of the table. A term '1' in the mode code of an ARR means the ARR is valid (under normal conditions) in the corresponding mode. Likewise, a term '1' in the component code indicates that the component appears in the ARR expression; only potential parametric faults take part in the code structure.

One more general code is defined; this code represents fault candidates. The structure of the fault candidates code is identical to the structure of the ARR component code and its initial value, after fault occurrence, is the fault isolation result. The fault candidates code could be refined by the proposed mode identification method.

In addition to ARR codes, each ARR is associated with two binary flags. The value of the flags is determined online, based on residuals evaluation. The ARR-certainty flag indicates whether the ARR is certain or not; the flag value is determined from the information given by two codes: (1) the fault candidates code and (2) the ARR component code. ARR-consistency flag indicates whether or not the ARR is consistent with the actual behavior of the monitored system; this flag value is determined online, based on a simple comparison of residuals to a threshold:

$$\text{ARR}_\text{i} \text{ consistency} = \begin{cases} 1 & |r_i(t)| < \text{threshold} \\ 0 & |r_i(t)| \geq \text{threshold} \end{cases} \quad (5.23)$$

Table 5.8 ARR-CODE TABLE

ARR No.	ARR	Mode-code [0 1 2 3 4 5 6 7]	Component-code-[q_{in} R_5 R_4 R_3 R_2 R_1]
1	$q_{in} - g(R_4, h_1) - A_1\dot{h}_1 = 0$	[1 0 1 0 0 0 0 0]	[1 0 1 0 0 0]
2	$q_{in} - g(R_1, h_1) - g(R_4, h_1) - A_1\dot{h}_1 = 0$	[0 1 0 1 0 0 0 0]	[1 0 1 0 0 1]
3	$q_{in} - g(R_4, h_1) - g(R_3, h_1 - h_2) - A_1\dot{h}_1 = 0$	[0 0 0 0 1 0 1 0]	[1 0 1 1 0 0]
4	$q_{in} - g(R_1, h_1) - g(R_4, h_1) - g(R_3, h_1 - h_2) - A_1\dot{h}_1 = 0$	[0 0 0 0 0 1 0 1]	[1 0 1 1 0 1]
5	$-g(R_5, h_1 - h_2) - A_2\dot{h}_2 = 0$	[1 1 0 0 0 0 0 0]	[0 1 0 0 0 0]
6	$-g(R_2, h_2) - g(R_5, h_2) - A_2\dot{h}_2 = 0$	[0 0 1 1 0 0 0 0]	[0 1 0 0 1 0]
7	$g(R_3, h_1 - h_2) - g(R_5, h_2) - A_2\dot{h}_2 = 0$	[0 0 0 0 1 1 0 0]	[0 1 0 1 0 0]
8	$g(R_3, h_1 - h_2) - g(R_2, h_2) - g(R_5, h_2) - A_2\dot{h}_2 = 0$	[0 0 0 0 0 0 1 1]	[0 1 0 1 1 0]
9	$q_{in} - g(R_4, h_1) - A_1\dot{h}_1 - g(R_5, h_2) - A_2\dot{h}_2 = 0$	[1 0 0 0 1 0 0 0]	[1 1 1 0 0 0]
10	$q_{in} - g(R_1, h_1) - g(R_4, h_1) - A_1\dot{h}_1 - g(R_5, h_2) - A_2\dot{h}_2 = 0$	[0 1 0 0 0 1 0 0]	[1 1 1 0 0 1]
11	$q_{in} - g(R_4, h_1) - A_1\dot{h}_1 - g(R_2, h_2) - g(R_5, h_2) - A_2\dot{h}_2 = 0$	[0 0 1 0 0 0 1 0]	[1 1 1 0 1 0]
12	$q_{in} - g(R_1, h_1) - g(R_4, h_1) - A_1\dot{h}_1 - g(R_2, h_2) - g(R_5, h_2) - A_2\dot{h}_2 = 0$	[0 0 0 1 0 0 0 1]	[1 1 1 0 1 1]

where $r_i(t)$ is the residual associated with ARR$_\text{i}$. More sophisticated methods for consistency-test can be utilized for real implementation which will be discussed later.

A consistent ARR is a powerful source of information. A consistent ARR indicates that all modes, given by its mode code, are possible modes, and all components given by its component code are normal components. Possible modes and normal components are represented by codes. If an ARR is consistent, then its possible modes code is equal to its mode code; this rule is independent of the ARR certainty. A possible modes code indicates that one of the modes represented by the code is definitely the actual mode of the monitored system. For normal components code generation, the method distinguishes between two ARR types, namely: (1) consistent-certain and (2) consistent-uncertain. For a consistent-uncertain ARR, its normal components code is equal to its component code. A normal components code indicates that all the components given by the code are definitely normal. This information is new, in the sense that it was unknown when the ARR was classified as uncertain. This information is utilized for fault candidates refinement. After the refinement, the status of consistent-uncertain ARRs is immediately changed to consistent-certain by the algorithm. For consistent-certain ARRs, all the components given in the ARR expression are definitely normal, but this information is known and represented by the ARR certainty. Hence, consistent-certain ARRs do not contribute new information for fault candidates refinement.

An inconsistent ARR yields weaker information than does a consistent ARR. If an ARR is inconsistent, it does not necessarily an indication of impossible modes.

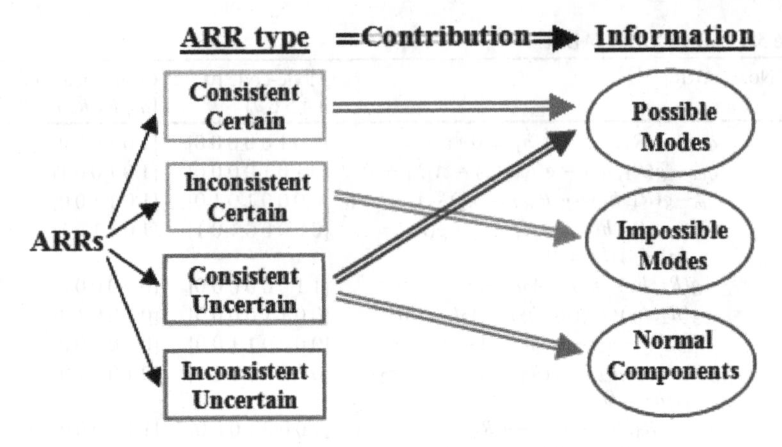

Fig. 5.17 ARR contribution to mode identification and fault candidates refinement

The ARR could be inconsistent due to the fault in the monitored system. Hence, the method distinguishes between two types of inconsistent ARRs, namely: (1) inconsistent-certain and (2) inconsistent-uncertain. For an inconsistent-certain ARR, it is known that all its components are normal; therefore, its inconsistency must be due to a mismatched mode (i.e., the ARR is not valid for the actual mode of the monitored system). For this inconsistent-certain ARR, its mode code indicates impossible modes, i.e., modes that are definitely not the actual mode of the monitored system. Inconsistent-uncertain ARRs do not contribute any information for mode identification or for fault candidates refinement; the reason for their inconsistency is unknown. The chart in Fig. 5.17 presents the four ARR types along with their contributions in the proposed method.

The method for mode identification and fault candidates refinement is based on nine basic rules. The first four concern individual ARRs and are given as follows:

1. If a certain ARR is consistent, then all its modes are possible modes.
2. If a certain ARR is inconsistent, then all its modes are impossible modes.
3. If an uncertain ARR is consistent, then all its modes are possible modes, and all its components are normal components.
4. If an ARR is consistent, then one of its modes is definitely the actual mode of the monitored system.

Individual ARRs generate information that is given by three codes, namely: (1) possible modes; (2) impossible modes; and (3) normal components. This information is merged into a unified result based on the following logic.

5. All possible modes codes from all ARRs are intersected to a unified possible modes code. This intersection is based on a logical AND.
6. All impossible modes codes from all ARRs are united to a unified impossible modes code. This union is based on a logical OR.

7. Modes given by the unified impossible modes code are removed from the modes given by the unified possible modes code. The result is a set of mode hypotheses represented by the mode hypotheses code.
8. All normal components codes from all ARRs are united to a unified normal components code. This union is based on a logical OR.
9. All normal components given by the unified normal components code are removed from the fault candidates code. This step is the fault candidates refinement.

The set of mode hypotheses given by rule (7) is the product of the proposed mode identification method. From rules (8) and (9), it is clear that under certain conditions, the fault candidates set is refined. This, of course, improves mode identification in the next running steps (because more ARRs become certain). The concept of individual ARR codes is summarized in Table 5.9. (Note that the possible modes code [1 1 ⋯ 1] has no influence on the result as a possible modes merger is based on logical AND; likewise, an impossible modes and normal components merger is based on logical OR and is not influenced by the code [0 0 ⋯ 0]).

Few simple examples, with regard to the two-tank system, are presented. The goal here is to illustrate the method logic. Therefore, a reliable consistency test is assumed (i.e., any valid ARR is found consistent and any invalid ARR is found inconsistent).

Case 1: The actual mode of the system is mode 6, and the faulty component is R_2; the fault information is unknown. From the GFSM in mode 6, the fault isolation result is $\{R_5, R_2\}$. All contributing ARRs along with their status and contribution are given in Table 5.10.

From the unified results, the following two possible modes are identified: (1) mode 4 and (2) mode 6. After removing impossible modes, this result remains unchanged; hence, the two modes (4 and 6) are the mode hypotheses. No normal components are identified, which can refine the fault candidates set.

Table 5.9 Information given by individual ARRS

ARR_i-certainty	ARR_i-consistency	ARR_i-possible-modes	ARR_i-impossible-modes	ARR_i-normal-components
0	0	[1 1 ⋯ 1]	[0 0 ⋯ 0]	[0 0 ⋯ 0]
0	1	ARR_i-mode-code	[0 0 ⋯ 0]	ARR_i-component-code
1	0	[1 1 ⋯ 1]	ARR_i-mode-code	[0 0 ⋯ 0]
1	1	ARR_i-mode-code	[0 0 ⋯ 0]	[0 0 ⋯ 0]

Table 5.10 Contributing ARRs-case 1

ARR no.	Status	Possible-modes [0 1 2 3 4 5 6 7]	Impossible-modes [0 1 2 3 4 5 6 7]	Normal-components [q_{in} R_5 R_4 R_3 R_2 R_1]
1	Certain inconsistent	[1 1 1 1 1 1 1 1]	[1 0 1 0 0 0 0 0]	[0 0 0 0 0 0]
2	Certain inconsistent	[1 1 1 1 1 1 1 1]	[0 1 0 0 0 0 0 0]	[0 0 0 0 0 0]
3	Certain consistent	[0 0 0 0 1 0 1 0]	[0 0 0 0 0 0 0 0]	[0 0 0 0 0 0]
4	Certain inconsistent	[1 1 1 1 1 1 1 1]	[0 0 0 0 0 1 0 1]	[0 0 0 0 0 0]
Unified results		[0 0 0 0 1 0 1 0]	[1 1 1 1 0 1 0 1]	[0 0 0 0 0 0]

Case 2: The same fault as in case 1 (i.e., R_2) is considered. The fault has been detected and isolated in mode 6; hence, the fault isolation result is $\{R_5, R_2\}$, and the fault candidates code is [0 1 0 0 1 0]. After the mode has changed, the actual mode of the monitored system is mode 4 (i.e., the faulty valve R_2 is now closed); the mode information is unknown. All contributing ARRs are given in the Table 5.11.

Now, a single possible mode is identified by the unified possible modes; this mode is mode 4. Considering also impossible modes, the mode hypothesis is mode 4. In case 2, few components are found normal by uncertain-consistent ARRs. After removing these normal components, the refined fault candidates code is [0 0 0 0 1 0], which means that R_2 is the faulty component. The fault candidates refinement improves mode identification of subsequent modes.

Case 3: In this case, the actual mode of the monitored system is mode 6, and the faulty component is R_3. From the GFSM in mode 6, the fault isolation result is R_3. All contributing ARRs are presented in Table 5.12.

From the unified results, the two possible modes are modes 2 and 6. After removing impossible modes, mode 6 is identified.

Table 5.11 Contributing ARRs-case 2

ARR no.	Status	Possible-modes [0 1 2 3 4 5 6 7]	Impossible-modes [0 1 2 3 4 5 6 7]	Normal-components [q_{in} R_5 R_4 R_3 R_2 R_1]
1	Certain inconsistent	[1 1 1 1 1 1 1 1]	[1 0 1 0 0 0 0 0]	[0 0 0 0 0 0]
2	Certain inconsistent	[1 1 1 1 1 1 1 1]	[0 1 0 0 0 0 0 0]	[0 0 0 0 0 0]
3	Certain consistent	[0 0 0 0 1 0 1 0]	[0 0 0 0 0 0 0 0]	[0 0 0 0 0 0]
4	Certain inconsistent	[1 1 1 1 1 1 1 1]	[0 0 0 0 0 1 0 1]	[0 0 0 0 0 0]
7	Uncertain consistent	[0 0 0 0 1 1 0 0]	[0 0 0 0 0 0 0 0]	[0 1 0 1 0 0]
9	Uncertain consistent	[1 0 0 0 1 0 0 0]	[0 0 0 0 0 0 0 0]	[1 1 1 0 0 0]
	Unified results	[0 0 0 0 1 0 0 0]	[1 1 1 1 0 1 0 1]	[1 1 1 1 0 0]

Table 5.12 Contributing ARRs-case 3

ARR no.	Status	Possible-modes [0 1 2 3 4 5 6 7]	Impossible-modes [0 1 2 3 4 5 6 7]	Normal-components [q_{in} R_5 R_4 R_3 R_2 R_1]
1	Certain inconsistent	[1 1 1 1 1 1 1 1]	[1 0 1 0 0 0 0 0]	[0 0 0 0 0 0]
2	Certain inconsistent	[1 1 1 1 1 1 1 1]	[0 1 0 1 0 0 0 0]	[0 0 0 0 0 0]
5	Certain inconsistent	[1 1 1 1 1 1 1 1]	[1 1 0 0 0 0 0 0]	[0 0 0 0 0 0]
6	Certain inconsistent	[1 1 1 1 1 1 1 1]	[0 0 1 1 0 0 0 0]	[0 0 0 0 0 0]
9	Certain inconsistent	[1 1 1 1 1 1 1 1]	[1 0 0 0 1 0 0 0]	[0 0 0 0 0 0]
10	Certain inconsistent	[1 1 1 1 1 1 1 1]	[0 1 0 0 0 1 0 0]	[0 0 0 0 0 0]
11	Certain consistent	[0 0 1 0 0 0 1 0]	[0 0 0 0 0 0 0 0]	[0 0 0 0 0 0]
12	Certain inconsistent	[1 1 1 1 1 1 1 1]	[0 0 0 1 0 0 0 1]	[0 0 0 0 0 0]
	Unified results	[0 0 1 0 0 0 1 0]	[1 1 1 1 1 1 0 1]	[0 0 0 0 0 0]

5.2.2 Implementation Schemes and Algorithms

The proposed method is based on nine basic rules [as given by rules (1)–(9)]. These rules are implemented by logic gates, and the implementation details are given in this section. The ARR-consistency flag principle is given in (5.23). The ARR-certainty flag is determined online by the logic scheme in Fig. 5.18.

In this logic diagram (and in all others presented in this section), bold lines stand for vector lines, and thin lines represent scalar variables. Logic operation between vectors must have inputs of the same length, and the operation is carried out in an elements-wise manner (i.e., the output is a vector of the same size of the inputs). When logic operation is carried out on a single input vector (e.g., the NOR gate in Fig. 5.18), the logic operation is carried out between all of the vector elements, and the output is a scalar. When logic operation is carried out between a vector and a scalar, the same logic operation is carried out between the scalar and each one of the vector elements (i.e., the output in this case is a vector). Thus, the ARR-certainty flag in Fig. 5.18 is equal to '1' only if the fault candidates code and the ARR-Components code have no elements in common.

For each ARR, the possible modes, impossible modes, and normal components codes are determined online based on Table 5.9. Using logic gates, this concept is implemented by the scheme presented in Fig. 5.19.

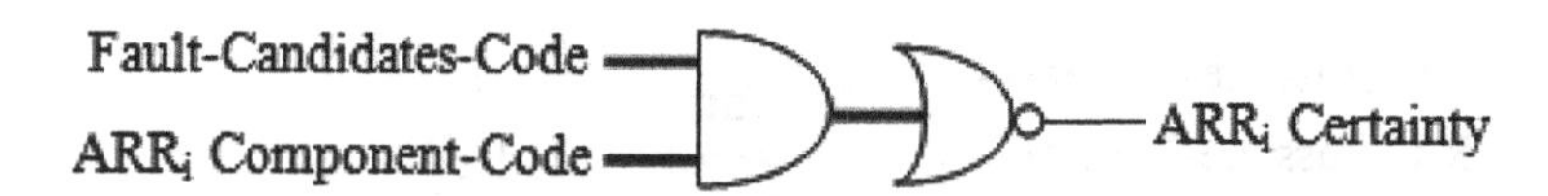

Fig. 5.18 Logic scheme ARR$_i$-certainty

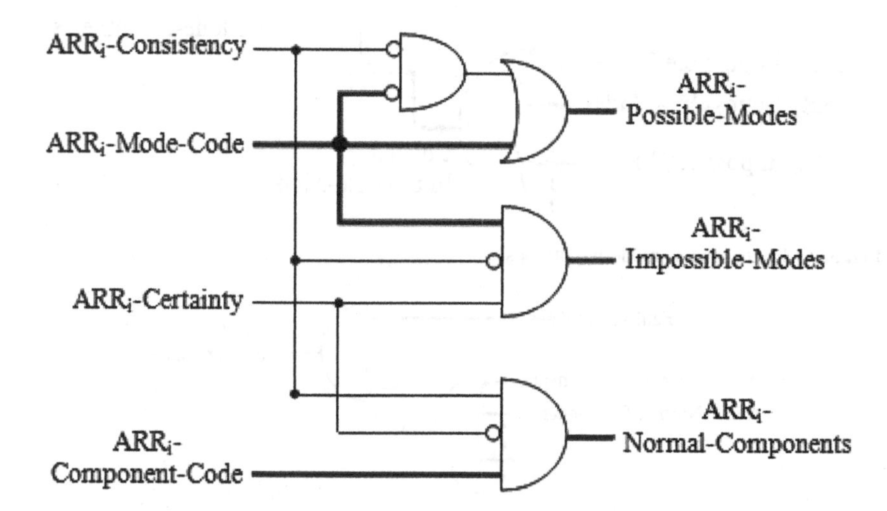

Fig. 5.19 Logic scheme the information given by ARR$_i$

The next step, after collecting the information from all ARRs, is information merger. This step is based on rules 5.-9. and yields the following two products: 1. mode hypotheses and 2. refined fault candidates. Rules 5.-7. are implemented by the logic scheme presented in Fig. 5.20, and rules 8. and 9. are carried out by the scheme in Fig. 5.21.

The refined fault candidates code replaces the fault candidates code in an iterative manner, i.e.,

$$\text{Fault\_Candidates}_{k+1} = \text{Refined\_Fault\_Candidates}_k \tag{5.24}$$

The iterations are carried out periodically, or at each time the fault candidates code is actually refined.

For large-scale systems with many sensors and modes, the size of the ARR set can be very large. For these systems, the parallel merger, as presented in Figs. 5.20 and 5.21, may demand large computational effort. Hence, in addition to the parallel merger given previously, an algorithm for an iterative merger is proposed. At first, the following abbreviations are defined:

Ce_i ARR_i-certainty;
Co_i ARR_i-consistency;
MC_i ARR_i-mode code;

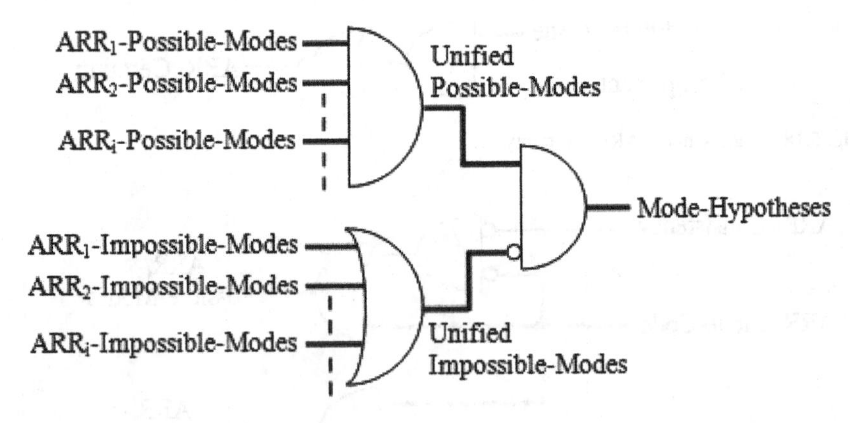

Fig. 5.20 Logic scheme mode-hypotheses

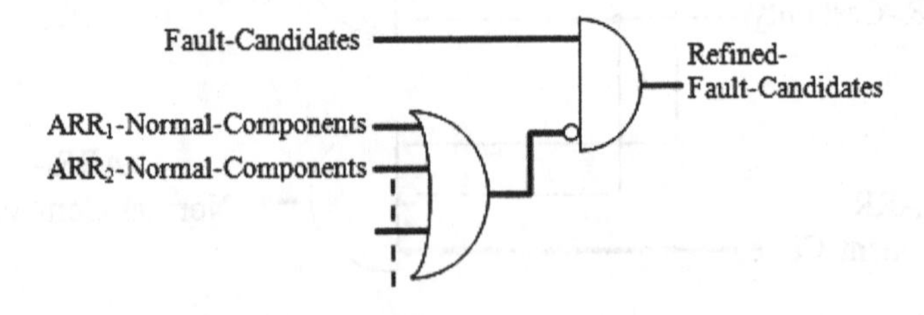

Fig. 5.21 Logic scheme refined fault-candidates

CC_i ARR_i-components-code;
PM_i ARR_i-possible modes code;
IM_i ARR_i-impossible modes code;
NC_i ARR_i-normal components code;
FC fault candidates code;
MH mode hypotheses code;
r_i residual that is associated to ARR_i;
N_{ARR} number of ARRs in the ARR set.

The iterative merger algorithm (pseudo-code) is given as follows

```
For i=1 to N_ARR
    r_i=evaluate(ARR_i)
    Co_i=|r_i| <Threshold
    Ce_i=/OR(AND(FC,CC_i))
    PM_i=OR(AND(/Co_i,/MC_i),MC_i)
    IM_i=AND(MC_i,/Co_i,Ce_i)
    NC_i=AND(Co_i,/Ce_i,CC_i)
    MH = AND(MH,PM_i,/IM_i)
    FC = AND(FC,/NC_i)
End
```

where the slash mark indicates logic complement.

The iterative algorithm loop is running online to process information received from individual ARRs. At each iteration, only one ARR is processed, and the following two products are updated: (1) MH and (2) FC. If the mode is consistent during N iterations (i.e., all ARRs are processed), the result of the iterative algorithm and the result of the parallel merger are identical.

The time response of the iterative algorithm to a mode change is longer (compared to the parallel algorithm); however, the computational effort needed at any iteration is lower. Another limitation is the algorithm convergence, which is promised only if all ARRs are processed within a single persistent mode ('too quick' mode changes could be missed).

An *all-mode all-fault certain* set of ARRs guarantees that under any condition of fault and mode, at least one ARR is valid and can be found consistent under a reliable consistency test. A consistent ARR contributes information of possible modes. On the other hand, an inconsistent ARR contributes information of impossible modes if the ARR is certain. It is clear that information of possible modes is stronger and more important than information of impossible modes. For example, if the monitored system consists of eight modes, and each ARR is valid for two modes (as the case of the two-tank system), then a single consistent ARR indicates two possible modes; whereas a single inconsistent-certain ARR contributes information of two impossible modes, which is equivalent to information of six possible modes. If N is the maximum number of valid modes for a single ARR (for the two-tank system, $N = 2$), then an all-mode all-fault certain set of ARRs guarantees that the number of mode hypotheses is never greater than N.

An *all-fault all-mode certain* set of ARRs is preferred, and it is achieved by $GARR_3$. If for a system an *all-fault all-mode* certain set of ARRs is unachievable, then the method is still feasible without any modifications. However, the result could be less precise, i.e., a larger number of mode hypotheses.

An *all-fault certain* set of ARRs guarantees that under any condition of fault and mode, at least one ARR is certain and contributes some information for the mode identification process. If the certain ARR is consistent, it contributes strong information of possible modes, and if the ARR is inconsistent, weaker information of impossible modes is achieved.

If the set of ARRs is *all-fault certain* (but is not *all-fault all-mode certain*), then under certain conditions of fault and mode, none of the ARRs is valid, and all ARRs are inconsistent. Under these conditions, the mode hypotheses set will consist of all modes that are not found impossible by certain-inconsistent ARRs. If n is the minimum number of valid modes for a single ARR (for the two-tank system, $n = 2$) and M is the number of systems modes, then an all-fault certain set of ARRs guarantees that the number of mode hypotheses is never greater than $M - n$. The two-tank system example with only two independent GARRs (i.e., $GARR_1$ and $GARR_2$) is presented to illustrate these insights.

Case 4: Consider the conditions of case 3, i.e., the actual mode is mode 6, the faulty component is R3, and the fault candidates code is [0 0 0 1 0 0]. If only $GARR_1$ and $GARR_2$ are considered, then none of the ARRs is consistent. On the other hand, four ARRs are certain and presented in Table 5.13.

Based on the unified results given in Table 5.13, the mode hypotheses code is [0 0 0 0 1 1 1 1]; this result includes all modes, which are not found impossible. This result has three more mode hypotheses than the result of case 3 and demonstrates the contribution of $GARR_3$ to the mode identification process.

An uninformative mode hypotheses set is defined as a mode hypotheses set of all modes. The uninformative mode hypotheses code is a vector full of "1" elements and provides no information. This result is unique to the case where all ARRs are uncertain and inconsistent. Under the single fault assumption, a reliable consistency test, and an *all-fault certain* set of ARRs, this result is impossible. If the set of ARRs is not *all-fault certain*, then an uninformative mode hypotheses set is possible.

Table 5.13 ARR code table

ARR no.	Status	Possible-modes [0 1 2 3 4 5 6 7]	Impossible-modes [0 1 2 3 4 5 6 7]	Normal-components [q_{in} R_5 R_4 R_3 R_2 R_1]
1	Certain inconsistent	[1 1 1 1 1 1 1 1]	[1 0 1 0 0 0 0 0]	[0 0 0 0 0 0]
2	Certain inconsistent	[1 1 1 1 1 1 1 1]	[0 1 0 1 0 0 0 0]	[0 0 0 0 0 0]
5	Certain inconsistent	[1 1 1 1 1 1 1 1]	[1 1 0 0 0 0 0 0]	[0 0 0 0 0 0]
6	Certain inconsistent	[1 1 1 1 1 1 1 1]	[0 0 1 1 0 0 0 0]	[0 0 0 0 0 0]
	Unified results	[1 1 1 1 1 1 1 1]	[1 1 1 1 0 0 0 0]	[0 0 0 0 0 0]

An empty mode hypotheses result is a vector full of zeros. Under the single fault assumption and a reliable consistency test, this result is impossible (even if the set of ARRs is not *all-fault certain*).

These constraints, on the scope of the mode hypotheses result (e.g., minimum and maximum of mode hypotheses), are important. Deviations are utilized to identify situations where faulty information is injected to the process, e.g., by an unreliable consistency test.

5.2.3 From Theory to Implementation

The theory presented in the previous section assumes a reliable consistency test. Any valid ARR is found consistent, and any invalid ARR is found inconsistent. In practice, such reliability is not trivial, or even impossible.

From a practical point of view, it is reasonable to assume that any valid ARR is found consistent. Difficulties here could be due to noise, disturbances, and model-plant mismatch. However, it is always possible to improve model accuracy, or to increase the threshold. Increased threshold may cause invalid ARRs to be found consistent; this problem is even more severe if the system is near steady state. Hence, for practical implementation, the assumption of reliable consistency test is removed. Instead, any invalid ARR (due to a mismatched mode or a fault) is assumed inconsistent. On the other hand, consistent ARRs are not necessarily valid. If an invalid ARR is found consistent, then this ARR is named a misleading ARR. A misleading ARR could have a destructive influence on the mode hypotheses result. In general, misleading ARRs are minor (i.e., exceptions), and for most ARRs, a reliable consistency test should be achieved. However, based on the theory presented in the previous section, and due to the merger of possible-modes (that is based on intersection), even a single misleading ARR may cause a completely incorrect result. Few examples are discussed.

Considering tank A_2 in mode 6, the faulty component is R_2 or R_3 (these conditions are consistent with cases 1 and 3 in previous section). If the difference between the inflow and outflow of tank A_2 is smaller than the threshold (i.e., $|q_{R_3} - q_{R_2}| <$ threshold), then from the system dynamical model, it is clear that $|A_2\dot{h}_2| <$ threshold. In this case, ARR$_5$ will be found consistent, and since its mode code is [1 1 0 0 0 0 0 0] (i.e., it is not valid for mode 6), ARR$_5$ is misleading. The same argument is also relevant to case 2 (where the actual mode is mode 4) if the two state variables $h_1(t)$ and $h_2(t)$ are very close to each other. In this condition, the inflow to tank A_2 is very low (i.e., $|q_{R_3}| <$ threshold) and ARR$_5$ is, again, misleading. Considering a misleading ARR$_5$ in Tables 5.10–5.12 (i.e., cases 1–3), the mode hypotheses result is empty (i.e., [0 0 0 0 0 0 0 0]). These simple examples show that the possible mode codes intersection, which is based on pure logic, is highly sensitive to misleading ARRs.

The following three solutions are proposed to deal with misleading ARRs:

1. ARR-reliability signal;

2. mode hypotheses latch;
3. algebraic merger.

The first solution utilizes an ARR-reliability signal. Here, it is proposed to identify, from the beginning, situations where ARRs may become misleading (i.e., although invalid, the absolute value of the residual will be low due to system dynamics). Around these states, the ARR-reliability signal is “0” (otherwise, the reliability signal is “1”).

ARR-reliability signal is considered in the merger process. The idea is to prevent the destructive influence of misleading ARRs. If the ARR-reliability signal is “0” and the ARR is found consistent, then the ARR is out of the game, and its consistency has no influence on the unified possible modes (and on unified normal components). If the ARR-reliability signal is “0” (i.e., unreliable) and the ARR is certain but inconsistent, then the information provided by the ARR is taken into account. In short, unreliable ARRs may contribute information of impossible modes but cannot contribute any information of possible modes. The extended logic is presented in Table 5.14, where the sign ϕ stands for the logic state “Don't-care” (either “0” or “1”). For reliable ARRs, Table 5.14 is identical to Table 5.9.

The second solution is more ad hoc but effective. Here, the problematic situation is identified online by the mode hypotheses result. If the mode hypotheses result becomes empty, it indicates that wrong information has been considered for the merger, e.g., information given by misleading ARRs. If the mode hypotheses result becomes empty, then the last reliable result (not empty) is locked until a new reliable result is achieved. When the mode hypotheses result is locked, so is the unified normal components. This way, misleading ARRs that cause empty result have no influence.

The third solution is based on an algebraic merger instead of the logic merger. Each ARR gives its vote to an algebraic unified result according to the scheme presented in Fig. 5.22.

The algebraic unified possible modes result is an algebraic sum of all possible modes codes from all ARRs. Similarly, the algebraic unified impossible modes result is an algebraic sum of all impossible modes codes from all ARRs. Then, the algebraic unified impossible modes are subtracted from the unified possible modes to give the algebraic merger result. This result contains the votes that each mode has achieved based on the information provided by online analysis of ARRs. The mode hypotheses are the modes that achieve the highest score; this principle is given by

Table 5.14 Information given by individual ARRS

ARR_i-reliability	ARR_i-certainty	ARR_i-consistency	ARR_i-possible-modes	ARR_i-impossible-modes	ARR_i-normal-components
0	0	0	$[1\,1 \cdots 1]$	$[0\,0 \cdots 0]$	$[0\,0 \cdots 0]$
0	ϕ	1	$[1\,1 \cdots 1]$	$[0\,0 \cdots 0]$	$[0\,0 \cdots 0]$
0	1	0	$[1\,1 \cdots 1]$	ARR_i-mode-code	$[0\,0 \cdots 0]$
1			Similar to Table 5.9		

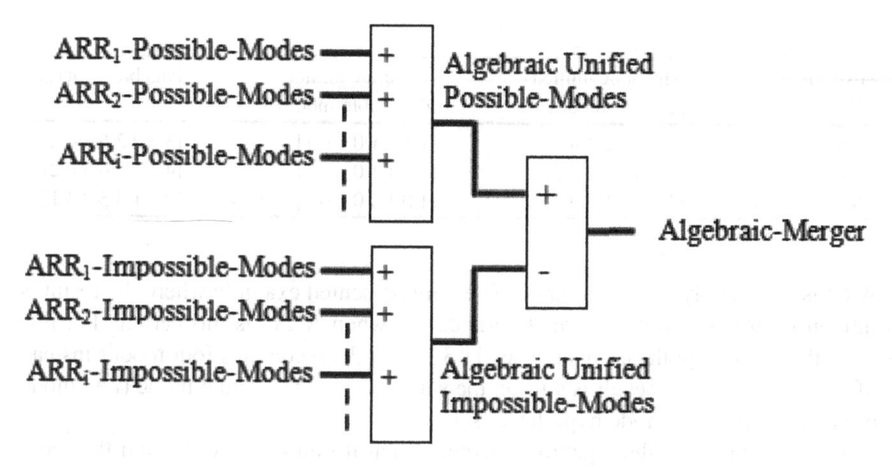

Fig. 5.22 Algebraic merger

$$\text{Mode\_Hypotheses} = \text{Max}(\text{Algebraic\_Merger}) \tag{5.25}$$

This way, the majority rules, and if the misleading ARRs are minor, the result is not damaged. It is easy to show that in terms of mode hypotheses, the result of the algebraic merger is always identical to the result of the logic merger if no misleading ARRs are presented. One problem of the algebraic merger is that some result is always achieved, even under very problematic conditions (an empty result is no longer an indication of a problem). On the other hand, a result of too many mode hypotheses can indicate a problem.

Few examples are presented to illustrate the algebraic merger and its ability of accommodating misleading ARRs. Consider, for example, cases 1 and 2 of previous section. If none of the ARRs is misleading, then the algebraic merger results are presented in the first lines of Tables 5.15 and 5.16 (codes of uncertain-inconsistent ARRs are not included in these results). It is clear that the mode hypotheses results are identical to the results of the logic merger (Tables 5.10 and 5.11). For example, the result of case 1 given in Table 5.15 is [2 2 2 2 4 2 4 2]; this result indicates that modes 4 and 6 are the mode hypotheses (in the tables, the maximum score is marked by an underline). Similarly, for case 2 in Table 5.16, the result is [3 2 2 2 6 3 4 2], which means that the mode hypothesis is mode 4. Even if ARR_2 or ARR_5 are misleading, it is clear from the two tables that the mode hypotheses result based on the algebraic merger is still correct, and in most cases, the misleading

Table 5.15 Algebraic merger-case 1

Misleading ARR	Algebraic unified possible-modes	Algebraic unified impossible-modes	Algebraic-merger
None	[3 3 3 3 4 3 4 3]	[1 1 1 1 0 1 0 1]	[2 2 2 2 4 2 4 2]
ARR_5	[4 4 3 3 4 3 4 3]	[1 1 1 1 0 1 0 1]	[3 3 2 2 4 2 4 2]
ARR_2	[2 3 2 3 3 2 3 2]	[1 0 1 0 0 1 0 1]	[1 3 1 3 3 1 3 1]

Table 5.16 Algebraic merger-case 2

Misleading ARR	Algebraic unified possible-modes	Algebraic unified impossible-modes	Algebraic-merger
None	[4 3 3 3 6 4 4 3]	[1 1 1 1 0 1 0 1]	[3 2 2 2 $\underline{6}$ 3 4 2]
ARR_5	[5 4 3 3 6 4 4 3]	[1 1 1 1 0 1 0 1]	[4 3 2 2 $\underline{6}$ 3 4 2]
ARR_2	[3 3 2 3 5 3 3 2]	[1 0 1 0 0 1 0 1]	[2 3 1 3 $\underline{5}$ 2 3 1]

ARR is completely accommodated. The only presented example where the result is influenced by the wrong information is case 1 when ARR_2 is misleading. For this case, the mode hypotheses result (i.e., [1 3 1 3 3 1 3 1]) contains four modes instead of two. However, the result is still on the safe side, and the actual mode (i.e., mode 6) is included in the mode hypotheses set.

One advantage of the algebraic merger is that it ranks all modes, and the score of each mode indicates something on the probability that this mode is the actual mode of the monitored system. The modes with the highest score are the most probable (and hence are chosen for the mode hypotheses result), and the modes with the lowest score are the least probable. For a large scale system (i.e., large number of modes and sensors), and if many misleading ARRs are expected, the mode hypotheses can be defined as all of those modes whose normalized score is higher than a certain threshold, and the normalized score is defined as the actual score divided by the score of the most probable mode (the threshold, in this case, is always smaller than one).

The consistency test described in (5.23) is not sufficient for practical implementation. If an ARR is valid, then the absolute value of its residual is lower than a threshold. The residual keeps a low value as long as the ARR is valid. Some residuals may have a low absolute value for a very short time, even though their corresponding ARR is not valid. One example is when the residual value crosses the zero line (i.e., its sign changes from positive to negative, and vice versa); another example is a short spike caused by noise or a disturbance. Therefore, in a more practical consistency test, an ARR is found consistent only if its absolute value is smaller than a threshold for at least a predefined period of time. Such decision-making mechanism is achieved by a *delayed-decision filter* (DDF), which is given in

$$Out(t) = \mathrm{DDF}(T_D, In(t)) \tag{5.26}$$

and defined in

$$\begin{aligned} \text{if} \quad & In(t) = In(t + T_D) \\ \text{then} \quad & Out(t) = In(t) \\ \text{otherwise} \quad & Out(t) = \text{unchanged} \end{aligned} \tag{5.27}$$

In practice, the consistency signal (5.23) is filtered by a DDF. One implication of the DDF is that any change is delayed by at least T_D seconds. Another implication is that very short changes (e.g., very short modes) are filtered. A mode is not identifiable if the monitored system does not stay in that mode for at least T_D seconds. Other methods, which cope with the problem of persistency of threshold violation, such as moving average, are discussed in [5] and [6].

5.2.4 Experimental Study

The signature-based mode tracking and the FDI module have been demonstrated in Sect. 5.1. These results in [1] presented a scenario of mode tracking under normal conditions and a single FDI. The next step is data collection for fault parameter estimation. These data are collected while the system is faulty, and its dynamical model is no longer valid. The fault parameter estimation requires information on the system mode evolution; this information comes from the method proposed in this section, namely, mode identification in the presence of fault. The experiments presented assume the system is faulty, and the fault isolation result is given. Based on the fault isolation result, an online evaluation of ARRs, and the methods presented in this section, a set of mode hypotheses is generated online. Even if the identified mode is not unique, this information is important and useful for the fault parameter estimation process.

The test-bed is the electric circuit presented in Fig. 5.23. The circuit dynamics, with little differences, is similar to the two-tank system dynamics. In the circuit, the two tanks are replaced by two electric capacitors, and the valves are represented by electric switches and resistors. The only differences to the DHBG are the two power variables that are now voltage and current (instead of liquid level and liquid flow) and the two voltage sensors v_1 and v_2 (instead of two level sensors h_1 and h_2). In addition, the modulated flow source $q_{in}(t)$ is replaced by a modulated voltage source $v_{in}(t)$ and resistor R. The signal $i_{in}(t)$ in the circuit is the analog of the signal $q_{in}(t)$ of the two-tank system; hence, $i_{in}(t)$ takes the place of the PI control signal (5.19) and is given as follows:

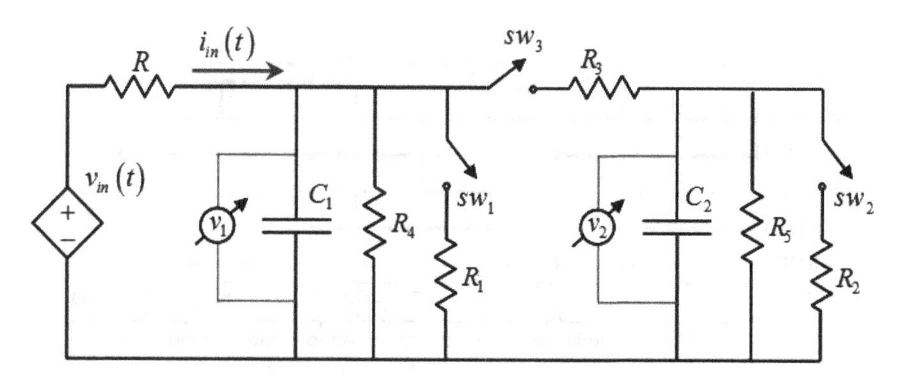

Fig. 5.23 Electric circuit with hybrid dynamics

$$i_{in}(t) = K_P(v_d - v_1(t)) + K_I \int_0^t (v_d - v_1(t))dt \tag{5.28}$$

Based on the circuit in Fig. 5.23 and the PI controller in (5.19), the modulated voltage source $v_{in}(t)$ is obtained in

$$v_{in}(t) = v_1(t) + R\, i_{in}(t) = v_1(t) + K_P\, R\, (v_d - v_1(t)) + K_I\, R \int_0^t (v_d - v_1(t))dt \tag{5.29}$$

One last difference between the circuit and the two-tank system is the constitutive relation of resistors, which is now linear and obeys Ohm's law.

The circuit in Fig. 5.23 has been chosen because of its similarity to the two-tank system. The mode of the circuit is given by the state of its three switches $[sw_3\ sw_2\ sw_1]$, and mode identification in the presence of fault is based on the ARR set given in Table 5.8 (with the differences mentioned above). The nominal physical parameters are given as follows: $R_1 = 80\,\text{k}\Omega$; $R_2 = 100\ \text{k}\Omega$; $R_3 = 30\,\text{k}\Omega$; $R_4 \to \infty$; $R_5 \to \infty$; $C_1 = 120\ \mu\text{F}$; $C_2 = 240\ \mu\text{F}$; $K_P = 1.5\,\text{S}$; $K_I = 0.01\ \text{S/s}$; and $R = 50\ \text{k}\Omega$ (where the infinity resistance means the resistor is cutoff).

The real-time implementation is based on the MATLAB® real-time windows target and two data-acquisition cards (i.e., NI PCI-6025E and NI PCI-6713). The PI-control law and the mode identification algorithm are implemented in the Simulink® environment. The PI-control goal is to maintain the voltage over C_1 close to a desired voltage $v_d = 6.5\,\text{V}$. The switches are implemented by three PhotoMOS relays (AQV251) and controlled electronically by the real-time software. The sampling time is 0.01 s, and the voltage source is modulated by a digital-to-analog converter channel.

Two fault scenarios are presented; in all scenarios, the initial condition is $v_1(0) = v_2(0) = 0$, and the three switches sw_1, sw_2, and sw_3 are programmed according to Fig. 5.24. The switches programming is only for demonstration, and it is unknown to the mode identification module (as the three valves in the two-tank system are

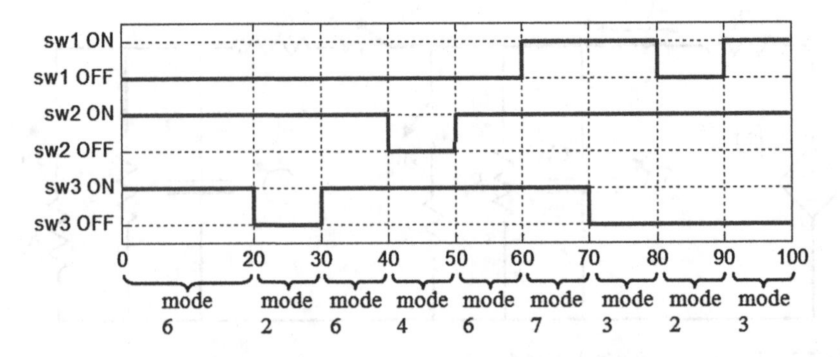

Fig. 5.24 Switches programming

managed by users). To cope with measurement noise, the pre-filter and derivative approximation in (5.17) are implemented in the real-time program.

Two scenarios are presented, and in each scenario, a different parametric fault is demonstrated. In the first scenario, the resistor R_2 is faulty, and its actual value is 180 kΩ (instead of 100 kΩ). This fault simulates a partial blockage at the outlet valve of tank A_2. In the second scenario, the fault is at R_3, and its faulty value is 100 kΩ (instead of 30 kΩ). This fault represents a partial blockage at the central tube connecting the two tanks. The two faults are detected and isolated in mode 6, and the fault isolation is based on the GFSM in Table 5.6, where $[a_3\, a_2\, a_1] = [1\,1\,0]$.

The experimental results of the first scenario are presented in Figs. 5.25, 5.26, and 5.27. The sensor measurements $v_1(t)$ and $v_2(t)$ are presented in the upper graph of Fig. 5.13. For reference, the first residual (ARR_1) is shown in the lower part of Fig. 5.13, along with its threshold. The threshold is 0.5e^{-5}, and it is identical for all residuals. In the time sections $t = 20$–30 s and $t = 80$–90 s, the system's actual mode is mode 2. Since ARR_1 is valid for mode 2, and since it is not sensitive to the fault (R_2), in these time sections, the residual value is below the threshold (it is shown by the graph). The graph also demonstrates the noisy signals that are found in the test-bed.

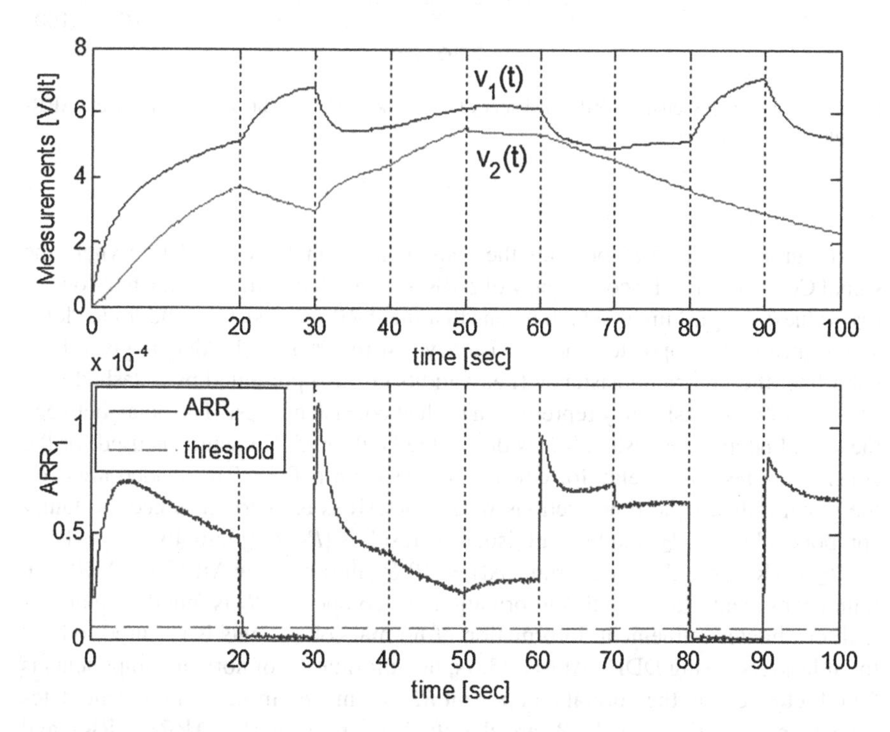

Fig. 5.25 *Upper graph* sensors measurement. Lower graph ARR_1

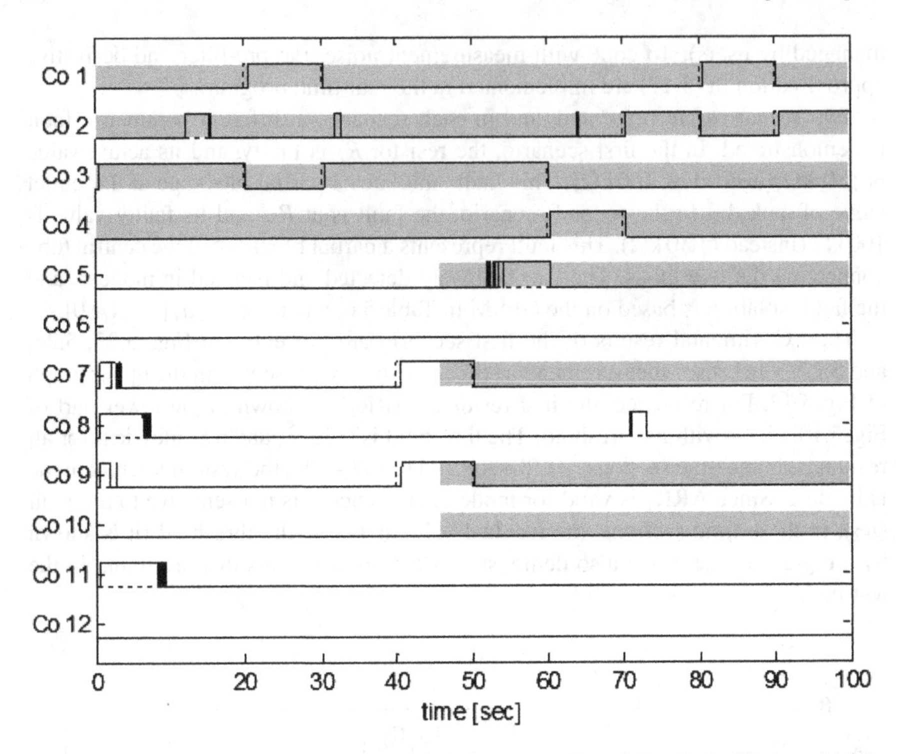

Fig. 5.26 First fault scenario ARR-consistency (*solid line*); ARR-validity (*dashed line*); and ARR-certainty (*shaded area*)

The graph in Fig. 5.26 presents the consistency signals of all ARRs (where the signal Co_i stands for the consistency of ARR_i). Consistency signals may have one of two values: a logic "high" which means that the ARR is consistent, and logic "low" which means the opposite. The consistency test result of each ARR is given by a solid line; theoretical consistency (i.e., validity) is also presented by a dashed line. The theoretical consistency represents a perfect consistency test. The delay between the solid line and the dashed line is due to the DDF in (5.27), which is used for the consistency test. The delay for consistency decision is 0.5 s. The shaded areas on the graph indicate the time sections where the ARRs are certain. Since the faulty component here is R_2, and the fault isolation result is $\{R_5, R_2\}$; initially, only ARR_1, ARR_2, ARR_3, and ARR_4 are certain. At $t = 40.6$ s, the uncertain ARR^5 and ARR^7 are found consistent. Based on this information, the component R_5 is found normal. For fault candidates refinement, information of normal components is required to hold for at least 5 s (using DDF). At $t = 45.6$ s, the information of normal components is found reliable, and the normal component $R5$ is removed from the fault candidates code. Consequently, more ARRs are classified as certain (ARR_5, ARR_7, ARR_9, and ARR_{10}), and the mode identification performances are improved. The graph shows

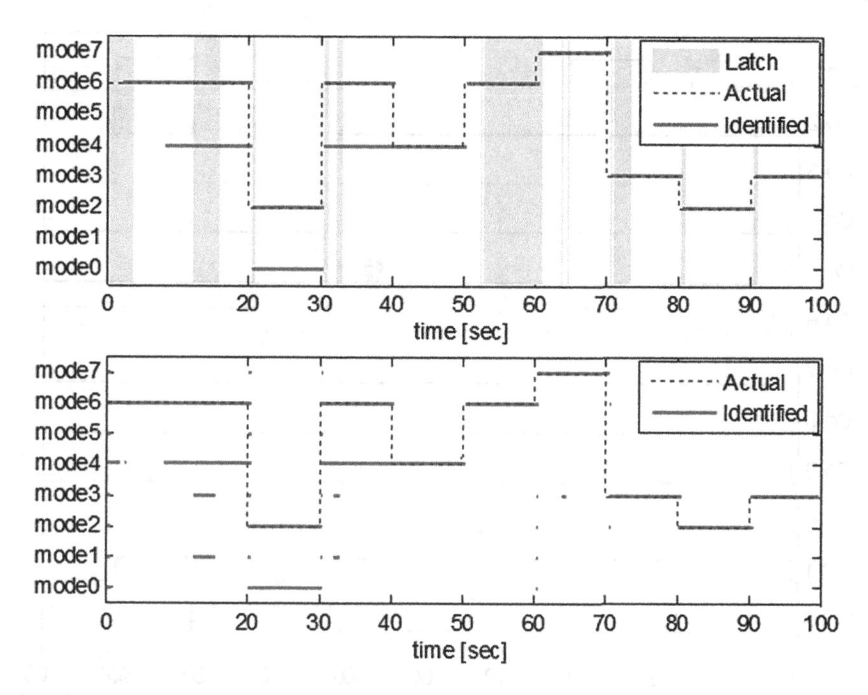

Fig. 5.27 Mode hypotheses of fault scenario 1. *Upper graph* logic merger and latch. *Lower graph* algebraic merger

that some ARRs are misleading for a short period of time. For example, ARR_2 is misleading in the time sections $t = 12.3-15.4$ s and $t = 31.9-32.8$ s, ARR_5 is misleading in $t = 52.1-60.5$ s, and ARR_8 is misleading in $t = 70.9-72.9$ s. In addition, some ARRs are misleading during the first few seconds, i.e., when the system is initiating.

The mode hypotheses result is presented in Fig. 5.27. The upper graph presents the result based on the logic merger and a latch. Until $t = 40$ s, the identified mode is not unique; for example, in the time section $t =$ 30–40 s, the actual mode is mode 6, but the mode hypotheses are modes 6 and 4 (the same conditions are presented in Table 5.10, case 1). At $t = 40$ s, the faulty valve is turned off, and the mode is uniquely identified (these conditions are also presented in Table 5.11, case 2). In addition, the component R^5 is found normal, and the fault candidates code is refined. Now, more certain ARRs are utilized for mode identification. In the time section $t =$ 50–60 s, the actual mode is mode 6. Based on the new conditions, this mode and all subsequent modes are uniquely identified. It is also clear from the upper graph that all misleading ARRs are accommodated (except in the first few seconds when no mode is identified). This is due to the latch that is integrated with the logic merger. The time sections where the latch is ON are presented on the graph by the shaded areas (due to the graph scale, not all details are visible).

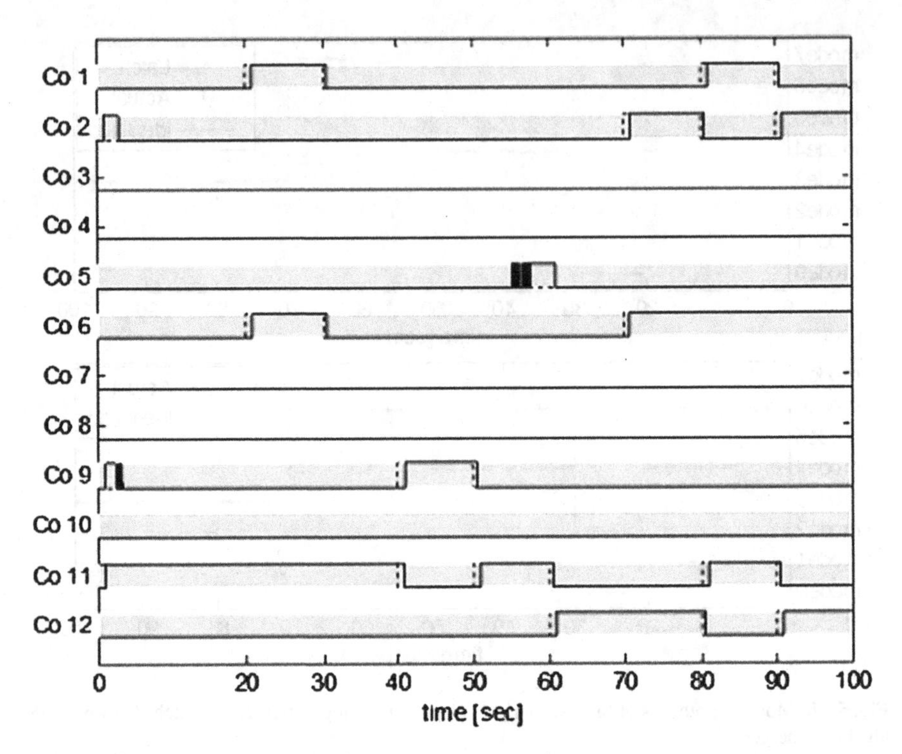

Fig. 5.28 Second fault scenario ARR Consistency (*solid line*), ARR Validity (*dashed line*), ARR Certainty (*shaded area*)

The mode hypotheses result of the algebraic merger is presented in the lower graph of Fig. 5.27. In general, the mode hypotheses here are identical to those presented in the upper graph. Not all misleading ARRs are accommodated, for example, when the actual mode is mode 6 and ARR_2 is misleading, the result includes four modes instead of two. The same conditions are also presented in Table 5.15 (case 1, R_2 is faulty). In addition, immediately after some mode changes, the result is "not clean" for a very short period of time. The reason is stated as follows: After a mode change, some ARRs become valid and others become invalid. Valid ARRs are found consistent, but the timing is not identical for all ARRs (e.g., one ARR may converge to the threshold faster than others). In the case of the logic merger, this phenomenon is filtered by the latch. For the algebraic merger, an additional DDF of 0.5 s acting on the mode hypotheses vector eliminates these undesired results (at the expense of additional delay). Note that the misleading ARR_5 (when $t = 52.1$–60.5 s) and the misleading ARR_8 ($t = 70.9$–72.9 s) are completely accommodated by the algebraic merger.

The experimental results of the second scenario are presented in Figs. 5.28 and 5.29. Here, the faulty component is R_3, which is uniquely isolable by the GFSM

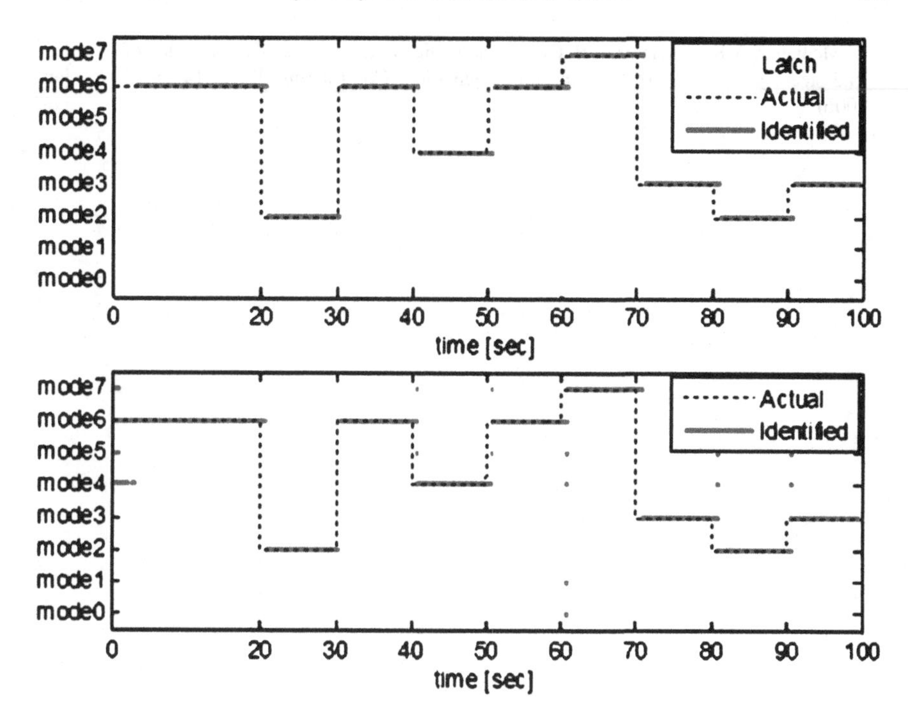

Fig. 5.29 Mode hypotheses of fault scenario 2. *Upper graph* logic merger and latch. *Lower graph* algebraic merger

at mode 6. From these graphs, it is clear that for the second scenario, all modes are uniquely identified, despite the faulty component (the value of R_3 is assumed unknown). The results of the logic merger and latch, and the results of the algebraic merger, are similar. The misleading ARR_5 ($t = 55.1$–61.0 s) is accommodated by the two algorithms.

References

1. S. Arogeti, D. Wang, C.B. Low, Mode tracking and FDI of hybrid systems, in *The 10th International Conference on Control, Automation, Robotics and Vision (ICARCV2008)* (Hanoi, 2008), pp. 892–897
2. C.B. Low, D. Wang, S. Arogeti, M. Luo, Fault parameter estimation for hybrid systems using hybrid bond graph, in *The 3rd IEEE multi-conference on systems and control (MSC 2009)* (Saint Petersburg, 2009), pp. 1338–1343
3. A.K. Samantaray, S.K. Ghoshal, Sensitivity bond graph approach to multiple fault isolation through parameter estimation. Proc. Inst. Mech, Eng. Part-I J. Syst. Control Eng. **221**, 577–587 (2007)
4. A.K. Samantaray, S.K. Ghoshal, S. Chakraborty, A. Mukherjee, Improvements to single-fault isolation using estimated parameters. Simulation **81**(12), 827–845 (2005)
5. A.K. Samantaray, B. Ould Bouamama, *Model-Based Process Supervision: A Bond Graph Approach* (Springer, London, 2008)

6. K. Medjaher, A.K. Samantaray, B. Ould Bouamama, M. Staroswiecki, Supervision of an industrial steam generator-Part 2: online implementation. Control Eng. Pract. **14**(1), 85–96 (Jan. 2006)

Chapter 6
Application of Real Time FDI and Fault Estimation to a Vehicle Steering System

6.1 Introduction

Many present-day complex systems, such as manufacturing system, automotive engine control, and embedded control system, exhibit behavior that can be modeled as hybrid systems. Such systems consist of continuous behavior and discrete states represented by modes. The hybrid nature of the system can be inherent in its normal behavior, or it may reflect abnormal behavior, where some modes represent faults (fault modes). In hybrid system diagnosis, two different types of faults are defined. The first is a parametric fault, where one or more model parameters are deviating from their nominal value to an unknown value. Practical examples of parametric faults in electro-hydraulic systems are described in [1]. The second type is fault mode; in this case, the faulty state is known a priori and can be modeled by known parameters only. Two examples are short- and open-circuit faults of power switches [2]. Fault modes can be detected and isolated by mode identification techniques.

There are several existing efforts made in FDI of electric vehicles. In [3], an autonomous vehicle called RobuCar is considered for demonstration of a fault tolerant control strategy. The dynamic modeling of the vehicle is developed using the bond graph method, and it is utilized for systematic generation of ARRs. These relations are then evaluated with actual measurements to generate residuals and to perform structural fault isolation. At the case of fault detection, an automaton is utilized to provide the best option to reconfigure the system such that the given control objective is achieved. Residual generation for actuators fault FDI of an electric vehicle is discussed in [4]. The modeling step is accomplished using the bond graph theory. Unmeasured flow variables and system nonlinearities which are considered as unknown inputs are estimated by a nonlinear observer with finite time convergence. The result is a linear time invariant model where the parity space approach is implemented for fault detection and isolation of the vehicle actuators. Experimental results are demonstrated on a test bed, based on the RobuCar electric vehicle.

Robust fault detection and isolation of an electric vehicle with structured and unstructured uncertainties is reported in [5]. This work focuses on the vehicle traction

D. Wang et al., *Model-based Health Monitoring of Hybrid Systems*,
DOI: 10.1007/978-1-4614-7369-5_6,

actuators, and utilizes the bond graph tool for model development in the linear fractional transformation form. Unstructured uncertainties are considered as modulated sources of unknown effort or flow, and are being estimated. The method facilitates systematic generation of robust ARRs and adaptive thresholds. The FDI methods presented in [3–5] are established upon continues dynamics principles. Temporal causal graph (TCG)-based methods for multiple fault diagnosis and distributed diagnosis of mobile robots are reported in [6] and [7], respectively. TCG is a qualitative model which is derived systematically from a BG model. These methods are demonstrated using a BG model of a wheeled mobile robot with differentially driven wheels.

In this chapter, the QHBG-based FDI framework presented in the previous chapters is studied in detail using a complex mechatronic system, namely, the steering system of an electric vehicle. Both incipient and abrupt parametric faults are considered, and modes may represent faults. The presented case studies reveal the complexity and difficulties of hybrid system diagnosis, where fault modes, sensor faults, and parametric faults are mixed in a single framework. The fault modes are detected, isolated and identified by the inherent mode tracking capability of the framework. Parametric faults, abrupt and incipient, are detected, isolated and an abrupt fault is estimated. All presented framework modules are based on the concept of GARRs.

6.2 Description of the Vehicle Steering System

6.2.1 The Electro-Hydraulic Steering System

The CyCab is an electric passenger transporter with four driving wheels and a dual steering system. The vehicle weight is about 350 kg and it can reach a maximum speed of 5 m/s. The focus of this study is FDI in the CyCab electro-hydraulic steering system as shown in Fig. 6.1. The steering system consists of motion controller, DC motor, gear, belt transmission, oil pump, oil tubs, hydraulic piston, steering mechanism, wheels, and tires.

The DC motor is controlled through a dc motor driver. A speed reducer (Gear) transfers the motor power to a belt, and the motor speed is measured by an incremental encoder, at the gear output shaft. The belt transfers the motor power to an oil pump, which generates an oil flow, and consequently, an oil pressure in one of two oil tubes. The pump is a bidirectional axial piston pump and converts mechanical speed to proportional oil flow. Two oil tubes connect the pump to a hydraulic actuator (an oil piston). The oil is pumped from an internal oil reservoir and returns to the pump through the low pressure tube. An internal nonreturn valve prevents oil from returning to the pump, at the high pressure tube. In addition, the pump is equipped with two pressure release valves, which protect the hydraulic components from over pressure. The oil piston drives an Ackerman steering mechanism, and the steering angle is measured by an absolute encoder. Two pressure sensors measure the pressure at the two cylinder sides, as shown in Fig. 6.2.

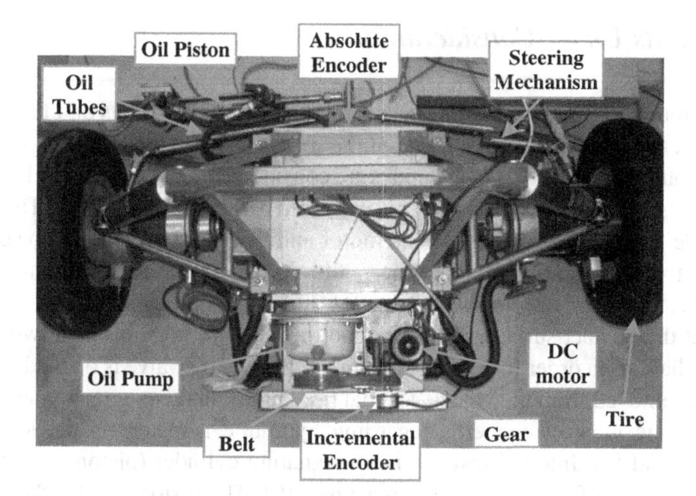

Fig. 6.1 The CyCab front steering system

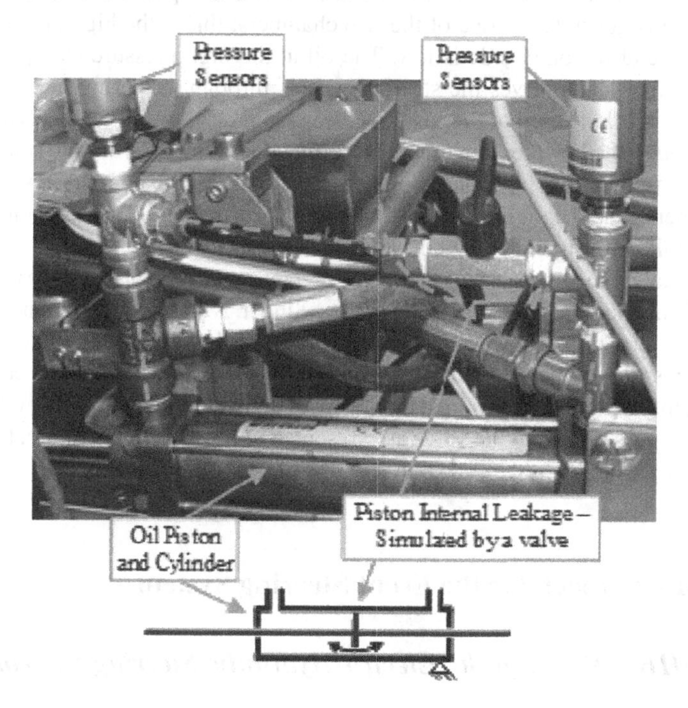

Fig. 6.2 Piston internal leakage—simulated by a valve

6.2.2 Faults Under Consideration

Seven different faults, representing sensor-faults, parametric-faults and fault-modes are considered in this study. Each one of the four sensors in the steering system is a possible source of fault. Two fault-modes are considered, the first is a burnt DC-motor or a burnt-driver; in this case no power is delivered to the steering system. The second fault-mode is a broken-belt, where the motor and gear are rotating but power is not delivered to the pump. Two parametric faults are discussed, a flat tire and a piston internal leakage. Flat tire fault is detected if the model-parameters representing the friction at the contact surface of the tire and the road are not consistent with actual system behavior. In order to simulate this fault, the wheel's valve is opened and air is gradually removed from the wheel. The flat tire fault is an incipient parametric-fault.

One of the greatest concerns regarding hydraulic actuators is the leakage of hydraulic fluid [8]. Internal leakage in an actuating cylinder (piston), due to worn seal, is a common fault in hydraulic machines [9]. The piston divides the cylinder into two chambers, which are filled with oil. When the pump rotates in one direction, oil pressure is generated in one of the two chambers; this is the high pressure side. The pressure at the other side is low. The oil at the high pressure side pushes the piston and the piston drives the steering wheels through a steering mechanism as shown in Fig. 6.1. The oil at the low pressure side returns to the oil reservoir at the pump, through the low pressure tube. For normal operation of hydraulic actuator, oil is not supposed to flow from one chamber to the other. If the internal flow between the chambers is not negligible, the piston efficiency is decreased. The piston internal leakage fault is illustrated at the bottom part of Fig. 6.2.

To simulate the internal leakage fault, an oil bypass has been installed, to allow free oil flow from one side to the other side of the cylinder. A proportional (manual) valve is used to control the oil flow in the bypass tube. When the valve is completely closed, it represents the normal condition. An open valve represents abnormal condition, and the fault severity is determined by the opening level of the valve. The oil piston, cylinder, bypass tubes and the valve are presented in Fig. 6.2. In the figure, the valve is marked by a circle.

6.3 FDI Approach for the Front Steering System

6.3.1 DHBG Model of the Electro-Hydraulic Steering System

The DHBG of the CyCab electro-hydraulic steering system is presented in Figs. 6.3, 6.4 and 6.5. The DHBG of the DC-motor and the motor-driver is presented in Fig. 6.3. The signal u_{in} controls the steering system. The current in the motor is proportional to the voltage u_{in}, and the torque generated by the motor is proportional to the current. The voltage-to-current ratio is k_1, and the current-to-torque ratio is k_2. R_1 is electric resistance, J_1 is the motor inertia, and due to the very fast response of the motor

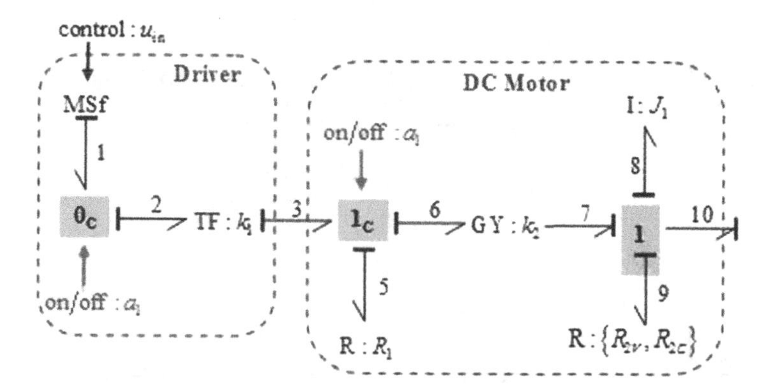

Fig. 6.3 DHBG: DC motor and driver

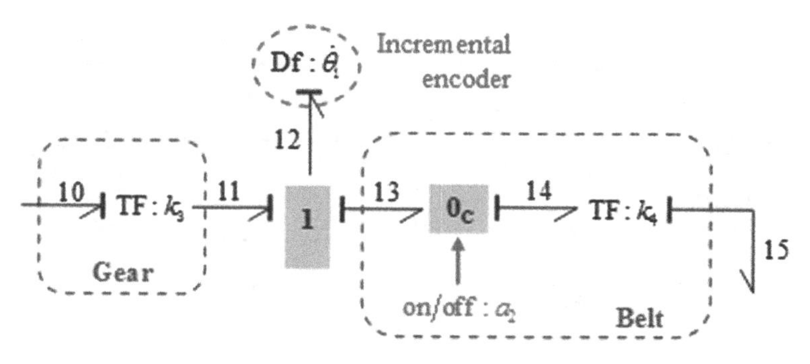

Fig. 6.4 DHBG: gear and belt

driver (in its current amplifier mode) the motor inductance is negligible. Mechanical friction, in the motor, is modeled by two parameters; R_{2V} represents viscous friction, and coulomb friction is given by R_{2C}. In this study, the relation between effort and flow, due to mechanical friction, is described by $g_i(\cdot)$, as follows

$$e = g_i(R_i, f) = R_{iV} f + R_{iC}\text{sign}(f), \quad R_i = \{R_{iV}, R_{iC}\} \tag{6.1}$$

The controlled-junction (CJ)'s state a_1, in Fig. 6.3, represents a burnt-driver or a burnt-motor fault. In case of fault, $a_1 = 0$, and the current through the motor is zero. The transmission from the motor to the pump is modeled in Fig. 6.4. The gear ratio is k_3, and the belt ratio is k_4. A sensor (incremental encoder) measures the shaft velocity at the gear output. The controlled-junction state a_2 is utilized to model the broken belt fault. The faulty state is represented by $a_2 = 0$, which disconnects the pump (load) from the motor (note that the pump velocity is zero if torque is not delivered to the pump).

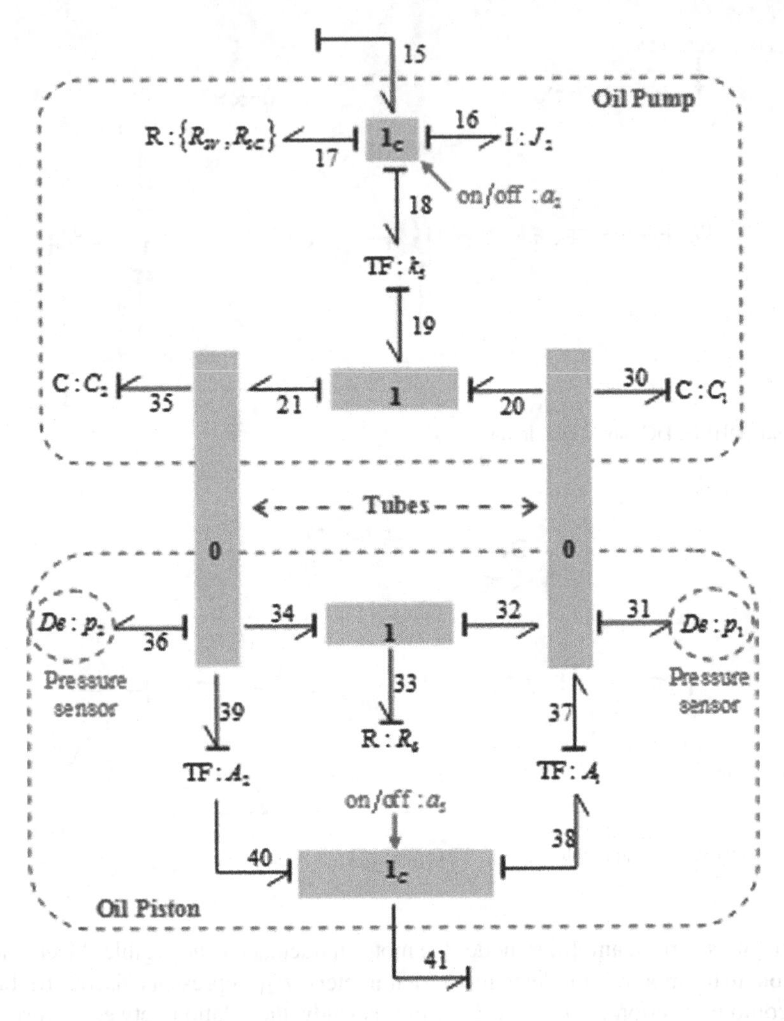

Fig. 6.5 DHBG: oil pump, tubes and piston

The hydraulic part is described in Fig. 6.5. This part includes the pump, tubes, and piston. The pump rotor is modeled by its mechanical friction $\{R_{3V}, R_{3C}\}$ and inertia J_2. The oil flow through the pump (and tubes) is proportional to the pump angular velocity, and the ratio is k_5. Two oil tubes connect the pump with an oil piston; these tubes are modeled by two 0-junctions (no pressure drop is assumed in the tubes). The pump is bidirectional and equipped with two internal pressure release valves. These are not given in the model and have no influence on the results presented in this chapter.

The oil compressibility and tubes elasticity is modeled by two storage components C_1 and C_2. The oil piston is modeled by two transformers A_1 and A_2, which represent the two effective cross section areas of the piston, from its both sides. The piston internal leakage is modeled by R_6, which represents internal resistance to oil flow. The relation between pressure difference Δe and oil flow f, through a valve or through a leak is described by $l_i(\cdot)$, as follows

$$f = l_i(R_i, \Delta e) = \frac{\text{sign}(\Delta e)\sqrt{|\Delta e|}}{R_i} \tag{6.2}$$

When the system is normal, $R_6 \to \infty$, and in case of internal leakage $R_6 \ll \infty$, and the value of R_6 is unknown. Two pressure sensors, p_1 and p_2, are installed at the two sides of the cylinder. The controlled-junction state a_5 represents the hybrid dynamical nature of the piston. The piston motion is limited by the two sides of the cylinder. Between these two limits, $a_5 = 1$, and the piston motion is smooth in two directions. When the piston position is in one of the two limits, the piston motion is restricted to only one direction. The variable a_5 represents an autonomous mode-change and its value is determined as follows:

$$\left.\begin{array}{l} x_{min} > x > x_{max} \\ x \le x_{min},\ p_{40} - p_{38} > 0 \\ x \ge x_{max},\ p_{40} - p_{38} < 0 \end{array}\right\} \Rightarrow a_5 = 1, \text{ else } a_5 = 0 \tag{6.3}$$

where x is the piston position and $x = 0$ when the steering angle is zero, $\{x_{min}, x_{max}\}$ are the two limits, and p_{38} and p_{40} are the two forces acting on the piston from its both sides due to oil pressure.

The DHBG of the steering mechanism and wheels is presented in Fig. 6.6. The steering mechanism is based on Ackerman's geometry and described in Fig. 6.7. The goal of this mechanism is to turn the two steering wheels in different steering angles

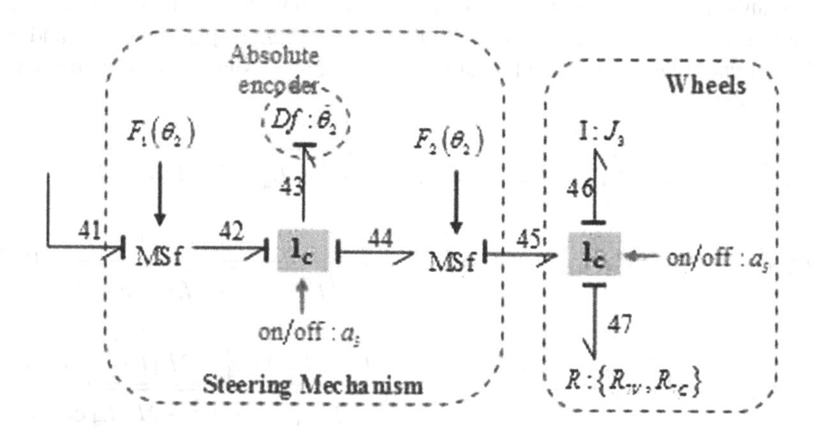

Fig. 6.6 DHBG: steering mechanism and wheels

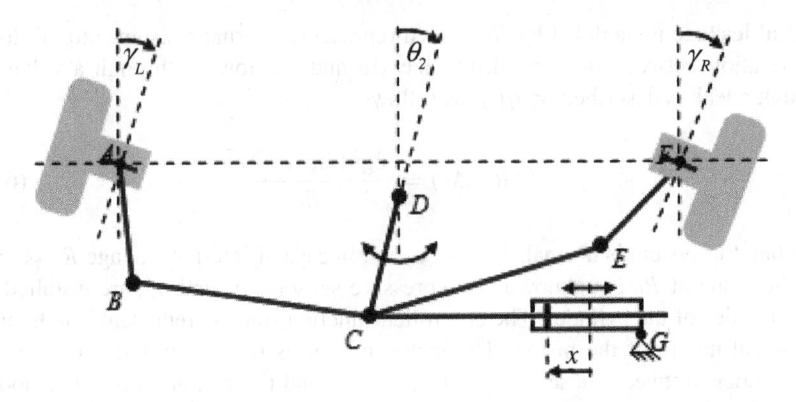

Fig. 6.7 The Ackerman's steering mechanism

(so that tire slippage is minimized). For modeling purpose, the two steering wheels are represented by a single virtual wheel, located at the center point between the two actual wheels. The steering angle of the virtual wheel is γ, and it is given as

$$\gamma = \frac{\gamma_L + \gamma_R}{2}, \ \dot{x} = F_1(\theta_2)\dot{\theta}_2, \ \dot{\gamma} = F_2(\theta_2)\dot{\theta}_2 \tag{6.4}$$

The steering angle is measured by an absolute encoder, located at point D in Fig. 6.7; this encoder measures the angle θ_2. The cylinder is fixed to the mobile robot frame at point G, and the piston position is given by x. Two nonlinear functions, $F_1(\theta_2)$ and $F_2(\theta_2)$, relate velocity variables in the steering mechanism, as described in (6.4).

In order to derive the two nonlinear functions $F_1(\theta_2)$ and $F_2(\theta_2)$, a detailed Ackerman's steering mechanism with geometry variables is depicted in Fig. 6.8. In the figure, $L_i, i = 1, 2, 3, 4, 6, 7, 8, 9$ and x_0 are length parameters, x and L_5 are length variables, β, δ, γ_1 and γ_2 are angle variables. Using the law of cosines in trigonometry,

$$L_1^2 + L_4^2 - 2L_1L_4\cos\beta = L_5^2 \Rightarrow L_5 = \sqrt{L_1^2 + L_4^2 - 2L_1L_4\cos\beta} \tag{6.5}$$

$$L_4^2 + L_5^2 - 2L_4L_5\cos\gamma_2 = L_1^2 \Rightarrow \cos\gamma_2 = \frac{L_4 - L_1\cos\beta}{\sqrt{L_1^2 + L_4^2 - 2L_1L_4\cos\beta}} \tag{6.6}$$

$$L_3^2 + L_5^2 - 2L_3L_5\cos\gamma_1 = L_2^2 \Rightarrow \cos\gamma_1 = \frac{L_3^2 + L_1^2 + L_4^2 - 2L_1L_4\cos\beta - L_2^2}{2L_3\sqrt{L_1^2 + L_4^2 - 2L_1L_4\cos\beta}} \tag{6.7}$$

Take derivative of (6.5) leads to

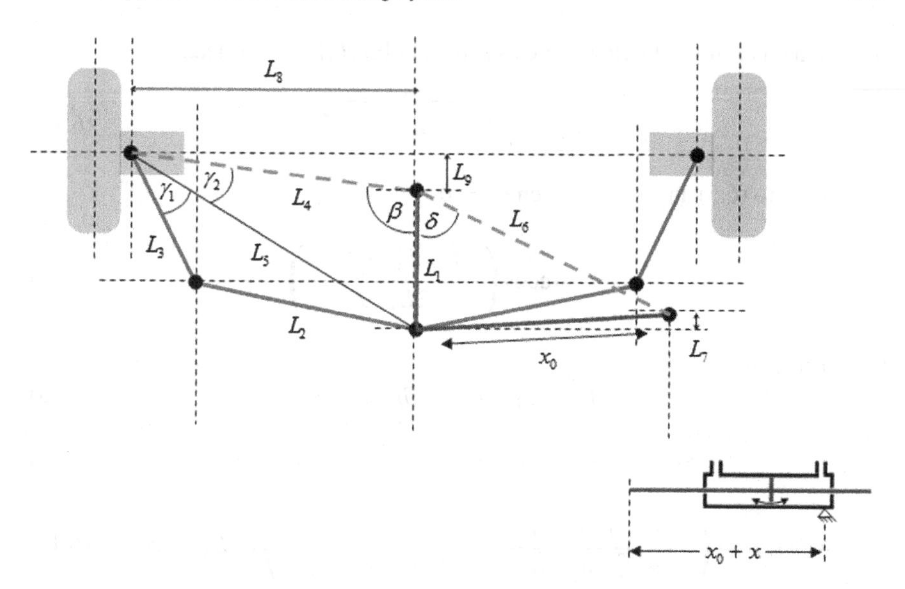

Fig. 6.8 The Ackerman's steering mechanism with geometry variables

$$L_1 L_4 \sin\beta\dot{\beta} = L_5 \dot{L}_5 \Rightarrow \dot{L}_5 = \frac{L_1 L_4 \sin\beta}{L_5}\dot{\beta} \tag{6.8}$$

Similarly, from (6.6) and (6.7) we get

$$\begin{aligned} & 2L_5\dot{L}_5 - 2L_4\dot{L}_5\cos\gamma_2 + 2L_4L_5\sin\gamma_2\dot{\gamma}_2 \\ \Rightarrow\ & (L_5 - L_4\cos\gamma_2)\dot{L}_5 = -L_4L_5\sin\gamma_2\dot{\gamma}_2 \\ \Rightarrow\ & \dot{\gamma}_2 = -\frac{L_5 - L_4\cos\gamma_2}{L_4L_5\sin\gamma_2}\dot{L}_5 \end{aligned} \tag{6.9}$$

$$\begin{aligned} & 2L_5\dot{L}_5 - 2L_3\dot{L}_5\cos\gamma_1 + 2L_3L_5\sin\gamma_1\dot{\gamma}_1 \\ \Rightarrow\ & (L_5 - L_3\cos\gamma_1)\dot{L}_5 = -L_3L_5\sin\gamma_1\dot{\gamma}_1 \\ \Rightarrow\ & \dot{\gamma}_1 = -\frac{L_5 - L_3\cos\gamma_1}{L_3L_5\sin\gamma_1}\dot{L}_5 \end{aligned} \tag{6.10}$$

From (6.9) and (6.10)

$$\dot{\gamma}_1 + \dot{\gamma}_2 = -\left(\frac{L_5 - L_3\cos\gamma_1}{L_3L_5\sin\gamma_1} + \frac{L_5 - L_4\cos\gamma_2}{L_4L_5\sin\gamma_2}\right)\dot{L}_5 \tag{6.11}$$

Combining (6.8) and (6.11) leads to

$$\dot{\gamma}_1 + \dot{\gamma}_2 = -\left(\frac{L_5 - L_3\cos\gamma_1}{L_3L_5^2\sin\gamma_1} + \frac{L_5 - L_4\cos\gamma_2}{L_4L_5^2\sin\gamma_2}\right)L_1L_4\sin\beta\dot{\beta} \tag{6.12}$$

If θ_2 is the measured steering angle as shown in Fig. 6.7, $\theta_2 = 0$ leads to

$$L_{5\_0} = \sqrt{L_8^2 + (L_1 + L_9)^2} \tag{6.13}$$

According to (6.5) and (6.13), when $\theta_2 = 0$

$$\beta_0 = \arccos\left(\frac{L_1^2 + L_4^2 - L_{5\_0}^2}{2L_1L_4}\right) \tag{6.14}$$

For left wheel

$$\beta_L = \beta_0 - \theta_2 \Rightarrow \dot{\beta}_L = -\dot{\theta}_2 \tag{6.15}$$

$$\gamma_L = \gamma_0 - (\gamma_{1L} + \gamma_{2L}) \Rightarrow \dot{\gamma}_L = -\dot{\gamma}_{1L} - \dot{\gamma}_{2L} \tag{6.16}$$

$$\dot{\gamma}_{1L} + \dot{\gamma}_{2L} = -\left(\frac{L_{5L} - L_3\cos\gamma_{1L}}{L_3L_{5L}^2\sin\gamma_{1L}} + \frac{L_{5L} - L_4\cos\gamma_{2L}}{L_4L_{5L}^2\sin\gamma_{2L}}\right)L_1L_4\sin\beta_L\dot{\beta}_L \tag{6.17}$$

where γ_0 is the sum of γ_1 and γ_2 when $\theta_2 = 0$.

In (6.17)

$$L_{5L} = \sqrt{L_1^2 + L_4^2 - 2L_1L_4\cos(\beta_0 - \theta_2)} \tag{6.18}$$

$$\gamma_{2L} = \arccos\left(\frac{L_4^2 + L_{5L}^2 - L_1^2}{2L_4L_{5L}}\right) \tag{6.19}$$

$$\gamma_{1L} = \arccos\left(\frac{L_3^2 + L_{5L}^2 - L_2^2}{2L_3L_{5L}}\right) \tag{6.20}$$

For right wheel

$$\beta_R = \beta_0 + \theta_2 \Rightarrow \dot{\beta}_R = \dot{\theta}_2 \tag{6.21}$$

$$\gamma_R = (\gamma_{1R} + \gamma_{2R}) - \gamma_0 \Rightarrow \dot{\gamma}_R = \dot{\gamma}_{1R} + \dot{\gamma}_{2R} \tag{6.22}$$

$$\dot{\gamma}_{1R} + \dot{\gamma}_{2R} = -\left(\frac{L_{5R} - L_3\cos\gamma_{1R}}{L_3L_{5R}^2\sin\gamma_{1R}} + \frac{L_{5R} - L_4\cos\gamma_{2L}}{L_4L_{5R}^2\sin\gamma_{2R}}\right)L_1L_4\sin\beta_R\dot{\beta}_R \tag{6.23}$$

In (6.23)

$$L_{5R} = \sqrt{L_1^2 + L_4^2 - 2L_1L_4\cos(\beta_0 + \theta_2)} \tag{6.24}$$

$$\gamma_{2R} = \arccos\left(\frac{L_4^2 + L_{5R}^2 - L_1^2}{2L_4L_{5R}}\right) \tag{6.25}$$

$$\gamma_{1R} = \arccos\left(\frac{L_3^2 + L_{5R}^2 - L_2^2}{2L_3 L_{5R}}\right) \tag{6.26}$$

From (6.15)–(6.17)

$$\dot{\gamma}_L = -\left(\frac{L_{5L} - L_3 \cos\gamma_{1L}}{L_3 L_{5L}^2 \sin\gamma_{1L}} + \frac{L_{5L} - L_4 \cos\gamma_{2L}}{L_4 L_{5L}^2 \sin\gamma_{2L}}\right) L_1 L_4 \sin(\beta_0 - \theta_2)\dot{\theta}_2 \tag{6.27}$$

From (6.21)–(6.23)

$$\dot{\gamma}_R = -\left(\frac{L_{5R} - L_3 \cos\gamma_{1R}}{L_3 L_{5R}^2 \sin\gamma_{1R}} + \frac{L_{5R} - L_4 \cos\gamma_{2L}}{L_4 L_{5R}^2 \sin\gamma_{2R}}\right) L_1 L_4 \sin(\beta_0 + \theta_2)\dot{\theta}_2 \tag{6.28}$$

Then

$$\dot{\gamma} = \frac{\dot{\gamma}_L + \dot{\gamma}_R}{2} = F_2(\theta_2)\dot{\theta}_2 \tag{6.29}$$

Next, consider the cylinder link

$$L_1^2 + L_6^2 - 2L_1 L_6 \cos\delta = (x_0 + x)^2 \tag{6.30}$$

$$\delta = \delta_0 + \theta_2 \tag{6.31}$$

$$\delta_0 = \arccos\left(\frac{L_1 - L_7}{L_6}\right) \tag{6.32}$$

From (6.30) and (6.31)

$$L_1^2 + L_6^2 - 2L_1 L_6 \cos(\delta_0 + \theta_2) = (x_0 + x)^2 \Rightarrow x_0 + x = \sqrt{L_1^2 + L_6^2 - 2L_1 L_6 \cos(\delta_0 + \theta_2)} \tag{6.33}$$

Take derivative of (6.33) leads to

$$L_1 L_6 \sin(\delta_0 + \theta_2)\dot{\theta}_2 = (x_0 + x)\dot{x} \Rightarrow \dot{x} = \frac{L_1 L_6 \sin(\delta_0 + \theta_2)}{x_0 + x}\dot{\theta}_2 \tag{6.34}$$

According to (6.32)–(6.34)

$$\begin{aligned}\dot{x} &= \frac{L_1 L_6 \sin(\delta_0 + \theta_2)}{\sqrt{L_1^2 + L_6^2 - 2L_1 L_6 \cos(\delta_0 + \theta_2)}}\dot{\theta}_2 \\ &= \frac{L_1 L_6 \sin\left[\arccos\left(\frac{L_1 - L_7}{L_6}\right) + \theta_2\right]}{\sqrt{L_1^2 + L_6^2 - 2L_1 L_6 \cos\left[\arccos\left(\frac{L_1 - L_7}{L_6}\right) + \theta_2\right]}}\dot{\theta}_2 = F_1(\theta_2)\dot{\theta}_2 \end{aligned} \tag{6.35}$$

The inertia of the virtual steering wheel is J_3, and the friction between the wheel and the road is modeled by two coefficients $\{R_{7V}, R_{7C}\}$.

6.3.2 Development of GARRs

The DHBG of the CyCab electrohydraulic steering system is described in Sect. 6.3.1. All CJs and storage elements of the graph are assigned preferred causality for FDI. Causality reassignment after mode changes is not required, and unified relations for FDI, called GARRs, are derived in this part. Theoretically, the maximum number of independent GARRs is equal to the number of sensors. The sensors on the DHBG are assigned inverted causality, and the constitutive relation of each sensor junction is utilized to develop a GARR [14]. A GARR is achieved if all unknown variables in the junction's equation are solved using the causal paths of the DHBG.

The first GARR is developed using the constitutive relation of the sensor junction $\dot{\theta}_1$, given as

$$e_{12} = e_{11} - e_{13} = 0 \tag{6.36}$$

Based on the DHBG causal paths, the unknown variables are solved

$$\begin{aligned} e_{11} &= k_3^{-1} e_{10} = k_3^{-1}(e_7 - e_8 - e_9) \\ &= k_3^{-1}(k_2 f_6 - J_1 \dot{f}_8 - g_2(R_2, f_9)) \\ &= k_3^{-1}(a_1 k_1 k_2 u_{in} - J_1 k_3^{-1} \ddot{\theta}_1 - g_2(R_2, k_3^{-1} \dot{\theta}_1)) \end{aligned} \tag{6.37}$$

$$\begin{aligned} e_{13} &= a_2 e_{14} = a_2 k_4 e_{15} = a_2 k_4 (e_{16} + e_{17} + e_{18}) \\ &= a_2 k_4 (J_2 \dot{f}_{16} + g_3(R_3, f_{17}) + k_5 e_{19}) \\ &= a_2 k_4 (J_2 k_4 \ddot{\theta}_1 + g_3(R_3, k_4 \dot{\theta}_1) + k_5 (p_2 - p_1)) \end{aligned} \tag{6.38}$$

and GARR_1 is achieved

$$\begin{aligned} \text{GARR}_1 : k_3^{-1}(a_1 k_1 k_2 u_{in} - J_1 k_3^{-1} \ddot{\theta}_1 - g_2(R_2, k_3^{-1} \dot{\theta}_1) \\ - a_2 k_4 (J_2 k_4 \ddot{\theta}_1 + g_3(R_3, k_4 \dot{\theta}_1) + k_5 (p_2 - p_1)) = 0 \end{aligned} \tag{6.39}$$

The sensor junction p_1 is utilized to develop GARR_2. The junction equation is given as

$$f_{31} = -f_{20} + f_{32} + f_{37} - f_{30} = 0 \tag{6.40}$$

Using the DHBG, all unknown variables are solved

$$f_{20} = f_{19} = k_5 f_{18} = k_5 a_2 f_{15} = k_5 a_2 k_4 f_{14} = a_2 k_4 k_5 \dot{\theta}_1 \tag{6.41}$$

$$f_{37} = A_1 f_{38} = A_1 a_5 f_{41} = A_1 a_5 F_1 f_{42} = A_1 a_5 F_1 \dot{\theta}_2 \tag{6.42}$$

$$f_{32} = l_6(R_6, e_{33}) = l_6(R_6, e_{34} - e_{32}) = l_6(R_6, p_2 - p_1) \tag{6.43}$$

$$f_{30} = C_1 \dot{e}_{30} = C_1 \dot{p}_1 \tag{6.44}$$

and $GARR_2$ is given as follows:

$$-a_2 k_4 k_5 \dot{\theta}_1 + l_6(R_6, p_2 - p_1) + a_5 A_1 F_1 \dot{\theta}_2 - C_1 \dot{p}_1 = 0 \tag{6.45}$$

In a similar way, $GARR_3$ is developed from the sensor junction p_2 and is given as

$$a_2 k_4 k_5 \dot{\theta}_1 + l_6(R_6, p_2 - p_1) - a_5 A_2 F_1 \dot{\theta}_2 - C_2 \dot{p}_2 = 0 \tag{6.46}$$

The last GARR (i.e., $GARR_4$) is developed using the constitutive relation of the sensor junction $\dot{\theta}_2$, this GARR is valid only for modes where $a_5 = 1$

$$e_{43} = e_{42} - e_{44} = 0 \tag{6.47}$$

The unknown variables are solved as

$$e_{42} = F_1 e_{41} = F_1(e_{40} - e_{38}) = F_1(A_2 e_{39} - A_1 e_{37}) = F_1(A_2 p_2 - A_1 p_1) \tag{6.48}$$

$$\begin{aligned} e_{44} &= F_2 e_{45} = F_2(e_{46} + e_{47}) = F_2(J_3 \dot{f}_{46} + g_7(R_7, f_{47})) \\ &= F_2 J_3 \frac{d}{dt}(F_2 \dot{\theta}_2) + F_2 g_7(R_7, F_2 \dot{\theta}_2) \end{aligned} \tag{6.49}$$

and the GARR is given as

$$F_1(A_2 p_2 - A_1 p_1) - F_2 J_3 \frac{d}{dt}(F_2 \dot{\theta}_2) - F_2 g_7(R_7, F_2 \dot{\theta}_2) = 0 \tag{6.50}$$

Few simplifications, based on two assumptions, are made. First, it is assumed that the piston does not reach any of its two side limits ($a_5 = 1, \forall t$). Second, the oil compressibility is negligible ($C_i \dot{p}_i \cong 0$). It is also known that $A_1 = A_2 = A$. Using these assumptions, $GARR_2$ and $GARR_3$ are identical, and represented by

$$GARR_{2,3} : a_2 k_4 k_5 \dot{\theta}_1 - l_6(R_6, p_2 - p_1) - A_1 F_1 \dot{\theta}_2 = 0 \tag{6.51}$$

Lists of all considered faults, in terms of model parameters, are given in Tables 6.1 and 6.2.

6.3.3 FDI Approach

The FDI process of the electrohydraulic steering system is based on three GARRs: $GARR_1$, $GARR_{2,3}$, and $GARR_4$. The mode is defined by two variables, a_1 and

Table 6.1 MD-FSM of the steering system at mode $[a_2\ a_1] = [1\ 1]$

	$GARR_4$	$GARR_{2,3}$	$GARR_1$	D_b	I_b
R_6	0	1	0	1	1
$\{R_{7V}, R_{7C}\}$	1	0	0	1	1
$\dot{\theta}_1$	0	1	1	1	1
$\dot{\theta}_2$	1	1	0	1	1
$p_2 - p_1$	1	0	1	1	1

Table 6.2 MCSM of the steering system

CJ	$GARR_4$	$GARR_{2,3}$	$GARR_1$	D_b	I_b
a_1	0	0	1	1	1
a_2	1	1	0	1	1

a_2, and only three modes are considered: normal operation $[a_2\ a_1] = [1\ 1]$, burnt motor/driver $[a_2\ a_1] = [1\ 0]$, and broken belt $[a_2\ a_1] = [0\ 1]$. The wheel model in this paper does not include static friction, which could be dominant if the steering velocity is zero (i.e., a high oil pressure is required in order to start the steering motion). The wheel model is given in $GARR_4$, and it is found that reliable FDI requires $\dot{\theta}_2 > \dot{\theta}_{2\,\min}$. To eliminate misleading information given by $GARR_4$, it is reformulated in the following way ($\dot{\theta}_{2\,\min} = 0.001\ \mathrm{s}^{-1}$ in the presented results):

$$\overline{\mathrm{GARR}_4} = c_2\mathrm{GARR}_4,\ \text{if}\ \dot{\theta}_2 > \dot{\theta}_{2\,\min},\ \text{then}\ c_2 = 1,\ \text{else}\ c_2 = 0 \qquad (6.52)$$

In this work, the single fault assumption is adopted. This single fault could be a parametric fault or a fault mode. The mode $[a_2\ a_1] = [1\ 1]$ represents normal condition, and it is the only mode where a parametric fault is possible. Only the MD-FSM of the steering system, at mode $[a_2\ a_1] = [1\ 1]$, is needed for parametric-fault isolation. This matrix is given in Table 6.1. The flow resistance R_6 represents the piston internal leakage. When the system is normal, $R_6 \to \infty$, and in case of fault, the value of R_6 is unknown. The friction coefficients $\{R_{7V}, R_{7C}\}$ stand for flat-tire fault. If the tire is flat, the friction is higher than normal. The three sensor faults are represented by: $\dot{\theta}_1$, $\dot{\theta}_2$ and $(p_2 - p_1)$. If one of the pressure sensors is faulty, then it is only possible to isolate the difference $(p_2 - p_1)$, as an abnormal parameter. It is important to note that although the parameter $(p_2 - p_1)$ appears in all GARRs, not all GARRs are sensitive to this fault. When R_6 is normal, then its value is $R_6 \to \infty$, and $GARR_{2,3}$ is insensitive to the fault $(p_2 - p_1)$. Using the single fault assumption, when $(p_2 - p_1)$ is faulty, then R_6 is definitely normal, and only $GARR_1$ and $GARR_4$ are sensitive to the pressure-sensor fault as given in the table.

From Table 6.1, it is clear that all considered parametric faults are isolable in the mode $[a_2\ a_1] = [1\ 1]$. However, the steering system is modeled as a hybrid system with two fault modes, and their monitoring process is based on the unique mode tracking technique of the GARR framework.

The mode tracking process utilizes the MCSM [12]. This matrix represents cause-effect relations between mode changes (modeled by the change of state of CJ) and residuals (GARRS). The first column in the MCSM presents control junctions state variables (a_i). The steering system MCSM is presented in Table 6.2.

From the MCSM, it is clear that all mode changes are detectable and isolable. The initial mode of the steering system is known; this is the normal mode $[a_2\ a_1] = [1\ 1]$. The initial mode is fed into the three GARRs, and as long as inconsistency in not detected by GARRs, it is clear that the system is normal, and its mode is $[a_2\ a_1] = [1\ 1]$. If inconsistency is detected by GARRs, then the detected signature (also known as coherence vector [10]) is utilized for mode-change isolation, or for parametric-fault isolation. If the detected signature is [0 0 1] or [0 1 1], then a mode change is hypothesized from Table 6.2, and a mode identification process is triggered. In the mode identification process, ARRs of suspected modes are derived from the GARRs and evaluated in real time. If all ARRs of a suspected mode are consistent with the monitored system, then, a new mode is identified. If new mode is not identified, then it is concluded that the system mode is $[a_2\ a_1] = [1\ 1]$ and the inconsistency is due to a parametric fault. The detected fault signature and the MD-FSM at Table 6.1 are utilized for parametric fault isolation. If the detected signature is not found in the MCSM, then it is clear that the inconsistency is due to a parametric fault, and the mode-change isolation process is skipped. On the other hand, since the signature [0 0 1] is unique in Table 6.2 (the MCSM) and does not appear in Table 6.1, then mode identification process is not necessary if this signature is detected (it is clear that the inconsistency is due to a burnt motor/driver fault).

The last stage in the quantitative health monitoring framework is fault parameter estimation. In addition to three sensor faults, the steering system may undergo two parametric faults, piston internal leakage with the fault parameter R_6, and flat tire described by the friction parameters $\{R_{7V}, R_{7C}\}$. The estimation process is based on the nonlinear least-square optimization method, where GARRs are utilized as residuals for the cost function, and mode changes are allowed when data are collected for the estimation process. The complete method is presented in Chap. 4. For the convenience of implementation, the nonlinear least-square problem is formulated in the discrete-time domain, and the cost function is based on N samples of data (collected from the faulty system after the detection of the fault). The targeted parameters for estimation are the fault candidates, as given by the fault isolation module. The GARRs selected to estimate the targeted parameters are those that contain the targeted parameter. For instance, if the fault R_6 has been isolated, then, the cost function is given as [13]

$$V(R_6) = \frac{1}{2} r^T r \tag{6.53}$$

where

$$r = \begin{bmatrix} \mathrm{GARR}_{23}(R_6, 1) \\ \mathrm{GARR}_{23}(R_6, 2) \\ \vdots \\ \mathrm{GARR}_{23}(R_6, N) \end{bmatrix} \tag{6.54}$$

For the normal system, (with $R_6 \to \infty$), $V(R_6)$ is theoretically zero (or below a threshold in practice). Hence, the estimated fault value is the value R_6 which minimizes the cost function $V(\cdot)$ [11]. The nonlinear least-square optimization problem can be solved iteratively by the Gauss-Newton method as described in Chap. 4. For the CyCab steering system, no mode change is expected after the occurrence of a parametric fault due to the single fault assumption, and the mode identification method in the presence of fault (as presented in Chap. 5) is not required here.

6.4 Experiment Study

The nominal physical parameters of the CyCab steering system are given in Table 6.3. Some of these parameters are taken from data sheets, and others were identified by the genetic algorithm technique [15].

Table 6.3 Values of the nominal physical parameters

Parameter	Value	Unit
R_{2V}	$1.017e^{-5}$	Nm/(rad/s)
R_{2C}	$6.20e^{-2}$	Nm
R_{3V}	$3.161e^{-1}$	Nm/(rad/s)
R_{3C}	$3.55e^{-2}$	Nm
R_{7V}	33.28	Nm/(rad/s)
R_{7C}	22.11	Nm
J_1	$3.727e^{-5}$	kg m^2
J_2	$5.627e^{-4}$	kg m^2
J_3	$9.78e^{-2}$	kg m^2
k_1	3	A/V
k_2	$5.627e^{-4}$	Nm/A
k_3	4/70	/
k_4	1/2	/
k_5	$1.024e^{-4}$	m^3
A	$4.46e^{-4}$	m^2
L_1	0.18	m
L_2	0.435	m
L_3	0.16	m
L_4	0.47	m
L_6	0.405	m
L_7	0.028	m
L_8	0.465	m
L_9	0.065	m
x_0	0.376	m

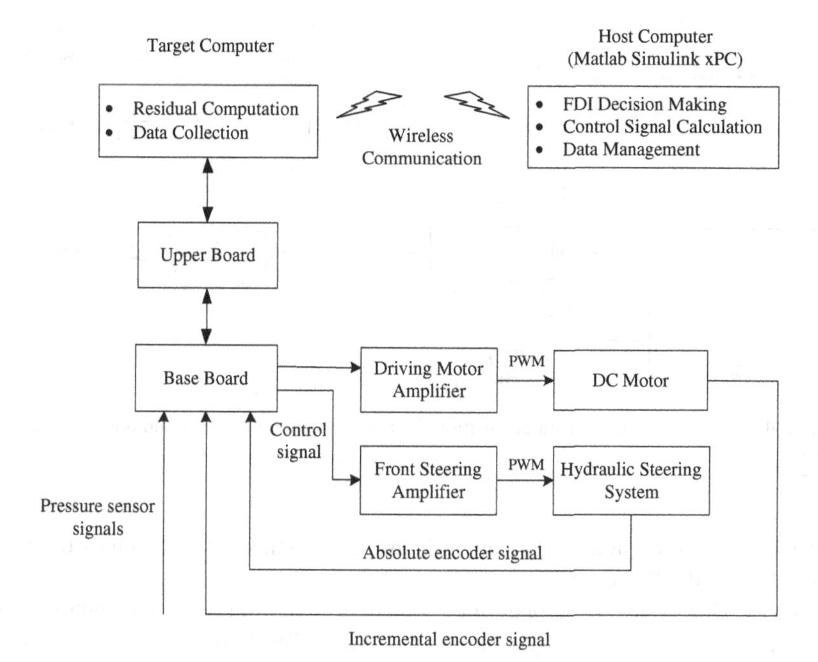

Fig. 6.9 Structure and hardware configuration of CyCab steering system

6.4.1 Experimental Hardware and Software

The mobile robot has four wheels which are driven by four independent DC motors. An incremental encoder is fitted on each DC motor to measure the wheel's angular velocity. An absolute encoder is installed on the robot's steering system to measure the steering angle. An incremental encoder mounted on the reducing output shaft of DC motor for steering is used to measure the angular velocity. Figure 6.9 depicts the overall structure and hardware configuration of CyCab steering system test bed. A notebook with Window operating system hosts the online FDI software. The software uses MatlabR2007B with Simulink® xPC Target to run and control real-time applications on the target PC. The host PC, i.e., the notebook, builds the FDI model using Simulink® blocks.

The Microsoft Visual C++ compiler in host PC creates executable code according to the Simulink® FDI model. The executable code is downloaded from the host PC to the target PC running the xPC Target real-time kernel. After downloading the executable code, one can run and test the target application in real time.

There are two circuit boards, namely base board and upper board, for connecting vehicle's I/O signals and sensor data to xPC target's PCI cards. The main purpose of base board is to connect all the sensors and I/Os into one place, and make a system more manageable and easy to understand. Signals connected to base board include, driving signal, steering signal, absolute encoder signal, incremental encoder signal

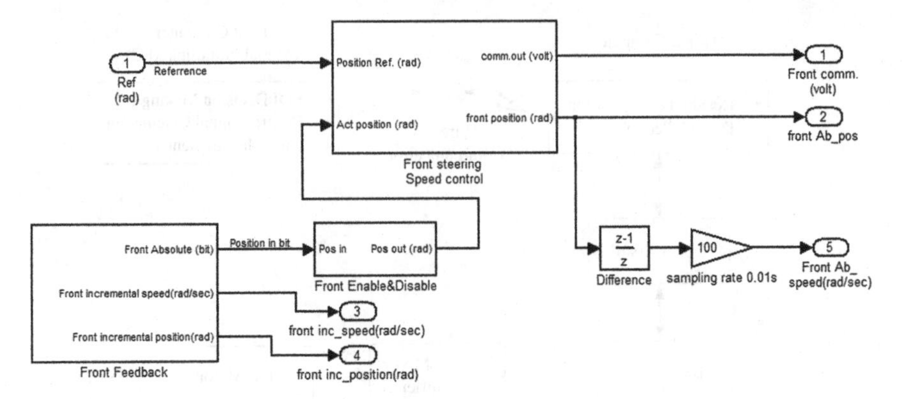

Fig. 6.10 Simulink model of data acquisition for the incremental encoder and absolute encoder signals

and pressure sensor signals. The upper board is to connect the all signals from the base board to the PCI cards.

The steering system is an open-loop system. The motion of the DC motor of the steering system is generated by a current-modulated PWM voltage from a DC motor amplifier. The translation of the piston is created under the pressure of the oil inside the cylinder, which in turn is pumped by a set of a DC motor and a hydraulic pump.

The target PC carries out the real time residual computation using measurements acquired from the sensors. All the measurements from the sensors are interfaced with the target PC via the PCI cards. The host PC performs the online FDI decision making using residual measurements acquired from the target PC. The host computer accesses the low-level target PC (onboard computer) through wireless communication using local wireless network.

Figure 6.10 describes the Simulink® model of data acquisition for the incremental encoder and absolute encoder signals. The signal from the absolute encoder is the steering angle. In order to obtain the steering angular velocity, the difference operation is used. Figure 6.11 depicts the Simulink® model of data acquisition for pressure sensor signals. The NI PCI-6025E data acquisition card is used for this purpose. All of the signals obtained are fed into the FDI block as shown in Fig. 6.12.

Figure 6.13 demonstrates the internal structure of the Simulink® model of the FDI process. In the model, signals from the pressure sensors are voltage signals and 1 Volt is equivalent to 5 bar. Thus these signals need to be multiplied by 5 before they are fed into the GARR equations. For ease of analysis, the absolute values of the residuals are used for final decision making process. The online FDI decision making is carried out using Matlab® m-file. In real experiment, the residual threshold selection is set by trial and error to reduce the possibilities of misdetection and false alarm. It is not reliable to consider the system as faulty if only one or two sampling residual values exceed the threshold during observation. So only if three or more successive residual values exceed the predefined threshold, the system is considered as faulty. After the fault occurrence, if three or more successive residual values return to threshold, then

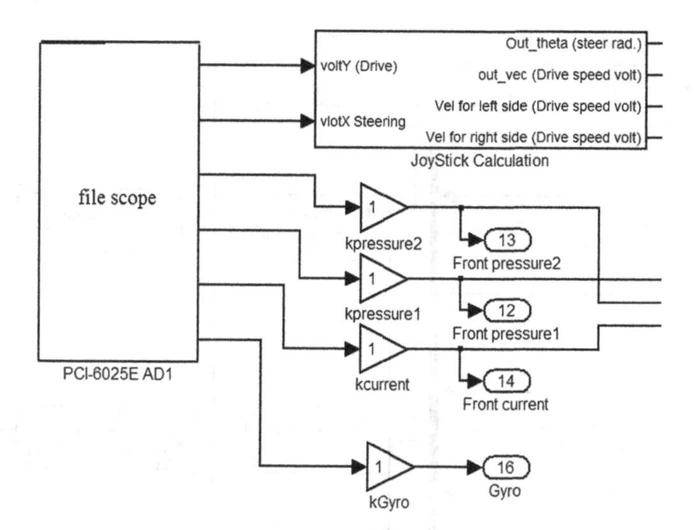

Fig. 6.11 Simulink model of data acquisition for the pressure sensor signals

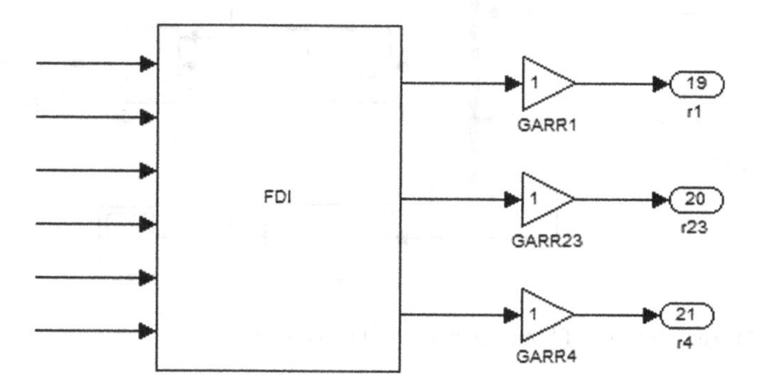

Fig. 6.12 Simulink model of FDI process

the system is considered as normal. Once a fault is detected, the fault signature is recorded and compared with the MD-FSM in Table 6.1 and the MCSM in Table 6.2 to carry out fault isolation.

6.4.2 Result and Analysis

The input signal of the CyCab steering system is defined as

$$u_{in} = 0.17(5 \sin 2t + 0.5 \sin t + 0.2 \sin 0.5t + \sin 2.5t) \tag{6.55}$$

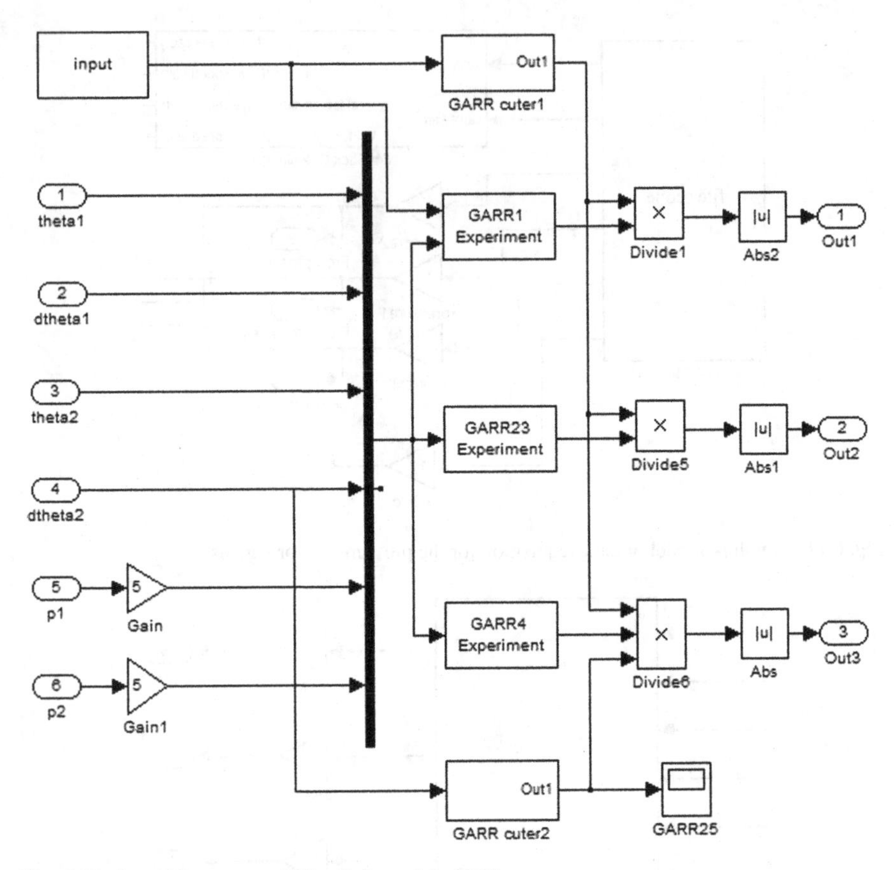

Fig. 6.13 Internal structure of Simulink model of FDI process

Six different fault scenarios are demonstrated. The first scenario is a sensor fault at the pressure sensor p_1. Due to a shortcut in the sensor electrical circuit, the sensor outputs is 5 V (which is equivalent to 25 bar, in a normal sensor). The fault is introduced at $t = 30$ s, and the three GARRs (absolute value) are presented in Fig. 6.14. From the graph, it is clear that the detected signature is [1 0 1] (the shaded areas on the graph represent thresholds). Using Table 6.1, a parametric fault in $(p_2 - p_1)$ is isolated.

The second fault scenario is the piston internal leakage. This fault is simulated by a special valve, which has been installed in the hydraulic system. At about $t = 30$ s, the valve is partially opened; the piston is still working, but its efficiency is decreased. The three GARRs are presented in Fig. 6.15. The detected signature is [0 1 0], and the parametric fault R_6 is isolated by Table 6.1.

After detection and isolation of the parametric fault R_6, 1000 samples of data are being collected for the fault estimation process. Figure 6.16 presents the cost function $V(R_6)$ in three cases. Case one is based on data, collected from the faulty system in

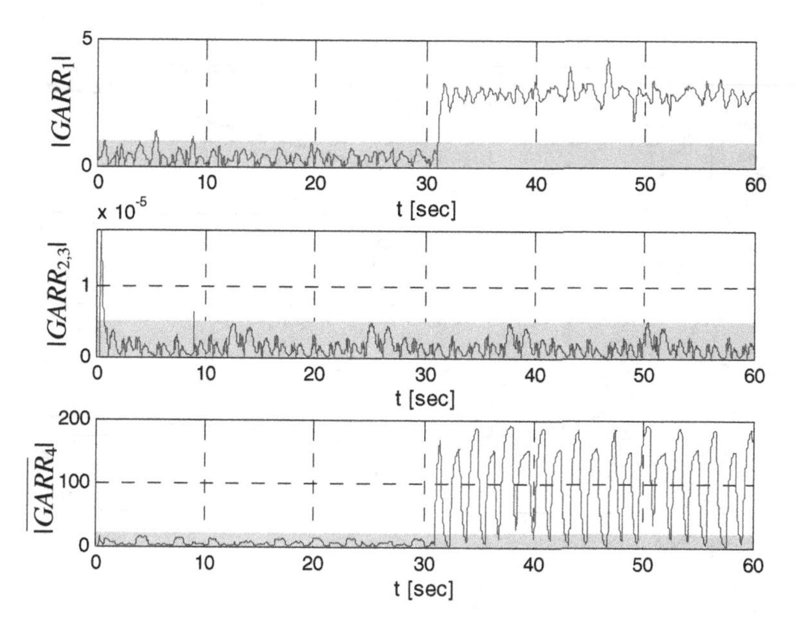

Fig. 6.14 GARRs: sensor-fault in p_1 (pressure sensor)

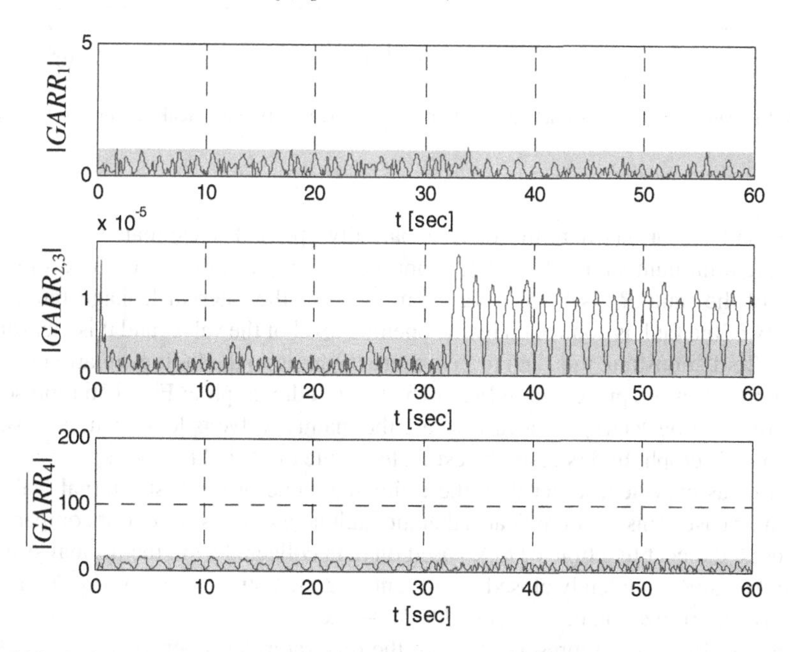

Fig. 6.15 GARRs: parametric fault in R_6 (internal leakage)

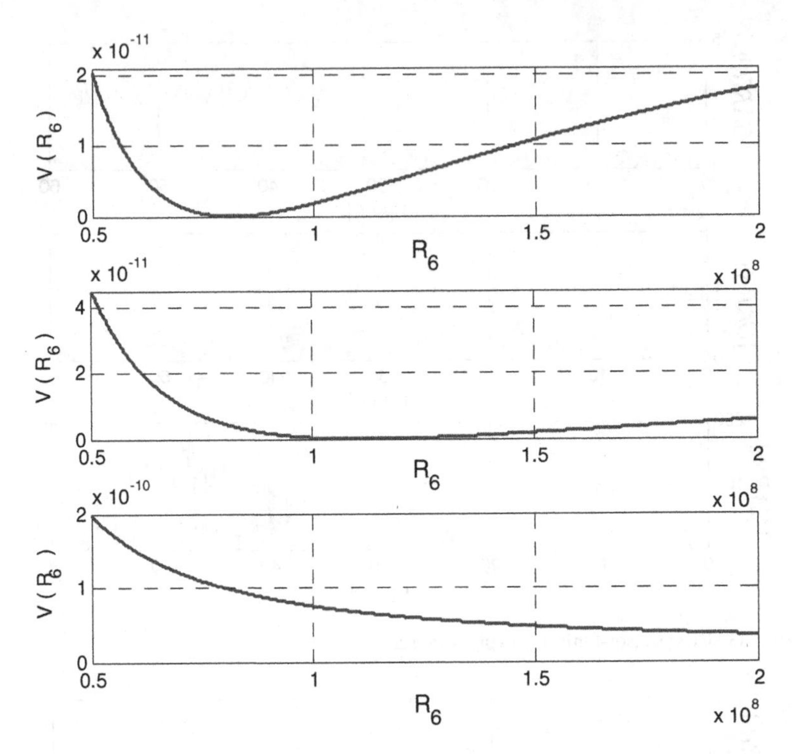

Fig. 6.16 Fault parameter estimation (R_6): internal leakage (*top*), minor leakage (*middle*), normal (*bottom*)

the second scenario (where the valve is partially open). The estimated R_6 is taken from the minimum value of $V(R_6)$. From the top graph in Fig. 6.16, the estimated value of the fault R_6 is $8.153\text{e}^7\ \text{kg}^{-0.5}\ \text{m}^{-0.5}$. For the internal leakage fault, the severity of the fault is controlled by the opening level of the valve, and it is measured by R_6. To demonstrate this, and to give more meaning to the fault estimation result, two more cases are presented in Fig. 6.16. The middle graph in Fig. 6.16 represents the cost function $V(R_6)$ for a case where the manual valve is less open, compared to the upper graph. In this case, the estimated value of R_6 is $11.216\text{e}^7\ \text{kg}^{-0.5}\ \text{m}^{-0.5}$, which is, as expected, larger than the estimated value of the first internal leakage presented case. This result indicates that the fault now is less severe. For comparison purposes, the cost function $V(R_6)$, based on data collected from the normal system (with the valve completely closed) is presented at the bottom graph of Fig. 6.16. For this case, minimization of $V(R_6)$ yields $R_6 \rightarrow \infty$.

The third scenario represents fault in the quadrature encoder $\dot{\theta}_1$. At about 30 s, the power supply to the sensor is suddenly stopped (due to a wire cutoff). Pulses are not generated by the sensor, any longer, and the counter reading (at the encoder interface card) remains constant. Due to the fault, the measured velocity $\dot{\theta}_1$ is read as zero, although the system is still running. The three GARRs are presented in

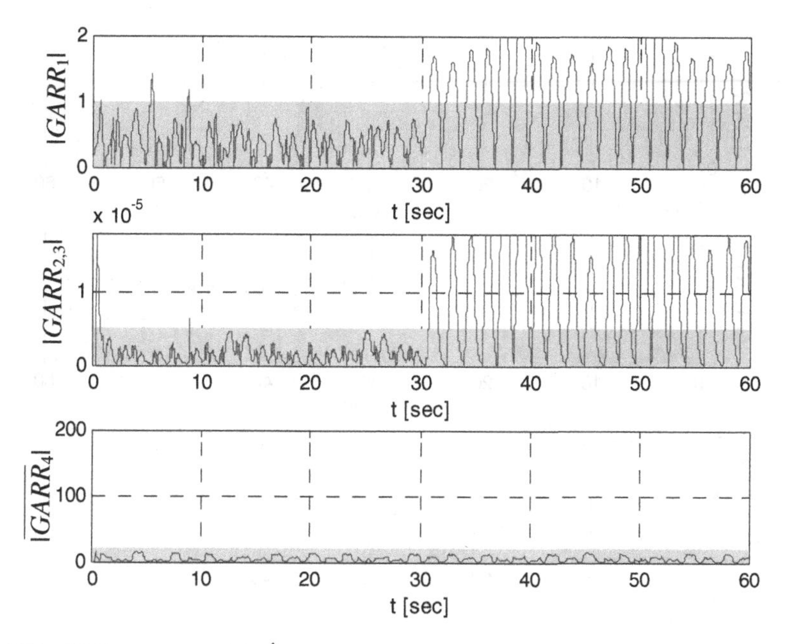

Fig. 6.17 GARRs: sensor fault in $\dot{\theta}_1$ (incremental encoder)

Fig. 6.17. From the graph, it is clear that the signature is [0 1 1]. This signature is also shared by the fault mode $[a_2\ a_1] = [0\ 1]$, and the signature itself is no sufficient for fault isolation. A further process of mode identification is required.

The fourth scenario demonstrates the broken belt fault; this fault is modeled as a fault mode. At about $t = 30$ s, the screw at the belt tension pulley is abruptly released, the belt is loose, and power is not delivered by the belt any more (this technique has been used instead of cutting the belt, although it remains some friction between the belt and the pulley, which makes the fault signature less clear). The three GARRs are presented in Fig. 6.18; from these results, the inconsistency is represented by the signature [0 1 1]. This signature may indicate a fault mode caused by a_2 (see the MCSM in Table 6.2 or a sensor fault represented by $\dot{\theta}_1$, as is demonstrated in scenario three. A distinction process is described in Fig. 6.19.

Figure 6.19 presents an evaluation of the two GARRs, $GARR_1$ and $GARR_{2,3}$, with the assumed mode $[a_2\ a_1] = [0\ 1]$. This evaluation is based on 1000 samples of data, collected from the system after the detection of the fault, and it is related to the third and fourth scenarios (1000 samples for each scenario). The two graphs at the left-hand side are related to the fault $\dot{\theta}_1$, and the two graphs at the right-hand side are corresponded to the fault a_2. From this evaluation, it is clear that for scenario four all ARRs of the mode $[a_2\ a_1] = [0\ 1]$ are consistent. Hence for scenario four, the broken belt fault is identified, likewise, in scenario three not all ARRs of the suspected mode show consistency, and the isolated fault is $\dot{\theta}_1$.

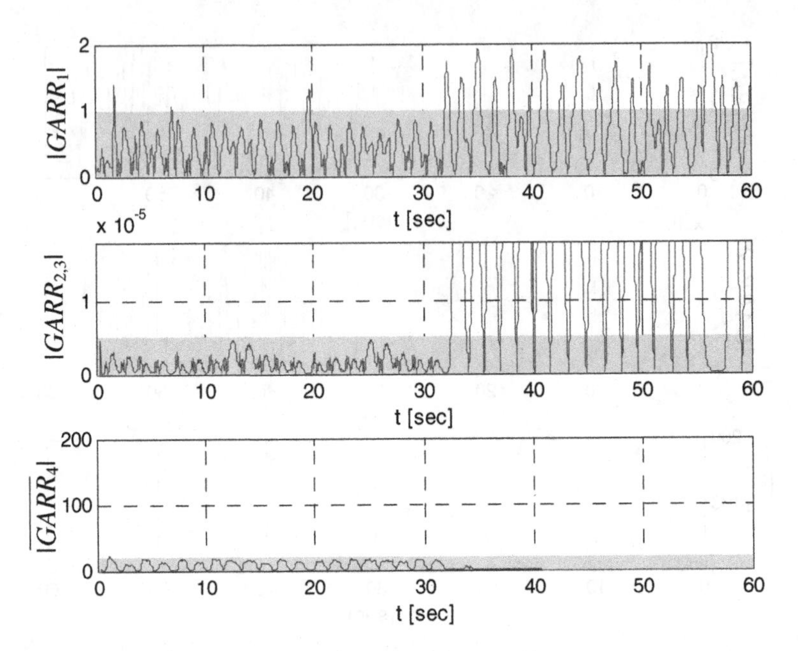

Fig. 6.18 GARRs: fault-mode related to a_2 (broken belt)

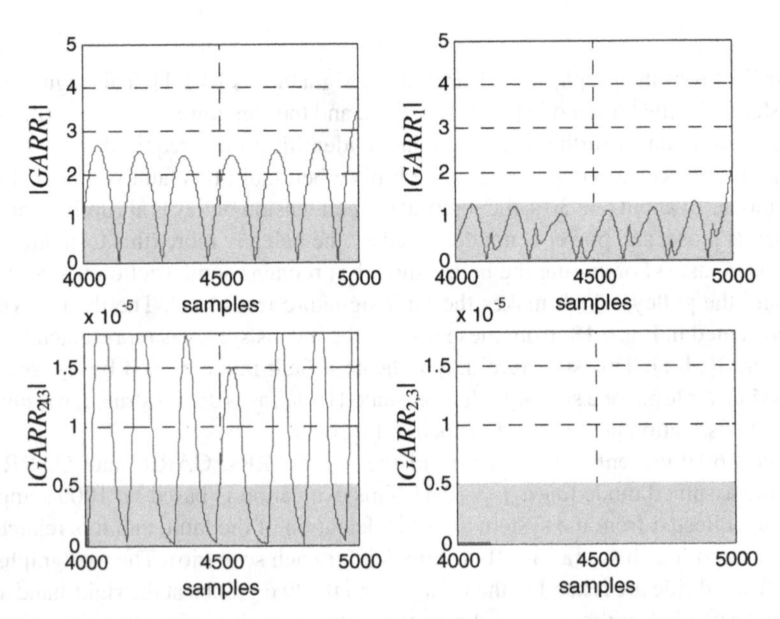

Fig. 6.19 GARR evaluation for mode identification and parametric fault isolation: sensor fault-$\dot{\theta}_1$ (*left*) versus a_2-Broken belt (*right*)

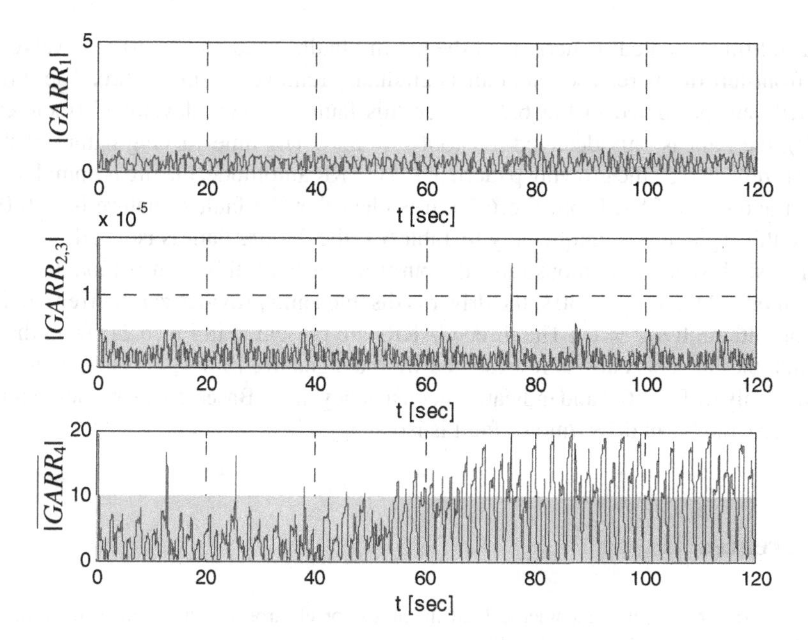

Fig. 6.20 GARRs: parametric fault in R_7 (flat tire)

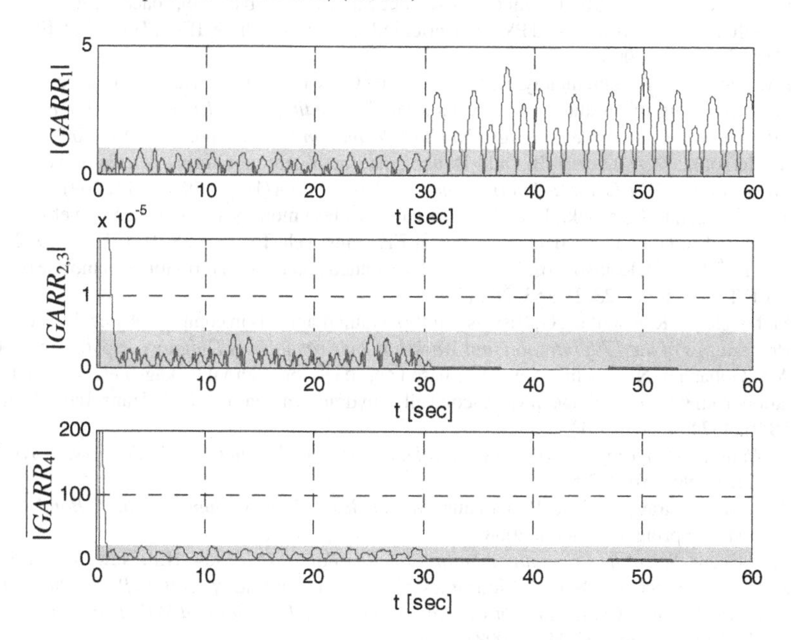

Fig. 6.21 GARRs: fault mode related to a_1 (burnt driver)

The fifth presented scenario is of the flat tire fault. At about $t = 30$ s the valve at the front-left tire is released, and air is gradually removed from the tire. The three GARRs are presented in Fig. 6.20. Since this fault is slowly developed (incipient fault), the fault is only detected at about $t = 55$ s. The time-varying nature of the fault is reflected by the growing pattern of the GARR amplitude; the tire is completely flat at about $t = 75$ s. From Fig. 6.20, it is clear that the fault signature is [1 0 0]. Since this signature is unique only to Table 6.1, the flat tire fault is isolated.

The sixth scenario demonstrates the burnt-driver fault; this fault is modeled as a fault mode. At about $t = 30$ s, the driver is disabled and provides zero current to the motor (although $u_{in} \neq 0$). The three GARRs are presented in Fig. 6.21; from these results, the inconsistency is represented by the signature [0 0 1]. This signature is unique only to Table 6.2 and indicates inconsistency in a_1. Based on the mode-change signature, the burnt driver/motor fault is isolated.

References

1. H.Z. Tan, N. Sepehri, Parametric fault diagnosis for electrohydraulic cylinder drive units. IEEE Trans. Ind. Electron. **49**(1), 96–106 (2002)
2. F. Zidani, D. Diallo, M.E.H. Benbouzid, R. Nait-Sait, A fuzzy-based approach for the diagnosis of fault modes in a voltage-fed PWM inverter induction motor drive. IEEE Trans. Ind. Electron. **55**(2), 586–593 (2008)
3. P.M. Pathak, A.K. Samantaray, R. Merzouki, B. Ould-Bouamama, Reconfiguration of directional handling of an autonomous vehicle, in *Proceedings of IEEE Region 10 and the 3rd International Conference on Industrial and Information Systems*, pp. 1–6, Dec. 2008
4. R. Merzouki, M.A. Djeziri, B. Ould-Bouamama, Intelligent monitoring of electric vehicle, in *Proceedings of IEEE/ASME International Conference on AIM*, pp. 797–804 (2009)
5. M.A. Djeziri, R. Merzouki, B. Ould-Bouamama, Robust monitoring of an electric vehicle with structured and unstructured uncertainties. IEEE Trans. Veh. Technol. **58**(9), 4710–4719 (2009)
6. M.J. Daigle, X.D. Koutsoukos, G. Biswas, Distributed diagnosis in formations of mobile robots. IEEE Trans. Robot. **23**(2), 353–369 (2007)
7. M. Daigle, X. Koutsoukos, G. Biswas, Multiple fault diagnosis in complex physical systems, in *Proceedings of the 17th International Workshop on Principles of Diagnosis*, pp. 69–76 (2006)
8. A.Y. Goharrizi, N. Sepehri, A wavelet-based approach for external leakage detection and isolation from internal leakage in valve-controlled hydraulic actuators. IEEE Trans. Ind. Electron. **58**(9), 4374–4384 (2011)
9. J. Watton, *Modelling, Monitoring and Diagnostic Techniques for Fluid Power Systems* (Springer, New York, 2007)
10. A.K. Samantaray, B. Ould Bouamama, *Model-Based Process Supervision: A Bond Graph Approach* (Springer, London, 2008)
11. M.T. Heath, *Scientific Computing: An Introductory Survey* (McGraw-Hill, New York, 1996)
12. S. Arogeti, D. Wang, C.B. Low, Mode tracking and FDI of hybrid systems, in *Proceedings of the 10th International Conference on Control, Automation, Robotics and Vision (ICARCV2008)*, Hanoi, Vietnam, pp. 892–897 (2008)
13. C.B. Low, D. Wang, S. Arogeti, M. Luo, Fault parameter estimation for hybrid systems using hybrid bond graph, in *The 3rd IEEE Multi-Conference on Systems and Control (MSC 2009)*, Saint Petersburg, pp. 1338–1343 (2009)
14. B. Ould-Bouamama, A.K. Samantary, M. Staroswiecki, G. Dauphin-Tanguy, Derivation of constraint relations from bond graph model for fault detection and isolation, in *International*

Conference on Bond Graph Modeling and Simulation (ICBGM'03), Simulation Series, vol. 35, number 2, pp. 104–109 (2003)
15. M. Yu, D.Wang, S. Arogeti, GA-based fault parameter identification for hybrid system with unknown mode changes, in *Proceedings of ISSCAA*, pp. 1–5 (2008)

Chapter 7
Multiple Failure Prognosis for Hybrid Systems

7.1 Prognosis of Multiple Incipient Faults

Recently, maintenance policies have evolved from early reactive maintenance, to age-based preventive maintenance, then to nowadays condition-based maintenance. Reactive maintenance is carried out usually after the system breakdown. In order to prevent emergency shutdowns and catastrophic failures, age-based preventive maintenance is introduced where maintenance is carried out based on system operating time regardless of the health condition of a machine. Age-based preventive maintenance may sometimes reduce unexpected failures, but it is not cost effective and cannot eliminate catastrophic failures. Consequently, these conventional maintenance strategies do not fulfill the needs of high reliability and better quality industrial equipments. Fortunately, CBM can be an effective alternative where it attempts to avoid unnecessary maintenance by taking maintenance actions only when there is evidence of abnormality in the monitored system [1].

Three key elements of effective CBM are data acquisition (i.e., the collection of machine health information), data processing (i.e., the conditioning and feature extraction/selection of acquired data) and decision making (i.e., the recommendation of maintenance actions through diagnosis and/or prognosis) [2]. Today's concept of machine diagnosis comprises the automated detection and classification of faults, whereas machine prognosis is the automated estimation of how soon and likely a failure will occur [3]. A CBM program can be used to do diagnostics or prognostics, or both. No matter what the objective of a CBM program is, the above three CBM steps are followed.

The word 'prognosis' indicates the foretelling of the probable course of a disease, a term widely used in medical practice. In the industrial and manufacturing arenas, prognosis is the capability to provide early detection and isolation of precursor and/or incipient fault condition to a component failure condition and to have the technology and means to manage and predict the progression of this fault condition to component failure [4].

D. Wang et al., *Model-based Health Monitoring of Hybrid Systems*,
DOI: 10.1007/978-1-4614-7369-5_7,

The time left before observing a failure is usually called remaining useful life (RUL). RUL, also called remaining service life, residual life or remnant life refers to the time left before observing a failure given the current machine age and condition, and the past operation profile. It is defined as the conditional random variable:

$$T - t|T > t,\ Z(t) \tag{7.1}$$

where T denotes a random variable of time to failure (TTF); t represents current age and $Z(t)$ denotes past condition profile up to the current time.

Many industrial systems exhibit increasing wear and tear of equipment during operation. For example, an automobile has many components, such as the suspension, gear box, and valves which exhibit different types of performance degradation due to erosion, friction, internal leakage, and cracks. Prognosis is viewed as an add-on capability to diagnosis. It is based on the analysis of failure modes, detection of early signs of wear and aging, and fault conditions. These signs are then correlated with a damage propagation model. The discipline that links studies of failure mechanisms to system life cycle management is often referred to as Prognostics and Health Management (PHM). In recent years, PHM has emerged as one of the key enablers for achieving efficient system-level maintenance and lowering life-cycle costs.

The most commonly used failure definition is: failure occurs when the fault reaches a predetermined level. Prognosis performs the important linkage of the diagnostic information with the maintenance scheduler. It is probably the least understood but most crucial part of the diagnostic/prognostic/CBM hierarchical architecture. Furthermore, it entails ambiguity and large-grain uncertainty. For example, it is difficult to model the degradation of a failure accurately due to the variations of operating conditions and environmental effects, or historical data is not readily available.

7.1.1 Augmented Global Analytical Redundancy Relations

In Chap. 3, the concept of GARRs has been introduced to describe the behavior of a hybrid system at all modes. However, these GARR equations can not be used directly for the identification of degradation of those components such as sensors and actuators which cannot be described by physical parameters. In this part, a new set of constraints, called Augmented Global Analytical Redundancy Relations, is developed to augment the capability of GARRs to identify the degradation of components which cannot be described by physical parameters. Without loss of generality, AGARR equations take the following form

$$\overline{F}_l(\theta, a, \beta, De, Df, u) = 0 \quad \text{for } l = 1, ..., m. \tag{7.2}$$

where m denotes the number of GARRs derived from the DHBG, $\theta = [\theta_1, ..., \theta_p]^T$ represents the parameters of the HBG components, where p represents the number of HBG parameters used to describe the hybrid system. $a = [a_1, ..., a_n]^T$ denotes the operating mode of the hybrid system. u denotes the system's inputs, De and Df denote the effort and flow sensors of the graph. $\beta = [\beta_1, ..., \beta_g]^T$, where g is the total number of nonparametric components, such as sensors and actuators.

In this part, the degradation of component is divided into two categories: parametric degradation and nonparametric degradation. Parametric degradation is characterized by the slowly changing of component parameter. For example, blockage of pipe is a typical example of incipient fault and the blockage, often modeled as resistance, usually increases over time. Nonparametric degradation refers to the deterioration of system components that are under monitoring but cannot be described by physical parameters. Examples include sensors, actuators, and subsystems. In order to quantify the severity of the fault, an efficiency factor β is introduced, where $\beta = 0$ indicates a total failure while $\beta = 1$ denotes no fault. $0 < \beta < 1$ represents a partial failure and $1 - \beta$ indicates the severity of the fault. When the system is under normal condition, all β are equal to 1, when a fault occurs, β related to the fault candidates drift away from value 1 and can be treated as unknown parameters in the dynamic model. In the HBG, actuator fault is modeled by multiplying the input by β as shown in Fig. 7.1, and sensor fault is modeled by multiplying the normal measurement by β as shown in Fig. 7.2. In normal condition, all β are known and equal to 1, when fault happens, those β related to the fault candidates become unknown parameters.

It is worth noting that in the AGARR equations described in (7.2), De and Df represent sensor measurements and can be faulty signals or normal signals. Since sensor faults are represented by multiplying normal measurement by β, normal measurement of sensors are represented by dividing De or Df by β in AGARR equations. In normal operation condition, all β are known and equal to 1. In a sensor total failure condition, De or Df is equal to 0 during observation window, all β related to the fault candidates (De or Df) are =0. For the other cases, i.e., partial failure with $0 < \beta < 1$, those β related to the fault candidates (De or Df) become unknown parameters.

Figure 7.3 shows the parametric degradation profile, in which curve A represents the gradual increasing of parameter due to parametric fault. For example, resistance value increases caused by pipe blockage. The curve B represents the gradual decreasing of parameter due to parametric fault. Figure 7.4 represents the nonparametric degradation profile, in which the efficiency factor β decreases gradually over time. In Fig. 7.3, t_0 is the time point at which the degradation first starts. Ideally,

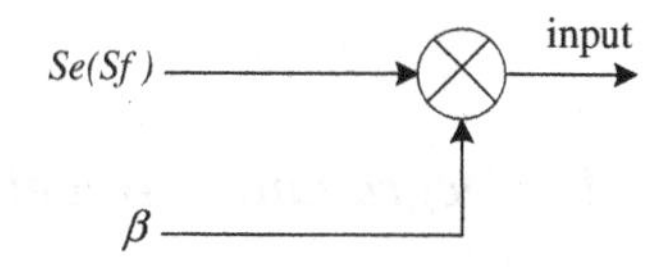

Fig. 7.1 Modeling of actuator fault in HBG

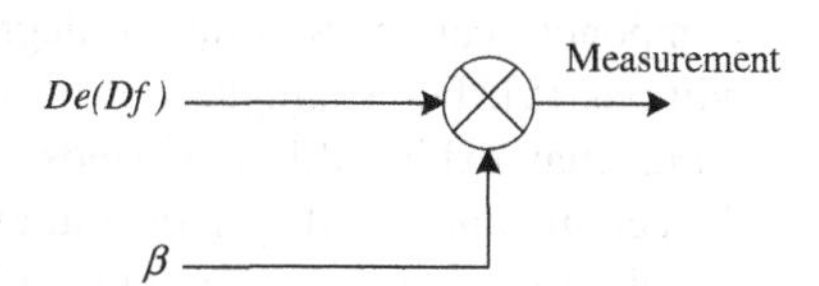

Fig. 7.2 Modeling of sensor fault in HBG

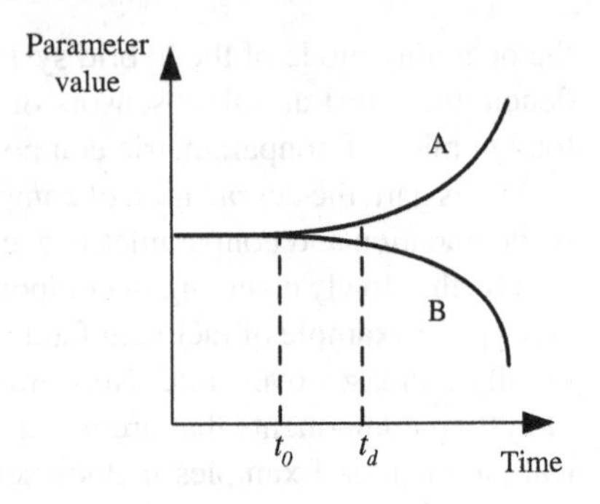

Fig. 7.3 Parametric degradation profile

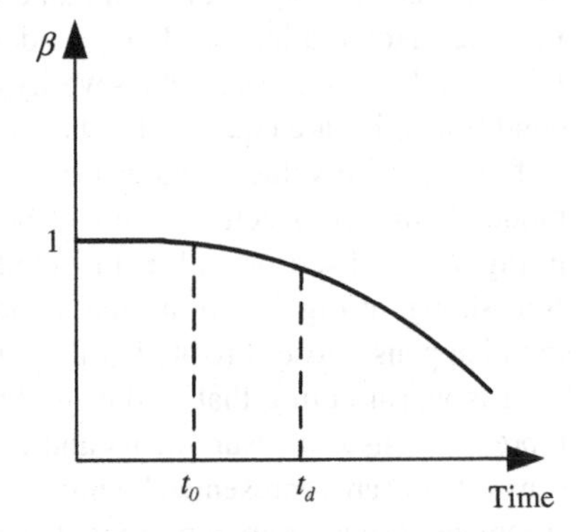

Fig. 7.4 Nonparametric degradation profile

deviations in residuals caused by faults and degradation should be detected very soon after fault occurrence. In practice, a certain delay between the time of fault occurrence and detection of fault can happen due to the fact that the degradation can occur at the non-detectable mode and is detected at a detectable mode, i.e., $t_d > t_0$. For example, in a hybrid system with $GARR_1 = V_1 - a_1 R_1 Df - De$, resistor R_1 degrades at a non-detectable mode, i.e., $a_1 = 0$, at t_0. Since the term $a_1 R_1 Df$ takes value 0 in at that mode, no matter what value R_1 takes, $a_1 R_1 Df$ is always equal to 0, $GARR_1$ is not sensitive to the fault in R_1 at $a_1 = 0$, so the fault cannot be detected by $GARR_1$. When the system enters a new mode, i.e., $a_1 = 1$, at t_d, the fault in R_1 can be detected by $GARR_1$.

7.1.2 Degradation Models

A component can show different degradation behaviors due to different operation conditions [5]. For example, a critical component of rotating machinery such as bearings may exhibit different kinds of incipient faults, such as corrosion, spalling and cracking, due to different operating conditions including load and environments. The fault progression of different incipient faults may be linear or nonlinear. In order to accurately predict the progression of incipient fault, the following first/second order ordinary differential equations (ODE) are used to describe the degradation progression of incipient faults

$$\begin{aligned}
\dot{F} &= b_1 F \\
\dot{F} &= b_2 F^2 \\
\ddot{F} &= b_3 \dot{F} + F + b_4 \\
\ddot{F} &= b_5 \dot{F}^2 + b_6 F + b_7
\end{aligned} \tag{7.3}$$

where F represents the fault component value, i.e., fault parameter value for parametric fault or β value for nonparametric fault; $b_1, b_2, b_3, b_4, b_5, b_6$ and b_7 are unknown degradation model coefficients.

Adams et al. [6] used dynamic degradation models with an additive term which varies over time. Here, some prescribed models in (7.3) are deployed for the behaviors of parameters and/or β evolving over time. These four degradation models only serve as examples to illustrate the proposed methodology and these prescribed degradation models should be chosen based on applications and understandings of the faulty components.

7.1.3 Particle Swarm Optimization for Prognosis

Let's consider a hybrid system with m AGARRs $\overline{F}_l(\theta, \beta, n)$ for $l = 1, 2, \ldots, m$ with $\theta \in \mathbb{R}^p$ and $\beta \in \mathbb{R}^g$. The variable n denotes the discrete sampling index. Once a fault is detected, N sample data are captured. Under no fault condition, $\theta = \overline{\theta}$ and $\beta = \overline{\beta}$, with $\overline{\theta}$ and $\overline{\beta}$ being the nominal values of parameters and efficiency factors. If a fault-detectable component i.e., $\theta_i \in \theta$ or $\beta_i \in \beta$, is faulty, the AGARRs $\overline{F}_l(\overline{\theta}, \overline{\beta}, n)$ that contains θ_i and/or β_i will be nonzero, and thus a fault is declared. After a fault is detected, the fault isolation module is activated to find a set of suspected fault candidates. If a fault is detected at the time of mode change, then the fault candidates include those parameters θ_i and/or β_i of such fault signature which is non-detectable in the previous mode. This indicates that if a fault occurs at the detectable mode, it will be detected before mode changes; on the contrary, if a fault starts at the non-detectable mode, this fault will be detected only when the system enters a new mode in which the fault is detectable. Consider Mode Dependent-FSM (MD-FSM) shown in Tables 7.1 and 7.2. In the tables, $a \in \{0, 1\}$ represents the operating mode of the hybrid system; θ_i, $i = 1, 2, 3, 4$ denote the parameters; β_i, $i = 1, 2, 3, 4$ represent the efficiency factors related to sensors or actuators; r_i, $i = 1, 2, 3$ denote the three residuals. The initial mode is $a = 0$, and the mode changes at 5 s. When residual r_3 becomes abnormal at 5 s, it leads to a coherence vector $C = [0\ 0\ 1]$ at mode $a = 1$. From Table 7.2, the fault could be due to changes of either θ_2, θ_3, β_1, or their combination. It is noted that the fault is detected at the time of mode change, and only

Table 7.1 MD-FSM at $a = 0$

	r_1	r_2	r_3	D_b
θ_1	1	0	0	1
β_1	0	0	0	0
θ_2	0	0	1	1
β_2	0	1	0	1
θ_3	0	0	0	0
β_3	0	1	0	1
θ_4	0	1	1	1
β_4	0	1	0	1

Table 7.2 MD-FSM at $a = 1$

	r_1	r_2	r_3	D_b
θ_1	1	0	0	1
β_1	0	0	1	1
θ_2	0	0	1	1
β_2	0	1	0	1
θ_3	0	0	1	1
β_3	0	1	0	1
θ_4	0	1	1	1
β_4	0	1	0	1

θ_3 and β_1 are non-detectable at previous mode $a = 0$. Therefore, the suspected fault candidates set could be $\delta = \{\theta_3, \beta_1, \theta_3\&\beta_1\}$. In the MD-FSM, the fault isolability (I_b) is not required. Unlike fault analysis under single fault assumption, the uniqueness of fault signature of a component can be due to one or multiple faults. For example, if both r_2 and r_3 are abnormal at mode $a = 1$, it leads to a coherence vector $C = [0\ 1\ 1]$. This fault signature is unique in Table 7.2, and is isolable under single fault assumption. However, for a system having multiple faults, the fault signature could be due to change of a single parameter θ_4 or due to simultaneous changes in more than one parameters, e.g., θ_2 and β_2, θ_3 and β_4, etc.

If the parameters θ and efficiency factors β are equal to the nominal values then $\overline{F}_l(\theta, \beta, n) = 0$ for $l = 1, 2, \ldots, m$. However, the parameters θ and efficiency factors β will change values over time due to degradation process. The identification of fault degradation requires matching the progressing profile from collected data to one of the prescribed degradation models. The matching is carried out through identifications of model coefficients. Once the model coefficients are determined, the identified model can be used to predict the RUL. For example, if θ_i is the true fault and the degradation profile matches the model of the first order nonlinear ODE, then the AGARR $\overline{F}_l(\theta, \beta, n) = 0$ that contains θ_i can be reformulated as

$$\overline{F}_l(sol, \theta_j, \beta, n) = 0$$

$$\text{with } sol = -\frac{1}{b_2 \cdot t - \frac{1}{\theta_i(0)}}$$
$$t = n \cdot t_s, j = 1, 2, \ldots, m, \text{and } j \neq i \tag{7.4}$$

where t_s is the sampling period. Note that in the solution, the initial condition $\theta_i(0)$ is not equal to the nominal value of θ_i. Since the fault initiates at the non-detectable mode, $\theta_i(0)$ is also an unknown coefficient.

With the identification process, an approximate model can be selected from a set of prescribed models. Here, a model selection scheme is developed as follows. A binary vector is defined as $\eta = [\eta_1 \quad \eta_2]$ and let the four binary values to associate with the four prescribed models given in (7.3), where $\eta = [0 \quad 0]$ means that the degradation model fulfills the first equation in (7.3), $\eta = [0 \quad 1]$ indicates that the degradation model matches the second equation in (7.3), $\eta = [1 \quad 0]$ means that the degradation model fulfills the third equation in (7.3) and $\eta = [1 \quad 1]$ indicates that the degradation model matches the last equation in (7.3). In this way, the model selection is represented by the binary selection in the identification process. The model selection scheme is shown in Fig. 7.5.

When a fault is detected and a set of suspected faults is isolated, the prognosis module is activated to predict the RUL of the faulty components. The first step of prognosis is to select a prescribed model and to identify the model coefficients. The next step is to predict the RUL of the faulty component(s) based on the identified degradation model. The identification aims to find continuous real numbers for model coefficients and discrete binary values for degradation models simultaneously. However, binary numbers representing different degradation behaviors are not involved in the

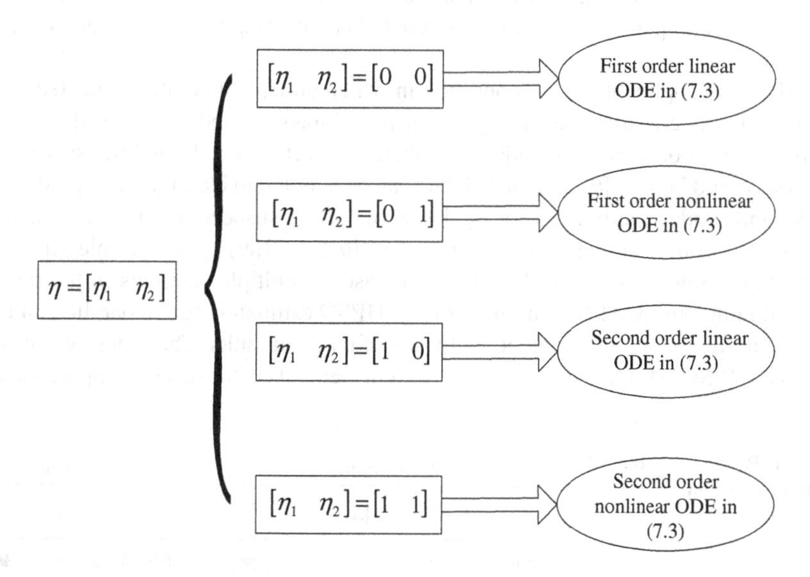

Fig. 7.5 A model selection scheme

fitness function, so traditional gradient based method cannot directly tackle this identification problem. In other words, gradient based optimization methods such as least square method and Gauss-Newton method, usually require the objective function to be differentiable with respect to the optimized parameters. In this work, the objective function to be minimized is not directly related to the binary vector η, thus no gradient information can be obtained regarding the binary numbers. Therefore, gradient-free stochastic optimization algorithms could be efficient alternatives. Stochastic optimization algorithms include ant colony optimization, GA, PSO, and so on. PSO does not have complex genetic operation and has fewer control parameters in PSO. All these features make PSO simple and easy to implement. As a result, PSO has been widely used in many areas, such as training neural networks [8, 9], control system design [7] and power system optimization [10].

The AHPSO developed in Chap. 4 is adopted to carry out the identification task, where RPSO is used to find continuous real numbers for model coefficients and initial values for certain degradation behavior, and BPSO is utilized to search the discrete binary values for degradation models. Figure 7.6 depicts the particle structure of AHPSO algorithm. In the figure, s is the number of coefficients for certain degradation model.

Both RPSO and BPSO evolve simultaneously and are coupled through the common fitness function defined as follows

$$F_{fitness} = \frac{1}{\sum_{l=1}^{m} \sum_{n=1}^{N} |\overline{F}_l^n| + \varepsilon} \tag{7.5}$$

with n represents the sampling time index, and ε is a small positive constant to avoid zero division during the search process. Since the GARRs are used to identify the targeted fault parameters, only those GARRs containing targeted parameters are selected.

AHPSO is applicable to the condition in which an isolable fault is detected. For instance, let us consider the example shown in Tables 7.1 and 7.2, if residual r_1 is abnormal at mode $a = 1$, it leads to a coherence vector [1 0 0], and the suspected faults set could be $\delta = \{\theta_1\}$. Then AHPSO can be adopted to identify the degradation model and model coefficients for θ_1. However, if the suspected faults set consists of more than one elements, for example, $\delta = \{\theta_3, \beta_1, \theta_3\&\beta_1\}$, a Multiple AHPSO (MAHPSO) algorithm is developed for the case of multiple elements in the set of targeted faults. In MAHPSO, number of P AHPSO estimators run in parallel, and P is equal to the number of elements in the set of targeted faults. The fitness values of various AHPSO estimators addressing different elements in the set are compared with

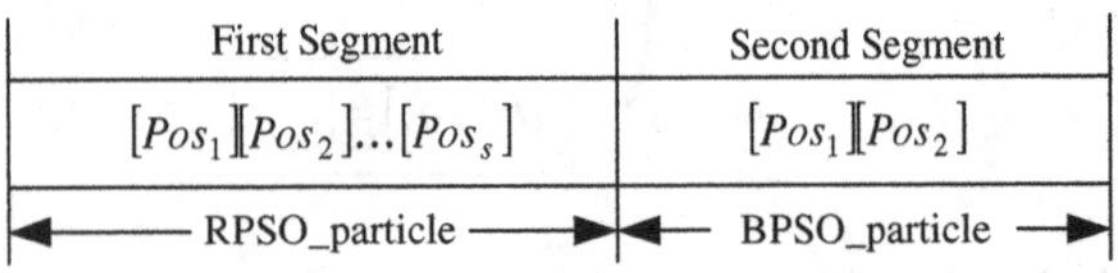

First Segment	Second Segment
$[Pos_1][Pos_2]...[Pos_s]$	$[Pos_1][Pos_2]$
← RPSO_particle →	← BPSO_particle →

Fig. 7.6 Particle structure of AHPSO algorithm

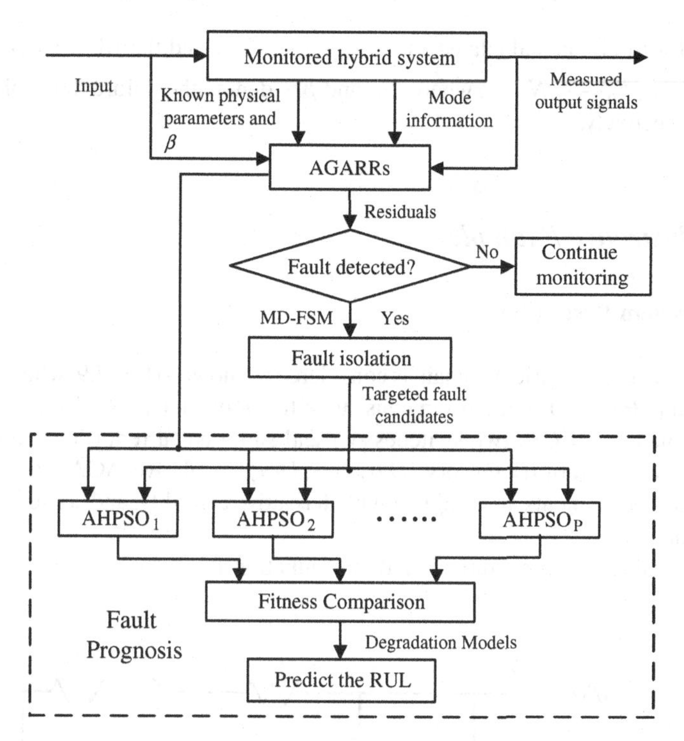

Fig. 7.7 Block diagram of the MAHPSO based prognosis method

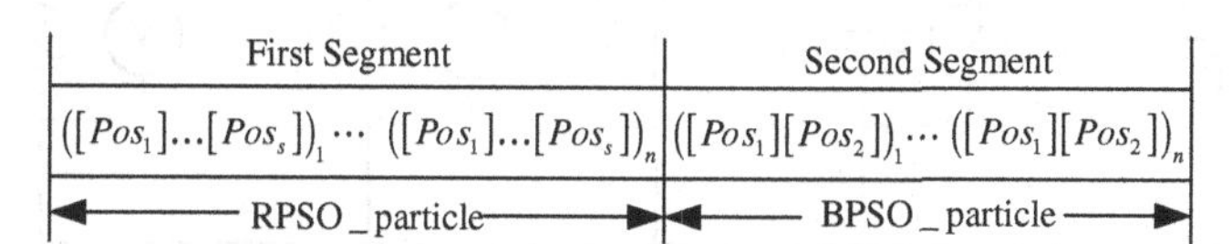

Fig. 7.8 Particle structure of AHPSO algorithm considering multiple faults

each other to choose the degradation model and model coefficients. MAHPSO based prognosis method is shown in Fig. 7.7. Moreover, if the element in the set of targeted faults includes more than one fault, for example, $\theta_3\&\beta_1$ in $\delta = \{\theta_3, \beta_1, \theta_3\&\beta_1\}$, the particle structure of corresponding AHPSO estimator needs to be modified, as shown in Fig. 7.8. In the figure, n is the number of faults in the element in the set of targeted faults.

According to (7.3), if the failure thresholds are given for different faults, then these thresholds are substituted into the corresponding solutions and the time to failure from the moment that faults are detected can be calculated. For example, consider a set of $\delta = \{\theta_3, \beta_1, \theta_3\&\beta_1\}$, where θ_3 and β_1 are true faults. The degradation profile of θ_3 matches the model of the first order nonlinear ODE, i.e., $\dot{\theta}_3 = b_2\theta_3^2$ and the degradation profile of β_1 matches the model of the first order linear ODE, i.e., $\dot{\beta}_1 = b_1\theta_1$. With the identified faults, estimated degradation models and coefficients, the RUL for θ_3 can be calculated as $\frac{\theta_{3F}-\theta_3(0)}{\theta_{3F}\theta_3(0)b_2} - N \cdot t_s$ and the RUL for β_1 can be computed as $\frac{\beta_{1F}}{\beta_1(0)b_1} - N \cdot t_s$, where θ_{3F} and β_{1F} denote the failure thresholds of θ_3 and β_1 respectively.

7.1.4 Illustrative Example

7.1.4.1 System Description

A hybrid system in electrical domain is considered as shown in Fig. 7.9, whose DHBG is depicted in Fig. 7.10. The system consists of five R elements $\{R_1, R_2, R_3, R_4, R_5\}$, two C elements $\{C_1, C_2\}$, two switches sw_1 and sw_2, one current (flow) sensor Df, and three voltage (effort) sensors $\{De_1, De_2, De_3\}$. To derive AGARRs from the DHBG, storage elements are assigned with derivative causality, and sensor causality is inverted.

First, consider the constitutive relation of junction 1_1

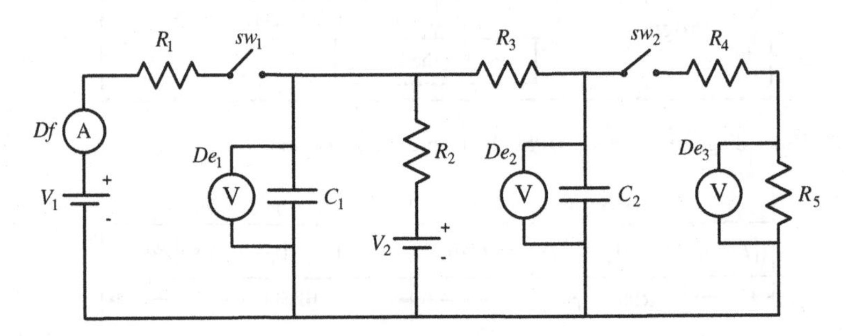

Fig. 7.9 A hybrid system: An electric circuit

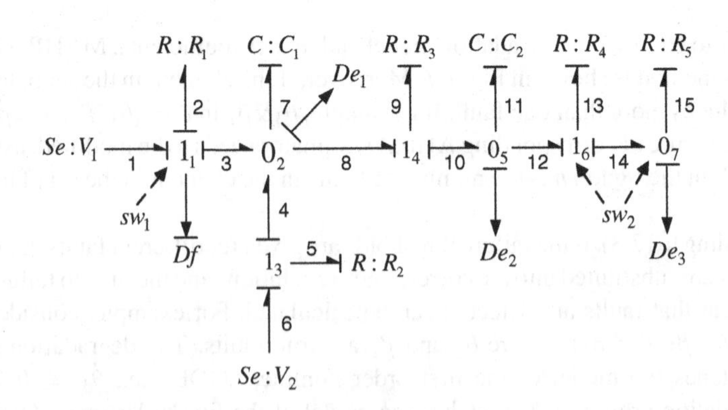

Fig. 7.10 DHBG of the electric circuit

$$e_1 - e_2 - e_3 = 0 \tag{7.6}$$

Since f_2 is measurable, $f_2 = Df$, and the causal paths lead e_2 to

$$e_2 = R_1 Df \tag{7.7}$$

Substituting (7.7) into (7.6), AGARR$_1$ is achieved

$$\text{AGARR}_1 = a_1(\beta_{V_1} \cdot V_1 - R_1 \frac{Df}{\beta_{Df}} - \frac{De_1}{\beta_{De_1}}) = 0 \tag{7.8}$$

Next, consider junction 0_2 attached to De_1, the constitutive relation of this junction can be expressed as

$$a_1 f_3 + f_4 - f_7 - f_8 = 0 \tag{7.9}$$

Then tracking back the causal paths yields

$$f_3 = Df \tag{7.10}$$

$$f_4 = f_5 = \frac{1}{R_2} e_5 = \frac{1}{R_2}(e_6 - e_4) = \frac{1}{R_2}(V_2 - De_1) \tag{7.11}$$

$$f_7 = C_1 \frac{d}{dt} e_7 = C_1 \frac{d}{dt} De_1 \tag{7.12}$$

$$f_8 = f_9 = \frac{1}{R_3} e_9 = \frac{1}{R_3}(e_8 - e_{10}) = \frac{1}{R_3}(De_1 - De_2) \tag{7.13}$$

Substituting (7.10)–(7.13) into (7.9) gives

$$\text{AGARR}_2 = a_1 \frac{Df}{\beta_{Df}} + \frac{1}{R_2}(\beta_{V_2} \cdot V_2 - \frac{De_1}{\beta_{De_1}}) - C_1 \frac{d}{dt}(\frac{De_1}{\beta_{De_1}}) - \frac{1}{R_3}(\frac{De_1}{\beta_{De_1}} - \frac{De_2}{\beta_{De_2}}) \tag{7.14}$$

Using junction 0_5 with De_2, yields

$$f_{10} - f_{11} - a_2 f_{12} = 0 \tag{7.15}$$

By following the causal paths, the flow variables f_{10}, f_{11} and f_{12} can be represented as

$$f_{10} = f_9 = \frac{1}{R_3} e_9 = \frac{1}{R_3}(e_8 - e_{10}) = \frac{1}{R_3}(De_1 - De_2) \tag{7.16}$$

$$f_{11} = C_2 \frac{d}{dt} e_{11} = C_2 \frac{d}{dt} De_2 \tag{7.17}$$

$$f_{12} = f_{13} = \frac{1}{R_4} e_{13} = \frac{1}{R_4}(e_{12} - e_{14}) = \frac{1}{R_4}(De_2 - De_3) \tag{7.18}$$

The third AGARR can be obtained by combining (7.15)–(7.18)

$$\mathrm{AGARR}_3 = \frac{1}{R_3}\left(\frac{De_1}{\beta_{De_1}} - \frac{De_2}{\beta_{De_2}}\right) - C_2 \frac{d}{dt}\frac{De_2}{\beta_{De_2}} - a_2 \frac{1}{R_4}\left(\frac{De_2}{\beta_{De_2}} - \frac{De_3}{\beta_{De_3}}\right) = 0 \tag{7.19}$$

Finally consider junction 0_7, with the constitutive relation

$$f_{14} - f_{15} = 0 \tag{7.20}$$

Tracking the casual paths gives

$$f_{14} = f_{13} = \frac{1}{R_4} e_{13} = \frac{1}{R_4}(e_{12} - e_{14}) = \frac{1}{R_4}(De_2 - De_3) \tag{7.21}$$

Then AGARR_4 is given by

$$\mathrm{AGARR}_4 = a_2\left[\frac{1}{R_4}\left(\frac{De_2}{\beta_{De_2}} - \frac{De_3}{\beta_{De_3}}\right) - \frac{1}{R_5}\frac{De_3}{\beta_{De_3}}\right] \tag{7.22}$$

From the four AGARRs derived, the MD-FSM can be obtained in Tables 7.3, 7.4, 7.5, 7.6. Note that β is used to represent the extent of the nonparametric fault, then β always has the same fault signature as that of its counterpart, e.g., β_{V_1} and V_1.

Table 7.3 MD-FSM at $a_1 = 0, a_2 = 1$

	r_1	r_2	r_3	r_4	D_b
β_{V_1}	0	0	0	0	0
β_{V_2}	0	1	0	0	1
C_1	0	1	0	0	1
C_2	0	0	1	0	1
R_1	0	0	0	0	0
R_2	0	1	0	0	1
R_3	0	1	1	0	1
R_4	0	0	1	1	1
R_5	0	0	0	1	1
β_{Df}	0	0	0	0	0
β_{De_1}	0	1	1	0	1
β_{De_2}	0	1	1	1	1
β_{De_3}	0	0	1	1	1

Table 7.4 MD-FSM at $a_1 = 0, a_2 = 1$

	r_1	r_2	r_3	r_4	D_b
β_{V_1}	1	0	0	0	1
β_{V_2}	0	1	0	0	1
C_1	0	1	0	0	1
C_2	0	0	1	0	1
R_1	1	0	0	0	1
R_2	0	1	0	0	1
R_3	0	1	1	0	1
R_4	0	0	0	0	0
R_5	0	0	0	0	0
β_{Df}	1	1	0	0	1
β_{De_1}	1	1	1	0	1
β_{De_2}	0	1	1	0	1
β_{De_3}	0	0	0	0	0

Table 7.5 MD-FSM at $a_1 = 0, a_2 = 0$

	r_1	r_2	r_3	r_4	D_b
β_{V_1}	0	0	0	0	0
β_{V_2}	0	1	0	0	1
C_1	0	1	0	0	1
C_2	0	0	1	0	1
R_1	0	0	0	0	0
R_2	0	1	0	0	1
R_3	0	1	1	0	1
R_4	0	0	0	0	0
R_5	0	0	0	0	0
β_{Df}	0	0	0	0	0
β_{De_1}	0	1	1	0	1
β_{De_2}	0	1	1	0	1
β_{De_3}	0	0	0	0	0

Consequently, in the tables, source elements and sensors are replaced by their corresponding efficiency factors.

7.1.4.2 Simulation Results

The physical parameters of the circuit are: $V_1 = 7$ Volt, $V_1 = 1$ Volt, $R_1 = 670\ \Omega$, $R_2 = 215.6\ \Omega$, $R_3 = 67.5\ \Omega$, $R_4 = 215.4\ \Omega$, $R_5 = 509\ \Omega$, $C_1 = 10000\ \mu$F and $C_2 = 4700\ \mu$F, sampling time t_s is 0.05 s, $\varepsilon = 0.01$ in (7.5). Two incipient faults are introduced. The first one is a gradual buildup of R_4 at $t = 5$ s, the degradation profile matches the model of the first order linear ODE in (7.3) with coefficients $b_1 = 0.04$, failure value for R_4 is chosen as $R_{4F} = 800\ \Omega$. The other one is a gradual

Table 7.6 MD-FSM at $a_1 = 0, a_2 = 0$

	r_1	r_2	r_3	r_4	D_b
β_{V_1}	1	0	0	0	1
β_{V_2}	0	1	0	0	1
C_1	0	1	0	0	1
C_2	0	0	1	0	1
R_1	1	0	0	0	1
R_2	0	1	0	0	1
R_3	0	1	1	0	1
R_4	0	0	1	1	1
R_5	0	0	0	1	1
β_{Df}	1	1	0	0	1
β_{De_1}	1	1	1	0	1
β_{De_2}	0	1	1	1	1
β_{De_3}	0	0	1	1	1

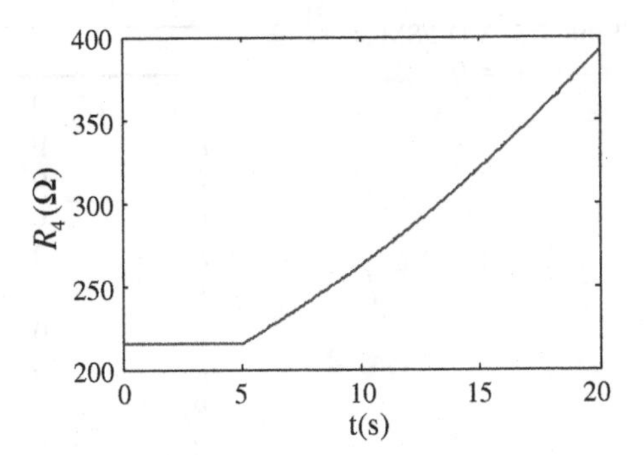

Fig. 7.11 Degradation profile in R_4

degradation in sensor De_3 introduced in β_{De_3} at $t = 6$ s, the degradation profile fulfills the model of the first order nonlinear ODE in (7.3) with coefficients $b_2 = -0.03$, failure value for β_{De_3} is set as $\beta_{De_3} = 0.4$. In simulation, the sensor fault in De_3 is realized by multiplying the normal sensor reading by β_{De_3} in Simulink®, where β_{De_3} is modeled as a function of time using function block in Simulink® library. The degradation profiles are shown in Figs. 7.11 and 7.12. Note that in real industrial applications it is commonplace for the fault to evolve slowly over a period of months, sometimes even over years. In order to avoid excessively long period simulations, the fault development rate has been increased so that significant effects are present after a few seconds [11]. The switches programming is shown in Fig. 7.13. Figure 7.14 presents the simulation responses of residuals due to the faults in R_4 and β_{De_3}, in the figure, dash lines denote the thresholds $\varepsilon_1 = 1e^{-6}$, $\varepsilon_2 = 0.03$, $\varepsilon_3 = 0.6e^{-3}$, $\varepsilon_4 = 5e^{-4}$. These thresholds are not pre-calculated, they are set by observing the normal residual responses. If the residual exceeds the predetermined threshold, then system is considered as faulty.

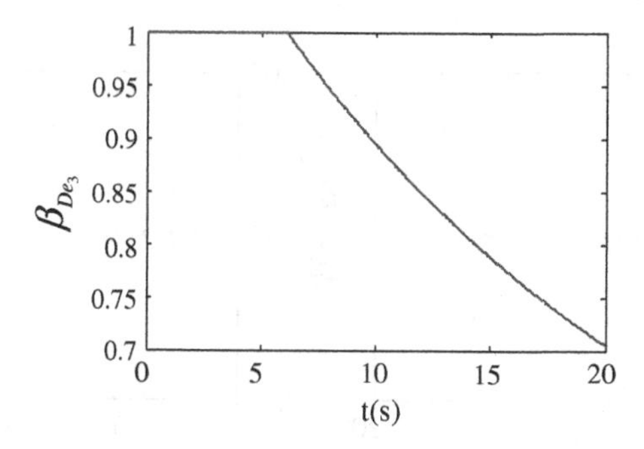

Fig. 7.12 Degradation profile in β_{De_3}

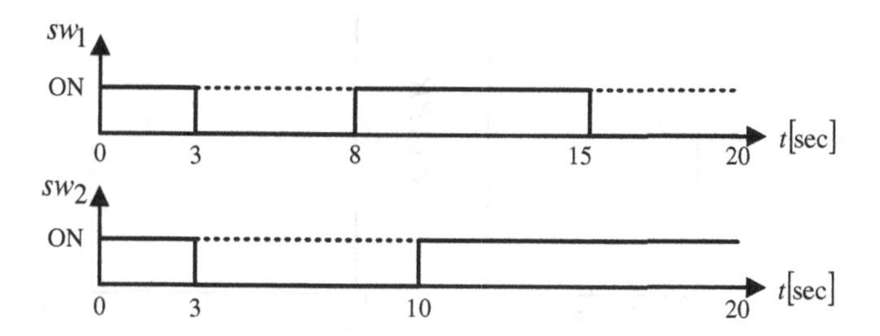

Fig. 7.13 Switches programming for simulation

According to the MD-FSM tables, it is known that the faults in R_4 and β_{De_3} initiate at a non-detectable mode, i.e., $a_1 = 0$ and $a_2 = 0$. The faults cannot be detected until the system enters a mode in which the faults are detectable, i.e., $a_1 = 1$ and $a_2 = 1$. Figure 7.14 reveals that residual r_3 and r_4 are sensitive to the faults at mode $a_1 = 1$ and $a_2 = 1$. From the MD-FSM table at that mode, R_4 and β_{De_3} have fault signature [0 0 1 1] which matches the simulation result. All observations have been captured from the moment that a fault is detected. With an observation window of 10 s, two hundreds of sample data ($N = 200$) are collected. According to the MD-FSM table at mode $a_1 = 1$ and $a_2 = 1$, R_4, R_5 and β_{De_3} are non-detectable at previous mode $a_1 = 1$ and $a_2 = 0$. The coherence vector [0 0 1 1] may result from one or more faults occurring simultaneously, and the faults are detected at the time of mode change, hence the targeted faults set is $\delta = \{R_4\&R_5\&\beta_{De_3}, R_4\&R_5, R_4, \beta_{De_3}, R_5\&\beta_{De_3}, R_4\&\beta_{De_3}\}$

The failure prognosis module is triggered and six AHPSO estimators in MAH-PSO run in parallel, each based on one element from the targeted faults set. The parameters associated with AHPSO are chosen as: Particle number = 350, Maximum iterations = 400, $\Psi_1 = 2$, $\Psi_2 = 2$, $w_r = 0.5$, $V_{max,r} = 4$. In order to eliminate the influence of randomness, each AHPSO is carried out 20 times from different initial seeds of the random number to ensure the repetitiveness of convergence. It has been observed that the final solutions obtained with different parameter settings do not differ much (standard deviation is less than 2 % of the mean value). Consequently,

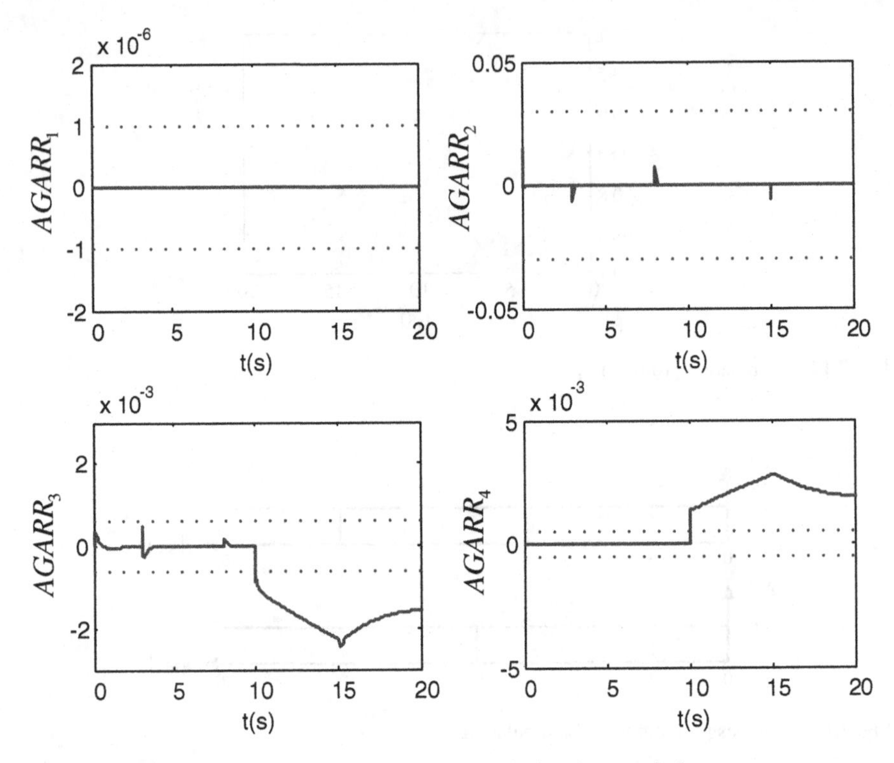

Fig. 7.14 Simulation residual responses for resistance fault and sensor fault

results presented are the mean of the 20 times. Figure 7.15 shows the evolution of mean fitness value versus iterations. It is obvious that the faults are in $R_4\&\beta_{De_3}$, not in the other elements in the targeted faults set.

Table 7.7 summarizes the prognosis results, in which $R_4(0)$ and $\beta_{De_3}(0)$ are the initial values of R_4 and β_{De_3} in the collected data. The identified degradation model for R_4 is the fisrt order linear ODE in (7.3) according to the binary vector $[\eta_1\ \eta_2] = [0\ 0]$ which matches the designed value. Similarly, binary vector $[\eta_1\ \eta_2] = [0\ 1]$ indicates that its degradation process fulfills the second equation in (7.3) and the result matches the designed value. The estimated RUL is 17.88 s for R_4 and 36.3 s for β_{De_3}, which are close to the designed values 17.85 s and 36.1 s. The simulation results show that the proposed prognosis method can accurately track the degradation dynamics and predict the RUL for hybrid systems.

7.1.4.3 Experiment Results

As shown in Fig. 7.16, a circuit test-bed is established to verify the proposed method. The real time FDI is based on the MATLAB® Real-Time Windows Target, and two data acquisition cards (NI PCI-6025E and NI PCI-6713). The switches are

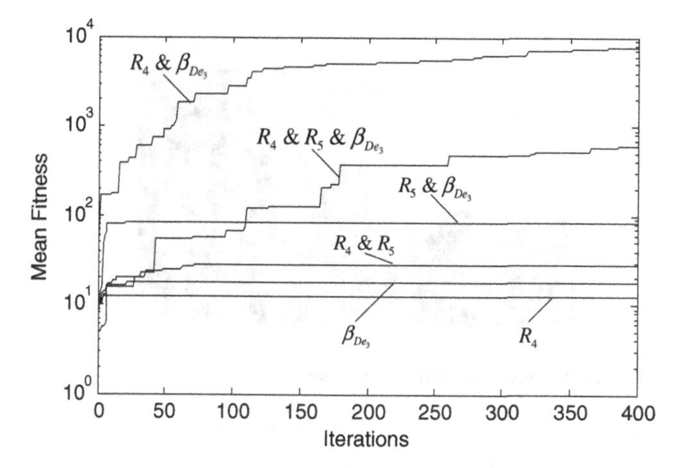

Fig. 7.15 Mean fitness evolution versus iterations for simulation

Table 7.7 Prognosis result at mode $a_1 = 0, a_2 = 1$

Fault in R_4	b_1	$R_4(0)$	η_1	η_2	RUL
Designed values	0.04	263.09	0	0	17.85 s
Estimated values	0.041	262.46	0	0	17.88 s
Fault in β_{De3}	b_1	$\beta_{De3}(0)$	η_1	η_2	RUL
Designed values	–0.03	0.893	0	1	36.1 s
Estimated values	–0.032	0.894	0	1	36.3 s

implemented by two PhotoMOS relays (AQV251) and controlled by the real time software. The physical parameters of circuit and sampling time are same as those used in simulation. Due to the numerical differentiation of noisy measurements in the AGARRs, the results for estimated variables are noisy. To get good estimates, a low pass filter is deployed to filter out the effect of measurement noise on residuals.

An actuator fault occurs when there is discrepancy between command input of an actuator and its actual output to system. It is introduced by reducing β_{V_1} at moment as shown in Fig. 7.17. The degradation profile of β_{V_1} matches the model of the second order linear ODE in (7.3), the failure value for β_{V_1} is chosen as $\beta_{V_{1F}} = 0.5$. In the experiment, the actuator fault is realized by multiplying input V_1 by β_{V_1} in Simulink® under Real-Time Windows Target environment, where β_{V_1} is modeled as a function of time using function block in Simulink® library and unknown for the designer. The switches programming is shown in Fig. 7.18. Since the fault is injected at mode $a_1 = 0$ and $a_2 = 0$, in which β_{V_1} is non-detectable, the fault cannot be detected by the residuals as shown in Fig. 7.19. Until $t = 10$ s the system enters a new mode $a_1 = 1$ and $a_2 = 1$, from MD-FSM tables, the fault signature of β_{V_1} is [1 0 0 0] which indicates it is detectable. The fault in β_{V_1} cannot be isolated because R_1 shares the same fault signature with the one of β_{V_1}, and both β_{V_1} and R_1 are non-detectable at previous mode $a_1 = 0$ and $a_2 = 1$, thus $\delta = \{R_1, \beta_{V_1}, R_1\&\beta_{V_1}\}$. Finally, the failure prognosis module is invoked in which three AHPSO estimators in MAHPSO run in

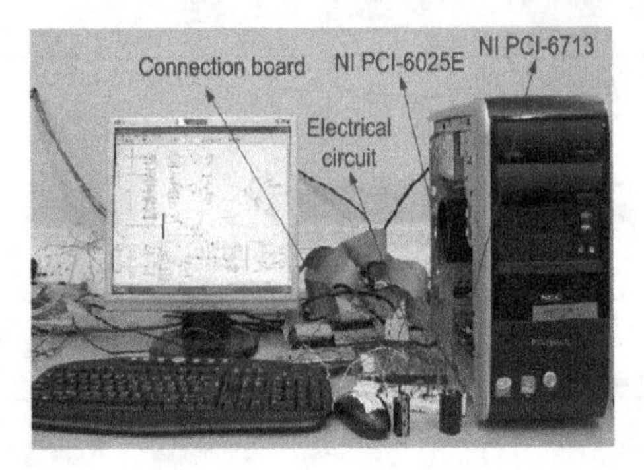

Fig. 7.16 An experimental setup for FDI and prognosis

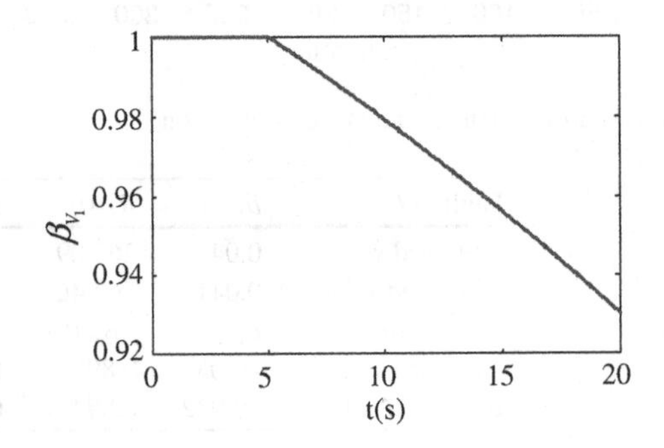

Fig. 7.17 Degradation profile in β_{V_1}

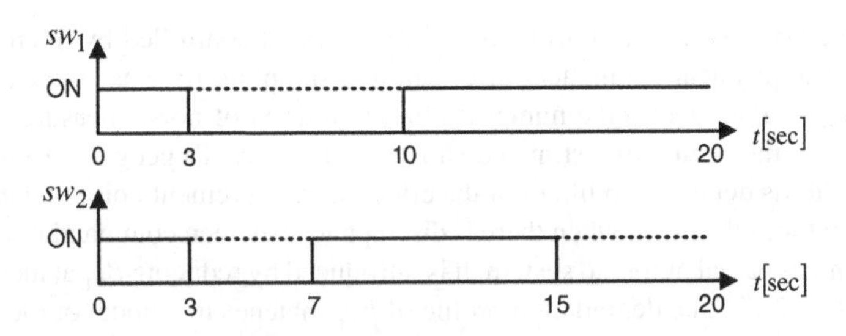

Fig. 7.18 Switches programming for experiment

parallel. Figure 7.20 depicts the evolution of mean fitness value versus iterations. It is obvious that the degradation is in β_{V_1}, not in R_1 and $R_1\&\beta_{V_1}$.

The prognosis results are listed in Table 7.8. The identified degradation model for β_{V_1} is the second order linear ODE in (7.3) according to the binary vector $[\eta_1\ \eta_2] = [1\ 0]$ which matches the designed value. The results confirm that the MAHPSO based prognosis method performs well in real experiment.

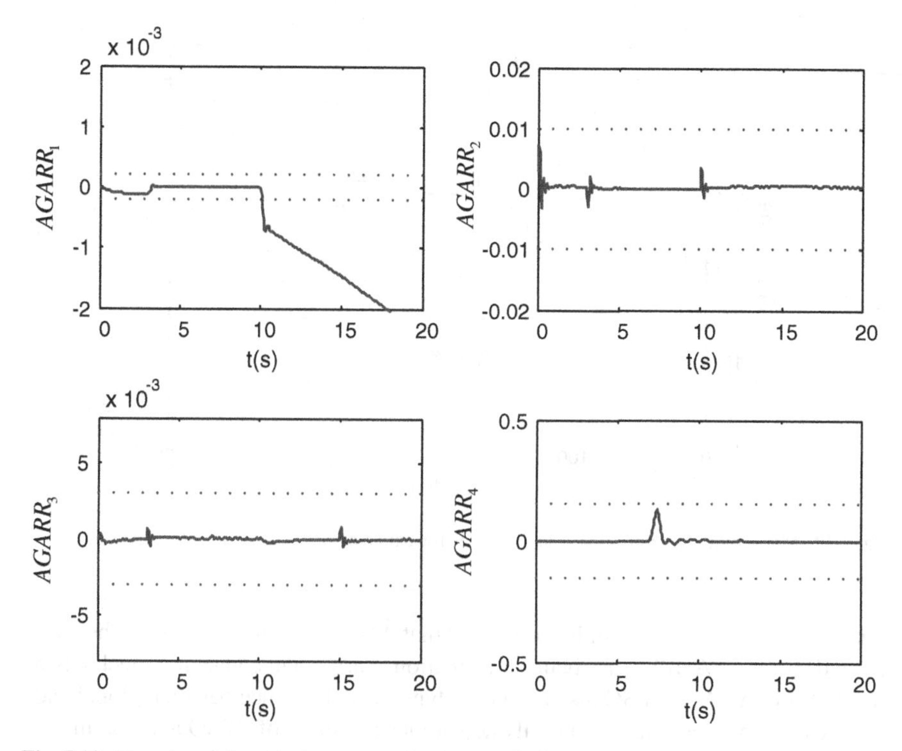

Fig. 7.19 Experimental residual responses for actuator fault

7.1.4.4 Comparison Study

For comparison purposes, AHPSO, GA and Adaptive GA (AGA) [12] approaches have been tested on the same experimental data collected. Since the true fault is β_{V_1}, then GA and AGA methods are carried out considering β_{V_1} as fault. The parameters of GA are set as: Population $= 350$, Maximum iterations $= 400$, $p_c = 0.6$ and $p_m = 0.04$. The parameters of AGA are the same as those of GA, the adaptive scheme of AGA consists of some rules to adaptively tune the crossover and mutation rates according to the performance of the current genetic operators. It tries to increase the probability of the genetic operator if it consistently produces a better offspring during the search process; on contrast, it tries to reduce the probability of the genetic operator if it always produces a poorer offspring. This method can adaptively regulate the balance between the exploration and exploitation of the solution space.

Comparison results are shown in Table 7.9 in which all methods identify that the degradation model for β_{v_1} is the second order linear ODE in (7.3) according to the

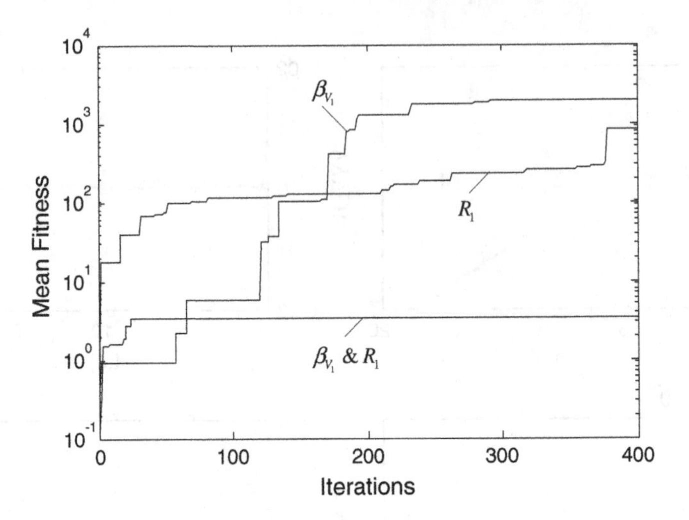

Fig. 7.20 Mean fitness evolution versus iterations for experiment

binary vector $[\eta_1\ \eta_2] = [1\ 0]$. It is observed that AHPSO is superior to the other two methods because AHPSO has better information sharing and conveying mechanism than GA and AGA. AHPSO also has good dynamics of balance between global and local search abilities, and it can easily adjust inertia weight of RPSO and maximum velocity of BPSO using simple "IF-THEN" rules.

7.2 Prognosis with Mode-Dependent Degradation Behaviors

Prognosis of hybrid systems is a challenging problem because multiple faults may happen simultaneously at a mode where these faults have different detectabilities. In other words, at the fault initiating mode, some of the faults are detectable while others are non-detectable. As a result, decision making based on only one observation of abnormal behavior is not reliable under this condition. This part focuses on the development of a model based prognosis framework for hybrid systems where a dynamic fault isolation scheme is proposed to facilitate the prognostic tasks. The degradation behavior of each faulty component is mode-dependent and can be estimated by a hybrid differential evolution algorithm. Thereafter, the remaining useful life of faulty component which varies with different operating modes is calculated by using both the estimated degradation model and the user-selected failure threshold. Experiments are carried out to validate the key concepts of the developed methods and results suggest the effectiveness.

Table 7.8 Summary for experiment results

Fault in β_{V_1}	b_3	b_4	$\beta_{V_1}(0)$	η_1	η_2	RUL
Designed values	−50	−1.2	0.979	1	0	47.7 s
Estimated values	−48.84	−1.19	0.972	1	0	47.3 s

Table 7.9 Comparison results using different methods

Methods	b_3	b_4	$\beta_{V_1}(0)$	η_1	η_2	RUL
AHPSO	−48.84	−1.19	0.972	1	0	47.3 s
GA	−47.93	−1.12	0.961	1	0	48.5 s
AGA	−48.56	−1.17	0.969	1	0	47.1 s

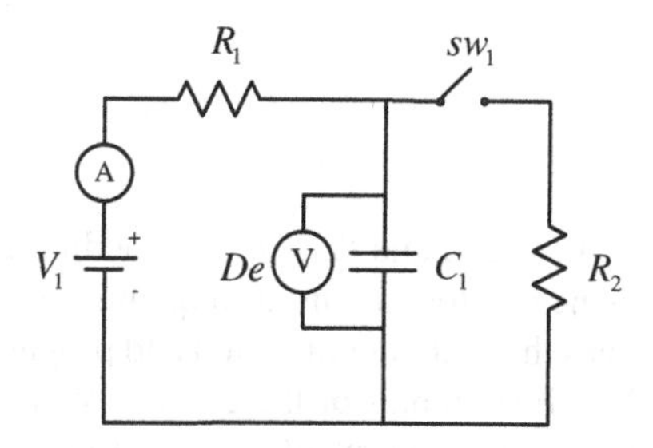

Fig. 7.21 An switched circuit

7.2.1 Dynamic Fault Isolation

As the decision making of traditional monitoring system cannot deal with the situation where the faults show different detectabilities at the fault initiating mode, prognostic task cannot be completed. In this section, a new prognosis framework for hybrid systems is developed.

An example of switched circuit with two modes is shown in Fig. 7.21. The system consists of one voltage source V_1, two resistors R_1 and R_2, one capacitor C_1 and one switch sw_1. Two sensors De and Df are mounted to measure current and voltage signals. The DHBG of the hybrid system is shown in Fig. 7.22. A controlled junction 1_3 models the switching behavior of the physical switch (sw_1) in HBG language. A DHBG is a HBG with a causality assignment that all its controlled junctions and storage components are assigned with preferred causalities. It is beneficial to design fault diagnosis algorithms based on a DHBG since the consistent causality of the graph provides a convenient way to describe the behavior of the hybrid system at all modes.

There are two AGARR equations that can be derived from the DHBG model as follows.

$$\mathrm{AGARR}_1 = \beta_{V_1} V_1 - R_1 \frac{D_f}{\beta_{D_f}} - \frac{D_e}{\beta_{D_e}} \tag{7.23}$$

$$\mathrm{AGARR}_2 = \frac{D_f}{\beta_{D_f}} - a\frac{1}{R_2}\frac{D_e}{\beta_{D_e}} - C_1\frac{d}{dt}(\frac{D_e}{\beta_{D_e}}) \tag{7.24}$$

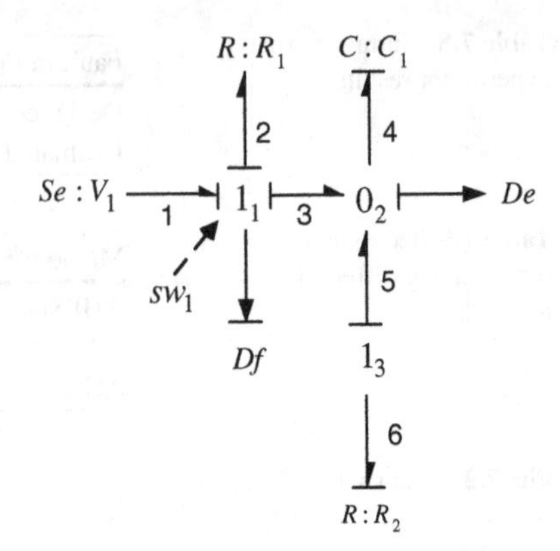

Fig. 7.22 DHBG of the circuit

In (7.23) and (7.24), three additional parameters (β_{V_1}, β_{D_f} and β_{D_e}) called efficiency factors are used to quantify the faults in sensors and actuator, which facilitates the fault identification and prognosis of these non-parametric components. The MD-FSM tables of the circuit are shown in Tables 7.10 and 7.11, where $a \in \{0,1\}$ represents the state of the switch and thus models the operating mode of the circuit.

The initial state of the switch is off ($a = 0$), and the switch is changed to on state ($a = 1$) at $t = 10$ s. If two incipient faults, i.e., one actuator fault in β_{V_1} and one resistor fault in R_2, occur simultaneously at $t = 6$ s, the detected coherence vector should be $C = [1\ 0]$. After comparing the detected coherence vector with the fault signature in Table 7.10, it is found that the possible set of fault candidates could be $\delta = \{\beta_{V_1}, R_1, \beta_{V_1}\&R_1\}$. However, since the fault in R_2 occurs at the mode $a = 0$ which is a non-detectable mode for R_2, the fault isolation result excludes

Table 7.10 MD-FSM at mode $a = 0$

	r_1	r_2	D_b
β_{D_f}	1	1	1
β_{D_e}	1	1	1
β_{V_1}	1	0	1
R_1	1	0	1
R_2	0	0	0
C_1	0	1	1

Table 7.11 MD-FSM at mode $a = 1$

	r_1	r_2	D_b
β_{D_f}	1	1	1
β_{D_e}	1	1	1
β_{V_1}	1	0	1
R_1	1	0	1
R_2	0	1	1
C_1	0	1	1

R_2 as a possible fault candidate. The decision making through only one observation of inconsistency is not reliable under this case and cannot identify R_2 as a fault candidate. As a result, the fault diagnosis task cannot be completed. In order to solve this problem during hybrid system monitoring, a dynamic fault isolation method is developed in which the system may set a waiting time (WT) to allow all faults to exhibit their symptoms on residuals. Assume that all faults could be detected within the predefined WT. In general, the WT takes the following form

$$T_{wt} = T_{rmct} + \sum_{k=0}^{n} T_{mct}^{k} \quad \text{with} \quad T_{mct}^{0} = 0 \tag{7.25}$$

where T_{wt} represents the WT, T_{rmct} denotes the remaining mode change time which means the time difference between the first observed inconsistency and the following mode change, $T_{mct}^{k}, k = 0, \ldots, n$ represents the successive mode change time interval with $T_{mct}^{0} = 0$.

It is worth to note that according to (7.25), the system is at least required to continuously monitor the process until the next mode change instance from the first observed fault condition. For the above example, if let $T_{wt} = T_{rmct} = 4s$ with $k = 0$ in (7.25), then at $t = 10$ s a new coherence vector $C = [1\ 1]$ could be detected after the first observed coherence vector. From Table 7.11 and previous fault identification result (i.e., identify β_{V_1} as the true fault), the possible set of fault candidates under this new fault condition could be $\delta = \{\beta_{V_1} \& R_2\}$. Since the new inconsistency is detected as the instance of mode change, then the set of fault candidates only considers the faults which are non-detectable at previous mode and detectable at current mode. Finally, the system is able to detect and isolate all faults using the dynamic fault isolation (DFI) scheme even if not all the faults are occurring at a detectable mode.

7.2.2 Mode-Dependent Degradation Behaviors

For a hybrid system, the same component will exhibit different degradation behaviors at different operating modes. For instance, the feed motor in a printer may be in the ramp-up, rotating with constant speed, ramp-down, and idle modes [13]. It is evident that the motor wear in the ramp-up mode is more severe than the wear in the idle mode. Under this condition, it would be desirable to describe the degradation process of the feed motor under various modes using different degradation models. The degradation behavior of a parameter or an efficiency factor is described as follows

$$\omega_1 \ddot{P} + (1 - \omega_1)\dot{P} = bP^{2\omega_2} + cP^{\omega_3} \tag{7.26}$$

where P represents the parameter value or efficiency factor value; b and c denote the degradation model coefficients; $\omega_i \in \{0,1\}$ for $i = 1, 2, 3$. Define vector $\Omega = [\omega_1, \omega_2, \omega_3]$ as a Degradation Model Structure Vector (DMSV).

It is evident that with different values of DMSV, Eq. (7.26) is able to describe various degradation behaviors. For example, if $\Omega = [\omega_1, \omega_2, \omega_3] = [0, 1, 1]$ for the resistor fault in R_2 of Fig. 7.22 at mode $a = 0$, the corresponding degradation behavior can be described as a nonlinear process $\dot{R}_2 = bR_2^2 + cR_2$. As for mode $a = 1$, the resistor fault in R_2 will degrade according to a linear form as $\dot{R}_2 = b + cR_2$ with $\Omega = [\omega_1, \omega_2, \omega_3] = [0, 0, 1]$. As a result, the incipient fault in R_2 can be described by different behaviors (linear or nonlinear differential equation) at different operating modes. These degradation processes under different operating modes are called Mode-Dependent-Degradation Models (MD-DM). Since the degradation model is mode-dependent, RUL for each faulty component is also mode-dependent, referred to as Mode-Dependent-RUL (MD-RUL).

7.2.3 Sequential Prognosis

After fault isolation, a set of suspected faults is obtained. The prognosis module is activated to predict the RUL of the faulty component. In this work, a new sequential prognosis method based on DFI is introduced. This method mainly includes two parts: standard prognosis process and auxiliary prognosis process. The standard prognosis process is based on the set of suspected faults through fault isolation. The auxiliary prognosis process is to predict the MD-RUL if mode change happens since the degradation behavior varies with mode change. This process does not rely on the set of suspected faults, but is based on the true faults from previous standard prognosis process.

Figure 7.23 shows the block diagram of the DFI based sequential prognosis method. After the standard prognosis according to the first observed inconsistency, true fault together with corresponding MD-RUL can be obtained. The system will continue the monitoring process within the predefined WT after the first observed faulty symptom. When a new inconsistency is detected, the fault isolation module is enabled to find the set of suspected faults. During the fault isolation, the true fault from the previous standard prognosis module must be included in every element of the set of suspected faults. For example, if θ_1 is found to be the true fault from the previous standard prognosis, the current set of suspected faults should have the form like $\delta = \{\theta_1\&\theta_i, \theta_1\&\beta_j, \theta_1\&\theta_i\&\beta_j, \ldots\}$ with $i, j \neq 1$. In addition, if the new inconsistency is detected at the time of mode change, the added fault signature is only caused by components which are non-detectable at previous mode and are detectable at current mode. On the other hand, if a new mode change, not an inconsistency, is detected, the auxiliary prognosis process is enabled. The purpose of this process is to predict the MD-RUL of the true faults obtained from previous standard prognosis module under this new mode.

The first step of prognosis is to select a prescribed degradation model together with the model coefficients. The next step is to predict the RUL of the faulty component based on the identified degradation model provided that a failure threshold is set. The purpose of the identification is to find continuous real numbers for degradation

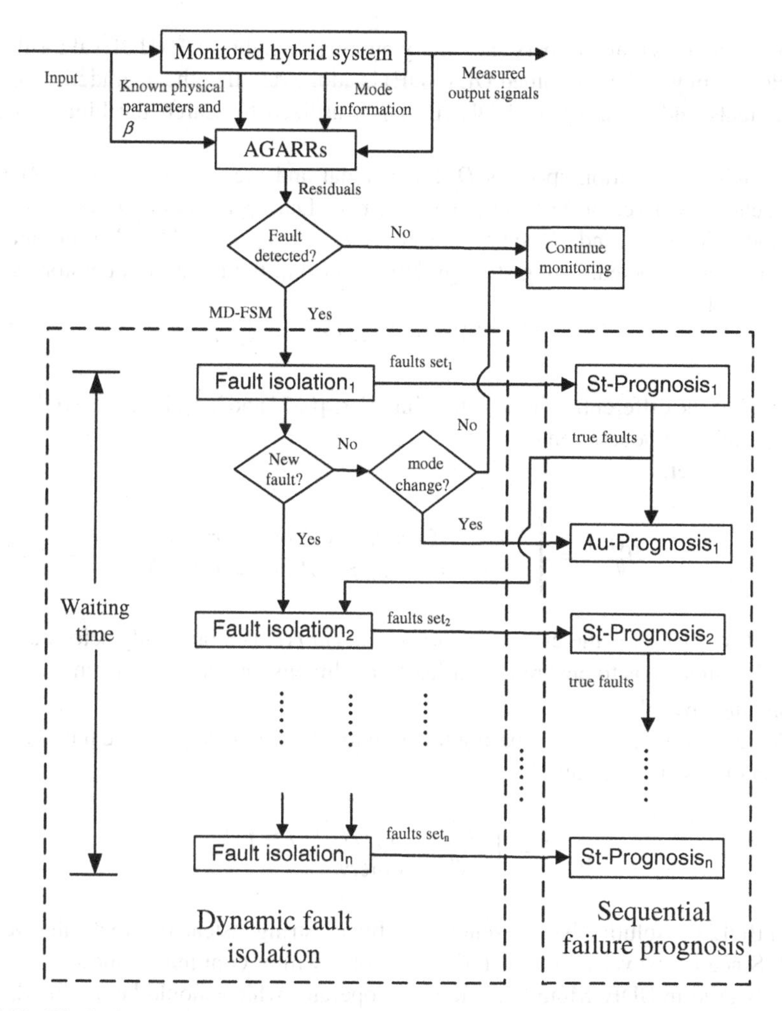

Fig. 7.23 Block diagram of the DFI based sequential prognosis method

model coefficients in (7.26) and discrete binary values for DMSV simultaneously. However, traditional gradient-based method cannot directly solve this identification problem because the objective function to be minimized is not directly related to the binary vector DMSV, thus no gradient information can be obtained. Therefore, heuristics-based evolutionary algorithms (EAs) can be efficient alternatives. EAs are a class of stochastic search methods that include genetic algorithm, evolutionary programming, evolution strategies, genetic programming, and their variants.

Differential evolution (DE) is originally due to Storn and Price and is able to deal with non-differentiable, non-linear, and multimodal objective functions [16]. Many published works demonstrated that DE converges fast and is robust, simple in implementation, and requires only a few control parameters [14, 15]. Since the solution includes real numbers and binary numbers, a hybrid DE (HDE) algorithm is developed, in which real-valued DE (RDE) is adopted to find the degradation model coefficients and binary-valued DE (BDE) is utilized to search the binary vector DMSV.

Assume the solution space is D-dimensional and the *ith* individual in RDE is represented by a vector $X_i = (x_{i1}, x_{i2}, \ldots, x_{iD})$. During mutation at generation G, two individuals $x_{r_1}^G$ and $x_{r_2}^G$ satisfying $r_1 \neq r_2$ and $r_1, r_2 \in [1, N]$ in population are randomly selected. Using strategy DE/best/1/bin, the mutation operator can be represented as

$$V_{ij}^{G+1} = X_{best}^G + F^r \times (X_{r_1j}^G - X_{r_2j}^G) \tag{7.27}$$

where F^r is the differential factor satisfying $F^r \in [0, 2]$ and X_{best}^G is the best individual in population at generation G.

In crossover,

$$H_{ij}^{G+1} = \begin{cases} V_{ij}^{G+1} & \text{if } rand_j \leq CR \text{ or } j = rnbr(i) \\ X_{ij}^G & \text{if } rand_j > CR \text{ or } j \neq rnbr(i) \end{cases} \tag{7.28}$$

where $CR \in [0, 1]$ is the crossover factor, and $rnbr(i)$ is a randomly selected index from D dimensions to ensure that at least one dimension parameter from V_i^{G+1} can be obtained by H_i^{G+1} [16].

Selection operates by comparing the individuals' fitness to generate the new population of next generation,

$$X_i^{G+1} = \begin{cases} H_i^{G+1} & \text{if } f(H_i^{G+1}) > f(X_i^G) \\ X_i^G & \text{otherwise} \end{cases} \tag{7.29}$$

In BDE, the solution is represented by a binary string instead of a real value vector [17]. Since crossover operates in discrete form within continuous space, it can be directly used in BDE. Mutation is the only operator which should be modified.

$$V_{ij}^{G+1} = X_{best}^G + F^b \bullet (X_{r_1j}^G \oplus X_{r_2j}^G) \tag{7.30}$$

where symbols ($\bullet$), (+) and ($\oplus$) denote Boolean algebra "AND", "OR" and "XOR", respectively. F^b is a random D-bit binary vector and it is not a control parameter.

The RDE and BDE evolve simultaneously and are coupled through the common fitness function

$$F_{fitness} = 1/(\sum_{l=1}^{m} \sum_{n=1}^{N} |\text{AGARR}_l^n| + \varepsilon) \tag{7.31}$$

where ε is a small positive constant which is used to avoid zero division during the search process, n is the discrete sampling index.

For each standard prognosis module in Fig. 7.23, if the set of suspected faults contains only one element, HDE is adopted to find the true degradation model with correct model coefficients. When the set consists of more than one element, several HDE estimators run in parallel, each estimator serves to identify one element in the set. The fitness values of various elements are compared with each other to choose the best degradation model and model coefficients. After that, with the identified MD-DM for each faulty component, and if the failure threshold is given, MD-RUL can be calculated by subtracting current time and data collection period from the time when the fault reaches the failure value. The objective of prognosis is to predict EOL at a certain time point t_f using the observation up to time $t_f + N \cdot t_s$, where N is the number of data collected and t_s is the sampling time. Let $T_{EOL} = 1$ denote the event that a failure threshold is exceeded, and 0 otherwise. EOL and RUL can be formulated as

$$EOL(t_f) = \inf\{t \in \mathbb{R} : t \geq (t_f + N \cdot t_s) \wedge T_{EOL} = 1\} \tag{7.32}$$

$$RUL(t_f) = EOL(t_f) - (t_f + N \cdot t_s) \tag{7.33}$$

In order to help better understand the Eqs. (7.32) and (7.33), let us consider a resistor fault in R_2 of Fig. 7.22 at mode $a = 1$, where the resistor fault in R_2 will degrade (slowly increase for this case) according to a linear form as $\dot{R}_2 = b + cR_2$. If the failure threshold is set as R_2^F, so T_{EOL} =1 if $R_2 > R_2^F$. With the estimated degradation model and coefficients, the EOL for this fault at mode $a = 1$ is calculated as $\log((\bar{b} + \bar{c} \times R_2^F)/(\bar{b} + R_2(0) \times \bar{c}))/\bar{c}$ and RUL at mode $a = 1$ is computed as $\log((\bar{b} + \bar{c} \times R_2^F)/(\bar{b} + R_2(0) \times \bar{c}))/\bar{c} - N \cdot t_s$, where $\bar{b}$ and $\bar{c}$ are estimated values of b and c, $R_2(0)$ is the value of R_2 at time point t_f. Note that the multiple HDE estimators are developed for the standard prognosis process, and the auxiliary prognosis process only employs single HDE estimator to predict the MD-RUL because the true faults are obtained from the previous standard prognosis module. In addition, the estimated MD-RUL of each prognosis module is independent of the MD-RUL for the same fault component of the previous prognosis module which means the estimated error of the previous prognosis module will not be propagated to the next prognosis module.

7.2.4 Experiment Result

Consider an electrical circuit as shown in Fig. 7.24. The circuit consists of five resistors R_1, R_2, R_3, R_4 and R_5, two voltage sources V_1 and V_2, two capacitors C_1 and C_2, one current sensor Df, two voltage sensors De_1 and De_2 and two switches sw_1 and sw_2. The DHBG of the system is shown in Fig. 7.25.

The next step is to derive AGARRs from the DHBG model, and only junctions with attached sensor are considered. Four structurally independent AGARRs are

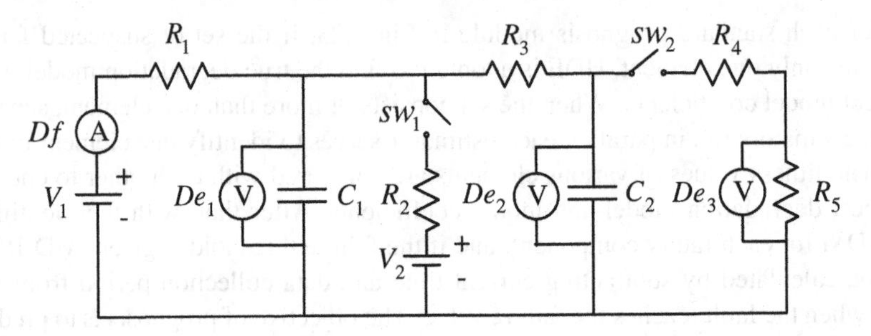

Fig. 7.24 An electric circuit with hybrid dynamics

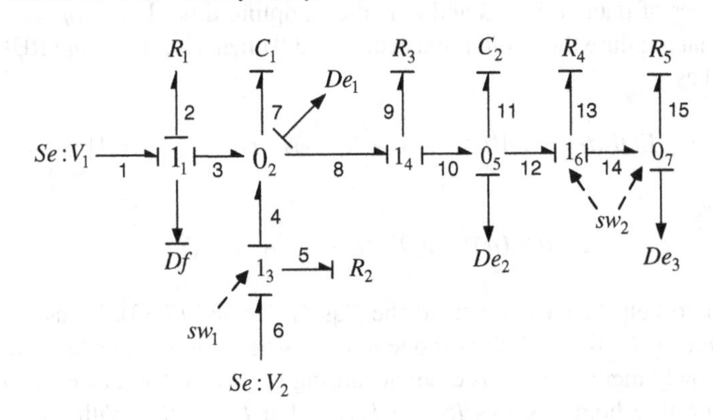

Fig. 7.25 DHBG of the circuit

obtained as follows

$$\mathrm{AGARR}_1 = \beta_{V_1} \cdot V_1 - R_1 \cdot \frac{Df}{\beta_{Df}} - \frac{De_1}{\beta_{De_1}} \tag{7.34}$$

$$\mathrm{AGARR}_2 = \frac{Df}{\beta_{Df}} - \frac{a_1}{R_2}(\beta_{V_2} \cdot V_2 - \frac{De_1}{\beta_{De_1}}) - C_1 \frac{d}{dt}(\frac{De_1}{\beta_{De_1}}) - \frac{1}{R_3}(\frac{De_1}{\beta_{De_1}} - \frac{De_2}{\beta_{De_2}}) \tag{7.35}$$

$$\mathrm{AGARR}_3 = \frac{1}{R_3}(\frac{De_1}{\beta_{De_1}} - \frac{De_2}{\beta_{De_2}}) - C_2 \frac{d}{dt}(\frac{De_2}{\beta_{De_2}}) - \frac{a_2}{R_4}(\frac{De_2}{\beta_{De_2}} - \frac{De_3}{\beta_{De_3}}) \tag{7.36}$$

$$\mathrm{AGARR}_4 = a_2[\frac{1}{R_4}(\frac{De_2}{\beta_{De_2}} - \frac{De_3}{\beta_{De_3}}) - \frac{1}{R_5}\frac{De_3}{\beta_{De_3}}] \tag{7.37}$$

where $a_1, a_2 \in \{0,1\}$ represents the states of the two switches (sw_1 and sw_2).

Table 7.12 MD-FSM at mode [0, 0]

	r_1	r_2	r_3	r_4	D_b
β_{D_f}	1	1	0	0	1
$\beta_{D_{e_1}}$	1	1	1	0	1
$\beta_{D_{e_2}}$	0	1	1	0	1
$\beta_{D_{e_3}}$	0	0	0	0	0
β_{V_1}	1	0	0	0	1
β_{V_2}	0	0	0	0	0
R_1	1	0	0	0	1
R_2	0	0	0	0	0
R_3	0	1	1	0	1
R_4	0	0	0	0	0
R_5	0	0	0	0	0
C_1	0	1	0	0	1
C_2	0	0	1	0	1

Table 7.13 MD-FSM at mode [1, 0]

	r_1	r_2	r_3	r_4	D_b
β_{D_f}	1	1	0	0	1
$\beta_{D_{e_1}}$	1	1	1	0	1
$\beta_{D_{e_2}}$	0	1	1	0	1
$\beta_{D_{e_3}}$	0	0	0	0	0
β_{V_1}	1	0	0	0	1
β_{V_2}	0	1	0	0	1
R_1	1	0	0	0	1
R_2	0	1	0	0	1
R_3	0	1	1	0	1
R_4	0	0	0	0	0
R_5	0	0	0	0	0
C_1	0	1	0	0	1
C_2	0	0	1	0	1

Tables 7.12, 7.13, 7.14 7.15 summarize the detectabilities of all components in the circuit under different operating modes ($[a_1, a_2]$). The nominal parameters of the circuit are: $V_1 = 9$ Vol, $V_1 = 0.5$ Vol, $R_1 = 670\ \Omega$, $R_2 = 215.6\ \Omega$, $R_3 = 67.5\ \Omega$, $R_4 = 215.4\ \Omega$, $R_5 = 509\ \Omega$, $C_1 = 1000\ \mu$F and $C_2 = 4700\ \mu$F. The sampling time is 0.05 s and the states of the two switches are shown in Fig. 7.26. Two incipient faults are considered, one is an actuator fault in V_1 and the other one is a sensor fault in De_3. The degradation behaviors of these two faults under different operating modes ($[a_1, a_2]$) are designed as

Table 7.14 MD-FSM at mode [0, 1]

	r_1	r_2	r_3	r_4	D_b
β_{D_f}	1	1	0	0	1
$\beta_{D_{e_1}}$	1	1	1	0	1
$\beta_{D_{e_2}}$	0	1	1	1	1
$\beta_{D_{e_3}}$	0	0	1	1	1
β_{V_1}	1	0	0	0	1
β_{V_2}	0	0	0	0	0
R_1	1	0	0	0	1
R_2	0	0	0	0	0
R_3	0	1	1	0	1
R_4	0	0	1	1	1
R_5	0	0	0	1	1
C_1	0	1	0	0	1
C_2	0	0	1	0	1

Table 7.15 MD-FSM at mode [1, 1]

	r_1	r_2	r_3	r_4	D_b
β_{D_f}	1	1	0	0	1
$\beta_{D_{e_1}}$	1	1	1	0	1
$\beta_{D_{e_2}}$	0	1	1	1	1
$\beta_{D_{e_3}}$	0	0	1	1	1
β_{V_1}	1	0	0	0	1
β_{V_2}	0	1	0	0	1
R_1	1	0	0	0	1
R_2	0	1	0	0	1
R_3	0	1	1	0	1
R_4	0	0	1	1	1
R_5	0	0	0	1	1
C_1	0	1	0	0	1
C_2	0	0	1	0	1

$$\begin{aligned}
\dot{\beta}_{V_1} &= -0.03\beta_{V_1}^2 + 0.01\beta_{V_1},\ \Omega = [0, 1, 1] \text{ at mode}[1, 0] \\
\dot{\beta}_{V_1} &= -0.02\beta_{V_1} + 0.005,\ \Omega = [0, 0, 1] \text{ at mode}[0, 0] \\
\dot{\beta}_{V_1} &= -0.05\beta_{V_1}^2 + 0.0035\beta_{V_1},\ \Omega = [0, 1, 1] \text{ at mode}[0, 1]
\end{aligned} \tag{7.38}$$

$$\begin{aligned}
\dot{\beta}_{De_3} &= -0.025\beta_{De_3} + 0.01,\ \Omega = [0, 0, 1] \text{ at mode}[1, 0] \\
\dot{\beta}_{De_3} &= -0.03\beta_{De_3}^2 + 0.003\beta_{De_3},\ \Omega = [0, 1, 1] \text{ at mode}[0, 0] \\
\dot{\beta}_{De_3} &= -0.06\beta_{De_3} + 0.015,\ \Omega = [0, 0, 1] \text{ at mode}[0, 1]
\end{aligned} \tag{7.39}$$

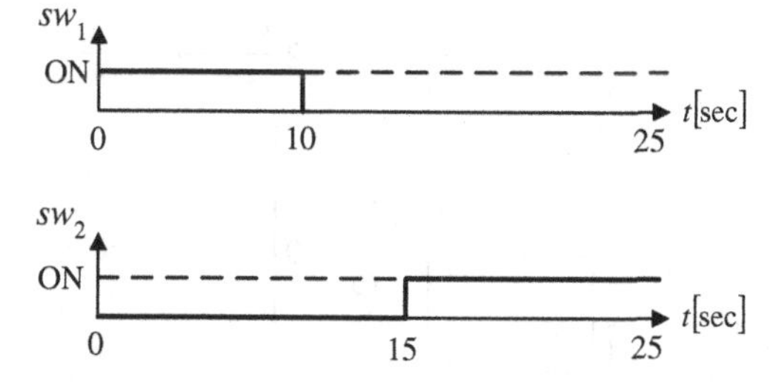

Fig. 7.26 States of the two switches

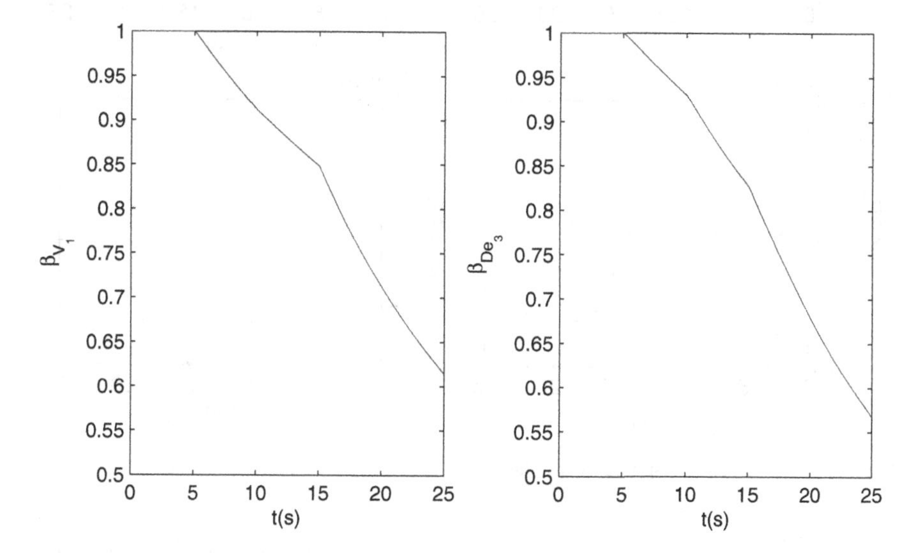

Fig. 7.27 Degradation process of two incipient faults

The parameters in (7.38) and (7.39) are selected to ensure monotonic decrease of two efficiency factors, i.e., β_{V_1} and β_{De_3}, under faulty condition. The failure values of these two faults are chosen as $\beta^F_{V_1} = 0.5$ and $\beta^F_{De_3} = 0.3$. Figure 7.27 demonstrates the degradation process of these two incipient faults. Figure 7.28 depicts the residual responses where the thresholds are set as $\varepsilon_1 = 5e^{-4}$, $\varepsilon_2 = 8e^{-4}$, $\varepsilon_3 = 5e^{-4}$, $\varepsilon_4 = 3e^{-4}$. These thresholds are set by observing residual responses under normal condition and they should be defined carefully to avoid false alarm. If any of the residual exceeds the corresponding threshold, the system is considered as faulty. The sensor and actuator faults are introduced at 5 s at mode $a_1 = 1$, $a_2 = 0$. Since the sensor fault in $\beta_{D_{e3}}$ is non-detectable at that mode according to Table 7.13, a deviation in residual of $AGARR_1$ is expected upon the fault occurrence in β_{V_1}. The experimental result confirms this behavior. After $t = 5$ s, a coherence binary vector is [1 0 0 0] which indicates the set of suspected faults is $\delta = [\beta_{V_1}, R_1, \beta_{V_1}\&R_1]$. One hundred sample data are collected. Based on the set of suspected faults, the

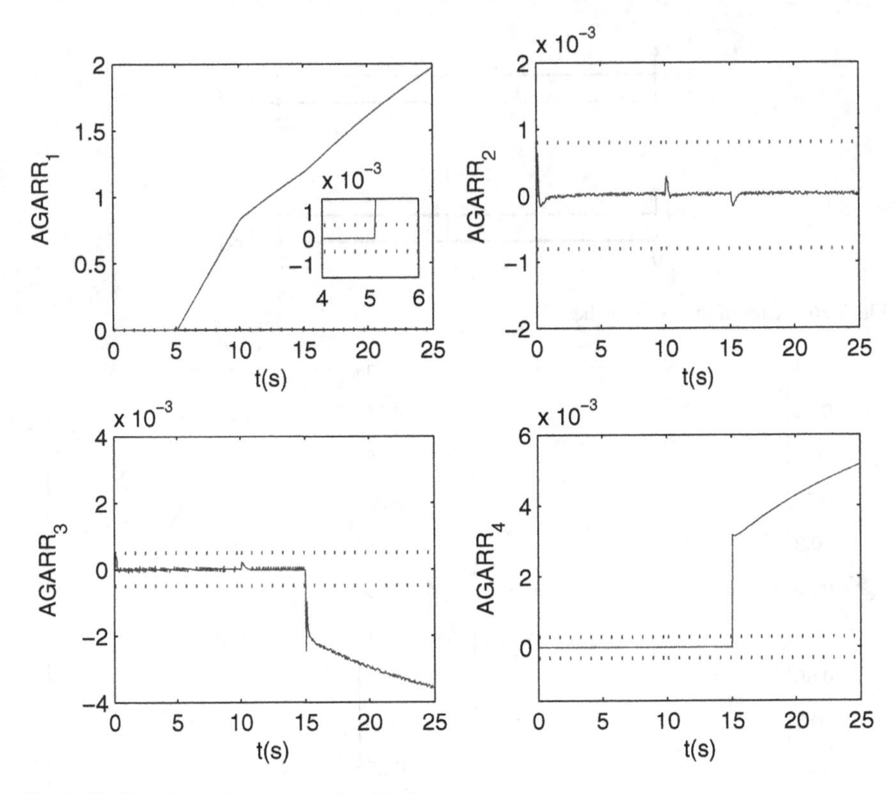

Fig. 7.28 Experimental response of residuals

standard prognosis module is triggered to find the true fault and then calculate the MD-RUL. The parameters associated with HDE are chosen as: Population size = 100, Maximum iterations = 400, $F^r = 0.4$, and $CR = 0.8$. Figure 7.29 shows the search process and results reveal that the true fault is in β_{V_1}. The calculated MD-RUL for this fault is summarized in Table 7.16.

In this experiment, the waiting time is defined as

$$T_{wt} = T_{rmct} + T_{mct}^{1} = 5 + 5 = 10\,s \quad \text{with} \quad k = 1 \tag{7.40}$$

The monitoring process is carried on and at $t = 10$ s, a new mode change occurs, and one hundred sample data are collected within 5 s upon the occurrence of mode change. After that, the auxiliary prognosis module is activated to predict the MD-RUL of β_{V_1} under this new mode, i.e., $a_1 = 0$, $a_2 = 0$. The prognosis results are put in Table 7.17. At $t = 15$ s, a new coherence vector [1 0 1 1] is observed. Since this fault is detected at the time of mode change, it is suspected that the mode change makes some non-detectable faults at previous mode to be detectable at this new mode, i.e., $a_1 = 0, a_2 = 1$. Hence, the suspected faults set under this new condition could be $\delta = [\beta_{V_1}\&R_4,\ \beta_{V_1}\&R_4\&R_5,\ \beta_{V_1}\&\beta_{D_{e_3}},\ \beta_{V_1}\&\beta_{D_{e_3}}\&R_4,\ \beta_{V_1}\&\beta_{D_{e_3}}\&R_5,\ \beta_{V_1}\&\beta_{D_{e_3}}\&R_4$

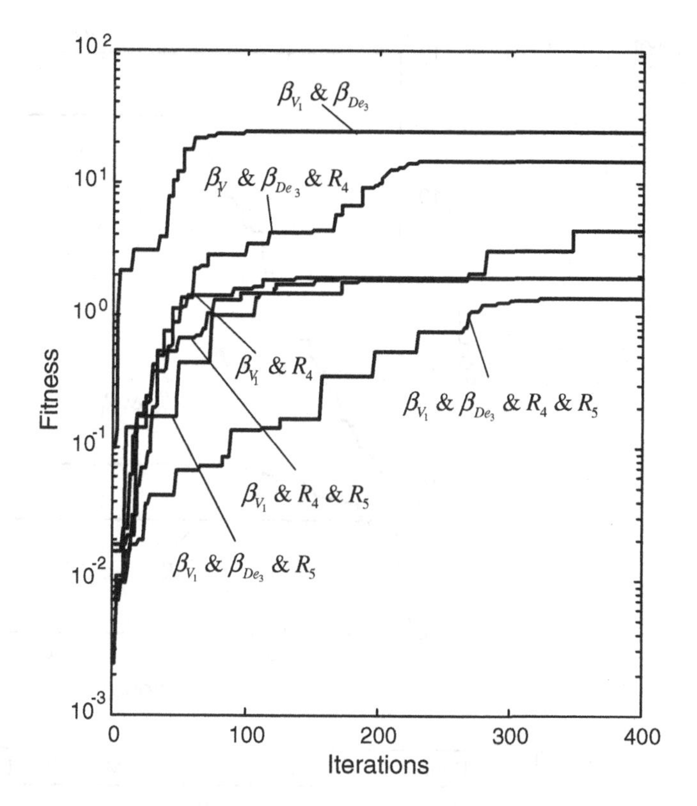

Fig. 7.29 Fitness evolution history of phase 1

Table 7.16 Prognosis result at mode $a_1 = 1, a_2 = 0$

Fault in β_{V_1}	b	c	MD-RUL	DMSV
Designed values	–0.03	0.01	64.35 s	[0,1,1]
Estimated values	–0.032	0.011	62.45 s	[0,1,1]

Table 7.17 Prognosis result at mode $a_1 = 0, a_2 = 0$

Fault in β_{V_1}	b	c	MD-RUL	DMSV
Designed values	0.005	–0.02	43.65 s	[0,0,1]
Estimated values	0.0052	–0.021	41.75 s	[0,0,1]

&R_5] based on the previous standard prognosis module result. One hundred sample data are gathered and then the standard prognosis module is activated where six HDE estimators run in parallel to find the true faults along with their DMSVs. Figure 7.30 demonstrates the search process and the results are listed in Table 7.18.

In order to demonstrate the effectiveness of the HDE in solving the prognosis problem under analysis, the HDE has been compared with GA and adaptive GA (AGA) [12]. The AGA can adaptively adjust the crossover and mutation rates based on the performance of the current genetic operators. It tries to increase the probability

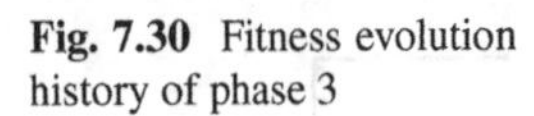

Fig. 7.30 Fitness evolution history of phase 3

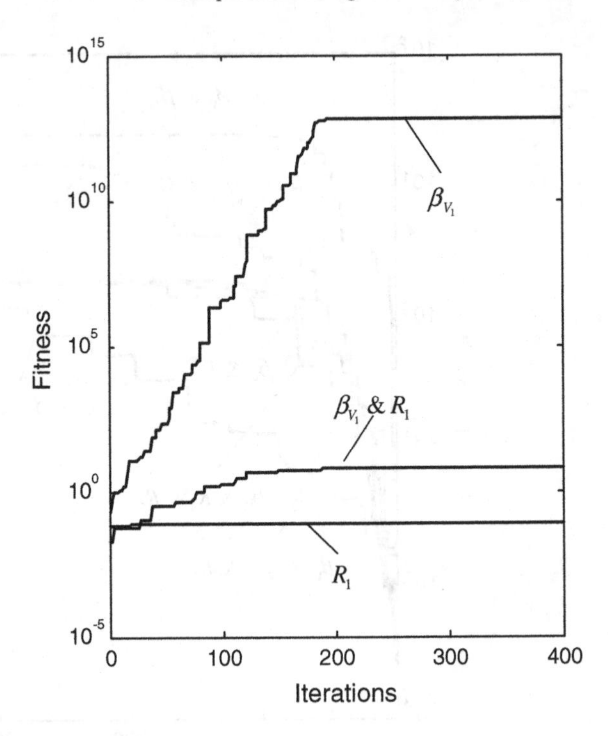

Table 7.18 Prognosis result at mode $a_1 = 0, a_2 = 1$

Fault in β_{V_1}	b	c	MD-RUL	DMSV
Designed values	–0.05	0.0035	13.5 s	[0,1,1]
Estimated values	–0.048	0.0033	13.85 s	[0,1,1]
Fault in $\beta_{D_{e3}}$	b	c	MD-RUL	DMSV
Designed values	0.015	–0.06	35.8 s	[0,0,1]
Estimated values	0.013	–0.056	34.1 s	[0,0,1]

Table 7.19 Performance comparison of different methods

	$F_{fitness}^{min}$	$F_{fitness}^{max}$	$F_{fitness}^{mean}$	σ
HDE	24.25	27.80	25.62	1.289
GA	19.75	23.74	21.88	3.435
AGA	22.35	25.6	24.01	2.023

of the genetic operator if it consistently produces a better offspring during the search process. On the contrary, it attempts to decrease the probability of the genetic operator if it always produces a poorer offspring. This approach can adaptively adjust the balance between the exploration and exploitation of the solution space [12].

All methods are tested on the same experimental data collected which are used for prognosis in phase 3. In order to perform a fair comparison, the population size and the maximum iteration of all approaches are set equal to 100 and 400, respectively. For each algorithm, 30 independent runs are carried out. Table 7.19 summarizes the minimum fitness $F_{fitness}^{min}$, the maximum fitness $F_{fitness}^{max}$ and the average fitness

$F_{fitness}^{mean}$ over the 30 runs. In the table, the standard deviation values σ corresponding to different approaches are also given. It is clear that HDE outperforms the other algorithms considered in terms of final solution and standard deviation.

References

1. J. Lee, J. Ni, D. Djurdjanovic, H. Qiu, Intelligent prognostics tools and e-maintenance. Comput. Ind. **57**(6), 476–489 (2006)
2. A.K.S. Jardine, D. Lin, D. Banjevic, A review on machinery diagnostics and prognostics implementing condition-based maintenance. Mech. Syst. Signal Process. **20**(7), 1483–1510 (2006)
3. A. Heng, S. Zhang, A.C.C. Tan, J. Mathew, Rotating machinery prognostics: State of the art, challenges and opportunities. Mech. Syst. Signal Process. **23**(3), 724–739 (2009)
4. G. Vachtsevanos, F. Lewis, M. Roemer, A. Hess, B. Wu, *Intelligent Fault Diagnosis and Prognosis for Engineering Systems* (Wiley, New Jersey, 2006)
5. B. Zhang, C. Sconyers, R. Patrick, G. Vachtsevanos, *A multi-fault modeling approach for fault diagnosis and failure prognosis of engineering systems, Annual Conference of the Prognostics and Health Management Society* (San Diego, USA, 2009)
6. D.E. Adams, M. Nataraju, A nonlinear dynamical systems framework for structural diagnosis and prognosis. Int. J. Eng. Sci. **40**(17), 1919–1941 (2002)
7. Z.L. Gaing, A particle swarm optimization approach for optimum design of PID controller in AVR system. IEEE Trans. Energy Convers. **19**(2), 384–391 (2004)
8. S.H. Ling, H.H.C. Iu, F.H.F. Leung, K.Y. Chan, Improved hybrid particle swarm optimized wavelet neural network for modeling the development of fluid dispensing for electronic packaging. IEEE Trans. Ind. Electron. **55**(9), 3447–3460 (2008)
9. A. Chatterjee, K. Pulasinghe, K. Watanabe, K. Izumi, A particle-swarm-optimized fuzzy-neural network for voice-controlled robot systems. IEEE Trans. Ind. Electron. **52**(6), 1478–1489 (2005)
10. Z.L. Gaing, Particle swarm optimization to solving the economic dispatch considering the generator constraints. IEEE Trans. Power Syst. **18**(3), 1187–1195 (2003)
11. S. Simania, R.J. Pattonb, Fault diagnosis of an industrial gas turbine prototype using a system identification approach. Control Eng. Pract. **16**(7), 769–786 (2008)
12. K.L. Mak, Y.S. Wong, X.X. Wang, An adaptive genetic algorithm for manufacturing cell formation. Int. J. Adv. Manuf. Technol. **16**(7), 491–497 (2000)
13. F. Zhao, X. Koutsoukos, H. Haussecker, J. Reich, P. Cheung, Monitoring and fault diagnosis of hybrid systems. IEEE Trans. Syst. Man, Cybern. Cybern. **35**(6), 1225–1240 (2005)
14. S. Das, P.N. Suganthan, Differential evolution: A survey of the state-of-the-art. IEEE Trans. Evol. Comput. **15**(1), 4–31 (2011)
15. F. Neri, V. Tirronen, Recent advances in differential evolution: a review and experimental analysis. Artif. Intell. Rev. **33**(1), 61–106 (2010)
16. R. Storn, K. Price, Differential evolution-a simple and efficient heuristic for global optimization over continuous spaces. J. Glob. Optim. **11**(4), 341–359 (1997)
17. L. Zhang, Y.C. Jiao, Z.B. Weng, F.S. Zhang, Design of planar thinned arrays using a Boolean differential evolution algorithm. IET Microw. Antennas Propag. **4**(12), 2172–2178 (2010)

Printed in the United States
by Bookmasters

Printed in the United States
By Bookmasters